Lecture Notes
in Control and Information Sciences 342

Editors: M. Thoma · M. Morari

Yasumichi Hasegawa · Tatsuo Suzuki

Realization Theory and Design of Digital Images

With 49 Figures

Springer

Authors

Prof. Yasumichi Hasegawa
Dr. Tatsuo Suzuki

Gifu University
Department of Electronics and Computer Engineering
Yanagido 1-1
501-11 Gifu
Japan

yhasega@gifu-u.ac.jp
tasuzuki@gifu-u.ac.jp

ISSN 0170-8643

ISBN-10 3-540-36115-4 **Springer Berlin Heidelberg New York**
ISBN-13 978-3-540-36115-2 **Springer Berlin Heidelberg New York**

Library of Congress Control Number: 2006928833

Springer is a part of Springer Science+Business Media

springer.com

Typesetting: Data conversion by authors.
Final processing by PTP-Berlin Protago-TEX-Production GmbH, Germany (www.ptp-berlin.com)
Cover-Design: WMXDesign, Heidelberg
Printed on acid-free paper 89/3141/Yu - 5 4 3 2 1 0

Preface

This monograph is concerned with description and design for two-dimensional and three-dimensional images; it will be of special interest to researchers and graduate students who specialized in image processing and system theory. From the data in digital images, mathematical models will be constructed. Then new systems which describe faithfully any two-dimensional or three-dimensional digital images will be proposed. Using the systems thus allows description to be treated as realization problem and design. By virtue of this approach, this monograph provides new results and their extensions which are designing of two-dimensional and three-dimensional images. Some actual design examples will be also shown.

In usual image processing today, two-dimensional images are transformed into one-dimensional signals, then which are analyzed by means of various established methods in signal processing theory. Likewise, three-dimensional images are transformed into two-dimensional signals and these signals are analyzed by established methods in two-dimensional signal processing theory. Another common processing procedure employs tree structures such as quad-trees for two-dimensional images and oct-trees for three-dimensional ones.

In this monograph, two-dimensional and three-dimensional images are viewed as input/output relations with special features, which are special cases of the behavior in Linear Representation Systems as discussed in our *Realization Theory of Discrete-Time Dynamical Systems* (T. Matsuo and Y. Hasegawa, Lecture Notes in Control and Information Science, Vol. 296, Springer, 2003). The processing method which we present in this monograph is built on this new insight and is very well-adapted to input/output relations. Capturing special features in two-dimensional and three-dimensional images and extending the Linear Representations Systems discussed in the previous book, this monograph will present new results for image processing by transforming a description problem into a realization problem. It will show how to design images by computer, and it will be clear that anyone can easily design images by using new method.

Realization and design problems for digital images can be roughly stated as follows:

A. For a given digital image, finding a mathematical model, equivalently a dynamical system, with the same image.
B. If possible, clarifying when the mathematical model can be actually embodied. In other words, investigating whether the mathematical model is finite dimensional.
C. Determining the mathematical model from a finite-sized digital image. This problem is called a partial realization problem.
D. If possible, finding the simplest mathematical model among all the models which describe the image. This problem is called a structure problem.
E. If possible, constructing the mathematical model for an image which exists wholly in our mind. This problem is very common, but at least so far has not been able to be managed using computers.

It is good to remember that the development of signal processing, including filtering, has been strongly stimulated by linear system theory and well-connected with the related mathematics. However, more such development of image processing has not occurred yet because there have been no suitable mathematical models for images.

In this respect, we submit that our new findings and proposals for digital images, fully discussed in this monograph, are important.

Image processing may have become a theme of scientific technology after the facsimile was invented in about 1840. In those days, the goal was only to transmit an image from one place to another by wire or wireless. Hence a method for converting an image, i.e., a two-dimensional signal, into a one-dimensional signal for instant transmission was developed. However, analysis of the image itself was never undertaken.

As already said, the usual method of treating digital images, whether two- or three-dimensional ones is almost the same as that for standard audio signal processing. There are also specialized methods such as one using quad-trees for two-dimensional images and oct-trees for three-dimensional images, or shape analysis. However, our method for processing two-dimensional or three dimensional images takes into account the image values which show the connections between horizontal and vertical and depth positions of the image's components.

Recently, computer graphics has been used to portray realities of natural objects or phenomena. Such portrayal has grown into an art genre called computer graphic arts and explosive works are shown in many museums. We also remember that such process of developments have originated in the pictorial art. Usually, generation algorithms for a work of graphic art are irregular and complex. On the contrary, design for traditional or folk handicraft articles such as furniture decoration or clothing are regular and simple. Such patterns may be more pleasing to human beings. This monograph treats geometrical patterns of the folk design type. They are generated by mathematical models which are suitable for computer programs. Our design problems are especially intended for regular patterns, that is, periodical designs.

It is also noteworthy that our method intensionally takes a positive attitude toward using computers. We will introduce a new system called a Commutative Linear Representation System which will realize, that is faithfully describe, any image.

We wish to acknowledge Professor Tsuyoshi Matsuo, who established the foundation for realization theory of two-dimensional images, and who taught us much on realization theory for discrete-time non-linear systems. He would be an author of this monograph, but in April thirteen years ago be sadly passed away. We greatfully consider him one of the authors of this manuscript in spirit.

We also wish to thank Professor R. E. Kalman for his suggestions. He stimulated us to research these realization problems directly as well as through his works.

Finally, Professor Kazuyasu Hamada pointed out some minor errors.

We also thank Special Lecturer M. L. Roecklein for making the first manuscript into a more elegant one.

May 2006

Yasumichi Hasegawa
Tatsuo Suzuki

Contents

1 Introduction

The realization problem for digital images that we will state here can be divided into the following three problems Ⓐ, Ⓑ and Ⓒ. The following notations are used in the problem description. D/I is the set of digital images, whether they are two-dimensional or three-dimensional. MM is the category of mathematical models with a behavior which is a digital image.

Ⓐ Existence and uniqueness in an algebraic sense. For any digital image $a \in D/I$, find at least one mathematical model $\sigma \in MM$ which has the behavior a. Also, prove that any two mathematical models that have the same behavior a are isomorphic in the sense of the category MM.

Ⓑ The finite dimensionality of the mathematical models. This problem is to clarify the conditions when a mathemmatical model $\sigma \in MM$ is finite-dimensional. Because finite- dimensional mathematical models are actually embodied by linear (or non-linear) circuits or computer programs, it is essential to make these conditions clear.

Ⓒ Deriving the mathematical models from finite data. This problem is regarded as a partial realization problem. The purpose of partial realization problem is to find the minimal mathematical model which fits the data of a given finite-sized digital image, and to clarify a condition under which the minimal mathematical models are isomorphic to each other.

In 1960, R. E. Kalman stated the realization problem for dynamical systems, that is, systems with input and output mechanisms, and he also established the realization theory for linear systems in algebraic sense. Based on his ideas, T. Matsuo and Y. Hasegawa established a realization problem for a very wide class of discrete-time non-linear systems [Matsuo and Hasegawa, 2003]. On the basis of these ideas, we propose a realization problem for digital images.

Discrete-time dynamical systems become ever more important syncronously with the development of computers and the establishment of mathematical programming. Discrete-time linear systems have provided the material of many fruitful contributions; as discrete-time non-linear dynamical systems as well. R. E. Kalman developed his linear system theory by using algebraic theory. Since then, algebraic theory provided significant resources for the developments of non-linear dynamical system theory [Matsuo and Hasegawa, 2003] as well. Our mathematical models for digital images are the first ones to be proposed in algebraic sense.

The usual treatments of images can be divided into the following four categories:

• Image transformation	:	image in	$\longrightarrow$	image out
• Image description	:	image in	$\longrightarrow$	model out
• Image understanding	:	image in	$\longrightarrow$	high-level model out
• Image generation	:	model in	$\longrightarrow$	image out

In current digital image processing, once the two-dimensional images are transformed into one-dimensional signals, the signals are analyzed by various methods established in one-dimensional signal processing theory. Similarly, three-dimensional images are transformed into two-dimensional signals, which are then analyzed by established methods in two-dimensional signal processing. Another common approach to digital image processing utilizes tree structures such as quad-trees for two-dimensional images or oct-trees for three-dimensional images. Digital images are also sometimes treated by shape analysis.

We wish to stress that in almost all cases, the methods used to treat three-dimensional images seek to display an image on a two-dimensional surface which looks as much as possible like the original three-dimensional object in the real world.

Our new approach states by considering two-dimensional images and three-dimensional images as input/output relations. These relations can be regarded as special cases of the behavior of Linear Representation Systems which were discussed in *Realization Theory of Discrete-Time Dynamical Systems* [Matsuo and Hasegawa, 2003]. On this basis, a new method of image processing is possible, based on and perfectly adapted to input/output relations.

A word about models, we will recall that there are two types. One is a mathematical model. Mathematical models are equations such as differential equations, partial differential equations, difference equations, and so on which describe observed objects. In physics, mathematical models have been used for describing observed physical phenomena by Newton, Leibnitz and so on. The other is a *Data – Model* which means something other than equations which describe observed objects. A typial example of a *Data – Model* is the relational model used in data-base systems.

The history of studies on phenomena illustrates the pattern of rigorous development in science. A typical example in physics is that of wave phenomena observed directly by our eyes. In the 18-th century, an equation describing one-dimensional wave phenomenon were found, and then an equation for two-dimensional wave phenomenon was considered, and the equation which describe it was obtained. Finally, in 19-th century, reseachers arrived at the equation which described three-dimensional waves, in other words, what we actually see. The analysis and expression of heat conduction phenomena followed a similar historical path.

From this point of view, present methods of image processing using transformed one-dimensional signals are not based on either mathematical models or *Data – Model*. We do not find scientific philosophy in these methods, that is to say, they are only a collection of various techniques. We venture to say that these methods are practical responses to the need to display something in print or on television which has been sent from other places. Methods based on tree-structure are regarded as *Data – Model*, but a shape-analysing method is not a method based on mathematical models.

Since our method for image processing is a method based on mathematical models, its potential must be considered to be the potential of logic.

The general plan of this monograph is to establish special features in digital images of two and three dimensions, with Linear Representation Systems (Matsuo and Hasegawa, 2003) in mind, and then to examine the new results for image processing, which have been obtained by transforming a description problem into a realization problem. It will also show how to design digital images, that is, how anyone can easily design digital images by using computer. Hence, this monograph is also a challenge to develop the potentials of our models for creation of new digital image processing and is yet untried digital image design.

The realization theory proposed here provides a new basis for treating both two- and three-dimensional digital images. Therefore, after two initial chapters concerning basic matters, this monograph is organized into two balanced sections of three chapters each.

Chapters 3 and 6 deal with the description, that is, the realization problem in the case of two-dimensional images and three-dimensional images, respectively. For each of these two dynamical systems, realization involves three issues:

1) The existence and uniqueness of our mathematical models for the digital images under consideration.
2) Their finite dimensionality.
3) Their partial realization problem.

Chapters 4 and 7 deal with the structure problems for two-dimensional images and three-dimensional images, respectively. Chapters 5 and 8 similarly provide design techniques for two-dimensional images and three-dimen-sional images, respectively. Finally, each chapter concludes with an Appendix containing the proofs for the results reached in each chapter.

Let us preview each chapter in somewhat more detail.

In Chapter 2, we clearly describe the features of two-dimensional images or three-dimensional images treated in this monograph. We have a clear connection between these images and input/output relations. We equivalently convert the usual description problem for image processing into our proposed realization problem for input/output relations. Finally, we propose a new theory of image processing.

In Chapter 3, we propose 2-Commutative Linear Representation Systems as mathematical models for any two-dimensional images, letting MM be the category of canonical (quasi-reachable and distinguishable) 2-Commutative Linear Representation Systems in detail. Next, we obtain the existence and uniqueness theorem. In addition, we investigate finite dimensional 2-Commutative Linear Representation Systems and posit a criterion for their canonicality. We show that in the isomorphic classes of finite dimensional canonical 2-Commutative Linear Representation Systems, there exists a unique Quasi-reachable Standard System and a unique Distinguishable Standard System. It is also shown that the following three conditions are equivalent:

1) A two-dimensional image is the behavior of a finite-dimensional 2-Commutative Linear Representation System;
2) The rank of infinite Hankel matrix is finite;
3) A two-dimensional image is rational.

A procedure to obtain the Quasi-reachable Standard System from a two-dimensional image is also given. Finally, the partial realization problem is discussed. For a partial two-dimensional image, there exists a minimum 2-Commutative Linear Representation System which exhibits the same behavior. Generally, minimum partial realizations are not unique up to isomorphism. To solve the uniqueness problem for a partial realization problem, we introduce the notion of natural partial realization, with the following results:

1) A criterion for the existence of natural partial realizations is given by the rank condition of the finite-sized Hankel matrix;
2) The existence condition of natural partial realizations is equivalent to the uniqueness condition of minimum partial realizations modulo isomorphism;
3) An algorithm to obtain the natural partial realization from a given two-dimensional image is given.

It is apparent that the results given by our method are the same as those obtained by linear system theory.

In Chapter 4, we propose a new structure problem for two-dimensional images. It can be stated as follows:

> Find a 2-Commutative Linear Representation System which has a simplest state space and a simplest state transition in the class of finite dimensional 2-Commutative Linear Representation Systems with the same behavior which is a two-dimensional image.

We will introduce a new 2-Commutative Linear Representation System called the Invariant Standard System, which will be used to discuss the structure problem. R. E. Kalman proposed a structure problem for linear systems, of which our structure problem for 2-Commutative Linear Representation Systems is the extension. Then applying this structure problem, we will propose a new coding problem for two-dimensional images and demonstrate a solution of effective coding.

In Chapter 5, the final chapter specially on two-dimensional images, we will propose a design problem. By the usual methods of treating a digital image, there is image generation. Various methods of image generation have been proposed intended mainly for presenting the reality of images. On the other hand, there is graphic generation by computer. Computer graphics have been used to present the realities of natural objects and phenomena. This chapter proposes a way to render a geometrical patterns imagined in the human mind on a computer display. The effectiveness of this method is illustrated by several examples.

With Chapter 6, we will take up discussion of the same matters in regard to three-dimensional images starting with the realization problem. We will propose new mathematical models called 3-Commutative Linear Representation Systems which are suitable for three-dimensional images. This is a direct extension from the models of two dimensional images. This method depends on how characteristic rules are extracted with preserving the connections in the positions, including three directions, of the pixels which show a three-dimensional image. The main realization theorem shown in this chapter is:

> For any three-dimensional image, there exist at least two canonical (quasi-reachable and distinguishable) 3-Commutative Linear Representation Systems which realize, that is, faithfully describe it and any two canonical 3-Commutative Linear Representation Systems which have the same behavior are isomorphic.

This chapter can be summarized as follows:

At first, the realization theory is stated. Next, finite-dimensional 3-Commutative Linear Representation Systems are investigated in detail. We give a criterion for the canonicality of finite dimensional 3-Commutative Linear Representation Systems, representation theorems of isomorphic classes for canonical 3-Commutative Linear Representation Systems and criteria for the behavior of finite dimensional 3-Commutative Linear Representation Systems. In addition, a procedure to obtain a canonical 3-Commutative Linear Representation System is outlined. Finally, partial realization for three-dimensional images is discussed, including the existence of minimum partial realization and natural partial realization. In summary, our discussion results in the following:

1) The necessary and sufficient condition for the existence of the natural partial realizations is given by the rank condition of the finite-sized Hankel matrix.
2) The existence condition of natural partial realizations is equivalent to the uniqueness condition of minimum partial realizations.
3) An algorithm to obtain the natural partial realization from a given finite-sized three-dimensional image is given.

It is evident that these results are the same as those obtained in linear system theory and in the theory for two-dimensional images presented in Chapter 3.

Here we want to mention for three-dimensional images:

Generally, human beings have always sought to express any phenomenon or subject of consideration as elegantly as possible, for example, the heat equation for heat conduction or the equation for electrical networks based on Kirchhoff's law and others. From the same point of view, Chapter 6 will propose a new mathematical model which describes any three-dimensional images faithfully. This idea is new for three-dimensional images. It is true that there are many methods of treating three-dimensional images such as shape analysis, techniques for computer graphics to be indicated on a screen and a method using oct-tree for encoding three-dimensional objects, but these methods seem to be technical, practical and very different from our method in neatness and purity. It is very important to know that the realization has yielded some new and neat scientific results and to understand what technical limitations it may have.

In Chapter 7, we will present a structure problem for three-dimensional images. This problem is an extension of a structure problem for linear systems as well as two-dimensional images. In fact, the structure problem for three-dimensional images can be the same as one for two-dimensional images. The problem can be stated as follows:

> Find a 3-Commutative Linear Representation System which has a simpler state space and a simpler state transition in the class of finite-dimensional 3-Commutative Linear Representation Systems which have the same behavior as a three-dimensional image.

With the application of this structure problem, we will propose a new coding problem for three-dimensional images and illustrate a solution of effective coding.

In Chapter 8, we will propose a design problem for three-dimensional images, where by the basic design method for two-dimensional images is extended to three-dimensional images. Therefore, this chapter will also propose how a pattern born in the human mind can be displayed on a computer-controlled screen.

Notations

N : the set of non-negative integers.

$N^2 := N \times N$: the product set in two sets of non negative integers.

$N^3 := N \times N \times N$: the product set of three sets of non negative integers.

N/pN : a finite field of residue class, where p is a prime number.

K : a field.

$K[z_\alpha, z_\beta]$: the commutative K-algebra of polynomials in two variables.

$K(z_\alpha, z_\beta)$: the field of rational function in two variables.

$K[[z_\alpha^{-1}, z_\beta^{-1}]]$: the K-linear space of formal power series in two variables.

$K((z_\alpha^{-1}, z_\beta^{-1}))$: the quotient field of $K[[z_\alpha^{-1}, z_\beta^{-1}]]$.

$K[z_\alpha, z_\beta, z_\gamma]$: the commutative K-algebra of polynomials in three variables.

$K(z_\alpha, z_\beta, z_\gamma)$: the field of rational function in three variables.

$K[[z_\alpha^{-1}, z_\beta^{-1}, z_\gamma^{-1}]]$: the K-linear space of formal power series in three variables.

$K((z_\alpha^{-1}, z_\beta^{-1}, z_\gamma^{-1}))$: the quotient field of $K[[z_\alpha^{-1}, z_\beta^{-1}, z_\gamma^{-1}]]$.

$F(X, Y)$: the set of all functions from X to Y.

$L(X, Y)$: the set of all linear maps from X to Y.

$L(X)$: the set of all linear maps from X to X.

K^n : the K-linear space of all n-vectors.

$K^{n\times n}$: the set of all $n \times n$-matrices.

im f : the image of a map f.

ker f : the kernel of a map f.

$\ll S \gg$: the smallest linear space which contains a set S.

Gr T : the graph of a relation T.

dom T : the domain of a relation T.

2 Two-Dimensional Images and Three-Dimensional Images

2.1 Two-Dimensional Images and Input/Output Relations

In this chapter, we will discuss the two-dimensional or three-dimensional digital images which are treated in this monograph.

A two-dimensional image can be considered as the following table:

$$\begin{array}{|llllll} \hline a(0,0) & a(0,1) & a(0,2) & \cdots & \cdots & \cdots \\ a(1,0) & a(1,1) & a(1,2) & \cdots & \cdots & \cdots \\ a(2,0) & a(2,1) & a(2,2) & \cdots & \cdots & \cdots \\ \vdots & \vdots & \vdots & \ddots & \vdots & \vdots \\ \vdots & \vdots & \vdots & \vdots & \ddots & \vdots \end{array}$$

$a(i,j) \in Y = K^p$ for any $i,j \in N$, where K is a field and p is a positive integer.

A two-dimensional image a can be expressed by $a \in F(N \times N, Y)$. Moreover, it can also be represented as the following formal power series in two variables, z_α and z_β:

$$a = \sum_{i=0}^{\infty} \sum_{j=0}^{\infty} a(i,j) z_\alpha^{-i} z_\beta^{-j} \in K^p[[z_\alpha^{-1}, z_\beta^{-1}]],$$

where $-i$ and $-j$ denote the position marker in two-directional axes for the position (i,j).

Consider a connection between this image and input/output relations.

Let U be a set of input values (alphabet; α and β), namely, let U be a set $\{\alpha, \beta\}$.

Let U^* be a set of words generated by the alphabet set U. Then any input/output map which satisfies causality condition and takes a value in Y can be expressed by an input response map $\underline{a} \in F(U^*, Y)$ which satisfies $\gamma(|\omega|) = \underline{a}(\omega)$, where $\gamma(|\omega|)$ denotes the value of the output when an input ω has been fed into the observed input/output map and $|\omega|$ denotes the length of input ω. See [Matsuo and Hasegawa, 2003] for input response maps.

Let us introduce a set $F_c(U^*,\ Y)$ of an input response map $\underline{a}$ which satisfies $\underline{a}(\omega_1|\omega_2) = \underline{a}(\omega_2|\omega_1) =$ for any $\omega_1,\ \omega_2 \in U^*$, where $|$ denotes the concatenation operator. Then an isomorphic relation between the set $F(N \times N, Y)$ and the set $F_c(U^*,\ Y)$ is obtained as follows:

$F(N \times N, Y) \to F_c(U^*,\ Y) : a \mapsto \underline{a}$ by setting $a(i,j) = \underline{a}(\overbrace{\alpha \cdots \alpha}^{i} \,|\, \overbrace{\beta \cdots \beta}^{j})$ for $i, j \in N$ and $\alpha, \beta \in U$.

Therefore, a two-dimensional image $a \in F(N \times N, Y)$ is equivalent to an input response map $\underline{a} \in F_c(U^*,\ Y)$.

Noting this equivalence relation between $F(N \times N, Y)$ and $F_c(U^*,\ Y)$, we will introduce a new problem for two-dimensional digital images, namely, a realization problem.

Roughly speaking, realization problems can be stated as follows:

≪ For given data which are typically considered as input/output data, find a mathematical model which realizes, that is, which faithfully describes them. ≫

It is understood that in image processing, description is the essential problem. Similarly, in input/output relation matters, realization is recognized as the most important problem.

In order to solve the problem, we will introduce a new special class of the mathematical models known as Linear Representation Systems [Matsuo and Hasegawa 2003]. These new mathematical models for digital images will be called 2-Commutative Linear Representation Systems.

2.2 Three-Dimensional Images

In the previous section, we noted a realization problem for two-dimensional images. Here, we will introduce a realization problem for three-dimensional images by analogy with the one for two-dimensional images.

A three-dimensional image a can be written as $a \in F(N \times N \times N, Y)$. The image a can also be represented as the following formal power series in three variables, z_α, z_β and z_γ:

$$a = \sum_{i=0}^{\infty} \sum_{j=0}^{\infty} \sum_{k=0}^{\infty} a(i,j,k) z_\alpha^{-i} z_\beta^{-j} z_\gamma^{-k} \in K^p[[z_\alpha^{-1}, z_\beta^{-1}, z_\gamma^{-1}]] ,$$

where $-i$, $-j$ and $-k$ denote the position marker in three-directional axes for the position (i,j,k).

We propose the same realization problem for three-dimensional images as for two-dimensional ones:

≪ For a given three-dimensional image, find a mathematical model which realizes, that is, which faithfully describes it. ≫

2.3 Historical Notes and Concluding Remarks

In the field of discrete-time systems, an infinite sequence or a one-dimensional signal $\acute{a}$ ($\acute{a}(i) \in Y$ for any $i \in N$) is considered as an impulse response sequence or a weighting pattern [Kalman, et al., 1969],[Willems, 1986].

In this monograph, it is shown for the first time that a close connection exists between two-dimensional images and input/output relations. Therefore, we can discuss a new realization problem for two-dimensional images.

This theory can then be extended to three-dimensional images as well.

It is also noteworthy that our methods are quite different from usual methods in image processing, for example, Fourier transformation, shape analysis, quad-trees, oct-trees and so on.

3 Realization Theory of Two-Dimensional Images

Let the set of output values Y be a linear space over the field K.

2-Commutative Linear Representation Systems present the following main theorem:

> For any two-dimensional image, there exist at least two canonical, that is, quasi-reachable and distinguishable 2-Commutative Linear Representation Systems which realize, that is, which faithfully describe it, and any two canonical 2-Commutative Linear Representation Systems with the same behavior are isomorphic.

As an application of non-linear realization theory [Matsuo and Hasegawa, 2003], we obtain a realization theory of two-dimensional images.

In Section 3.1, the realization theory is stated. Section 3.2 shows some examples of images generated by the finite-dimensional 2-Commutative Linear Representation Systems.

In Section 3.3, finite-dimensional 2-Commutative Linear Representation Systems are investigated in detail. We derive a criterion for the canonicality of finite-dimensional 2-Commutative Linear Representation Systems, the representation theorems of isomorphic classes for canonical 2-Commutative Linear Representation Systems and criteria for the behavior of finite-dimensional 2-Commutative Linear Representation Systems. In addition, a procedure to obtain a canonical 2-Commutative Linear Representation System is also given.

In Section 3.4, partial realization is discussed according to the results obtained in Section 3.3. Existence of minimum partial realization is clearly presented. Minimum partial realizations are rarely unique up to isomorphism. To solve the uniqueness problem, the notion of natural partial realizations is introduced. The main results for partial realization are the following three fold:

1) A necessary and sufficient condition for the existence of the natural partial realizations is given by the rank condition of the finite-sized Hankel matrix.
2) The existence condition for natural partial realization is equivalent to the uniqueness condition for minimum partial realizations.
3) An algorithm to obtain a natural partial realization from a given partial two-dimensional image is given.

It is evident that the results for our systems are the same as those obtained in linear system theory.

The main methods of processing two-dimensional images are treating one-dimensional signals which have either been transformed from the two-dimensional images or been passed through two-dimensional filters. Quad-trees is also used in a modeling of two-dimensional images.

On the contrary, our model of two-dimensional images is an application of realization problem in system theory. The unique feature of our model is a direct extension of describing natural phenomena, for example, wave equation, heat equation and a equation of motion. It has been a dream for human being to seek out equation which describes the given objects.

3.1 2-Commutative Linear Representation Systems

Definition 3.1.

(1) A system given by the following equations is written as a collection $\sigma = ((X, F_\alpha, F_\beta), x^0, h)$ and it is called a 2-Commutative Linear Representation System.

$$\begin{cases} x(i+1,j) = F_\alpha x(i,j) \\ x(i,j+1) = F_\beta x(i,j) \\ x(0,0) \quad\;\; = x^0 \\ \gamma(i,j) \qquad = hx(i,j) \end{cases}$$

for any $i, j \in N$, $x(i,j) \in X$, $\gamma(i,j) \in Y$, where X is a linear space over the field K. F_α and F_β are linear operators on X which satisfy $F_\alpha F_\beta = F_\beta F_\alpha$. $x^0 \in X$ is an initial state. $h : X \to Y$ is a linear operator.

(2) The two-dimensional image $a_\sigma : N \times N \to Y; (i,j) \mapsto hF_\alpha^i F_\beta^j x^0$ is called the behavior of σ.

(3) For a two-dimensional image $a \in F(N \times N, Y)$, σ which satisfies $a_\sigma = a$ is called a realization of a.

(4) A 2-Commutative Linear Representation System σ is said to be quasi-reachable if the linear hull of the reachable set $\{F_\alpha^i F_\beta^j x^0;\ i,j \in N\}$ equals X.

(5) A 2-Commutative Linear Representation System σ is called distinguishable if $hF_\alpha^i F_\beta^j x_1 = hF_\alpha^i F_\beta^j x_2$ for any $i,j \in N$ implies $x_1 = x_2$.

(6) A 2-Commutative Linear Representation System σ is called canonical if σ is quasi-reachable and distinguishable.

Remark 1: The $x(i,j)$ in the system equation of σ is the state that produces output value of a_σ at the place (i,j), while linear operator $h : X \to Y$ generates the output value $a_\sigma(i,j)$ at the place (i,j).

Remark 2: σ realizes a two-dimensional image a implies that σ is a faithful model for a.

Remark 3: Notice that a canonical 2-Commutative Linear Representation System $\sigma = ((X, F_\alpha, F_\beta), x^0, h)$ is a system with the most reduced space X among systems that have the behavior a_σ. See Definition (3-A.18), Proposition (3-A.23), Definition (3-A.24), Definition (3-A.29), Proposition (3-A.31) and Corollary (3-A.33) in Appendix 3-A.

Example 3.2.

(1) Let $K[z_\alpha, z_\beta]$ be a set of K-valued polynomials in two variables z_α, z_β. Let a linear operator z_α be $K[z_\alpha, z_\beta] \to K[z_\alpha, z_\beta]$; $\lambda \mapsto z_\alpha\lambda$, and let a linear operator z_β be $K[z_\alpha, z_\beta] \to K[z_\alpha, z_\beta]$; $\lambda \mapsto z_\beta\lambda$. For any two-dimensional image a and the unit element $\mathbf{1} \in K[z_\alpha, z_\beta]$, $((K[z_\alpha, z_\beta], z_\alpha, z_\beta), \mathbf{1}, a)$ is a quasi-reachable 2-Commutative Linear Representation System which realizes a. See Proposition (3-A.17).

(2) Let $F(N \times N, Y)$ be a set of any two-dimensional images. For any two-dimensional image $a \in F(N \times N, Y)$, let $S_\alpha a : N \times N \to Y; (i,j) \mapsto a(i+1,j)$ and $S_\beta a : N \times N \to Y; (i,j) \mapsto a(i,j+1)$. Then $S_\alpha, S_\beta \in L(F(N \times N, Y))$ and $S_\alpha S_\beta = S_\beta S_\alpha$ hold.

Let $(0,0) : F(N \times N, Y) \to Y; a \mapsto a(0,0)$ be a linear operator, and let a be any two-dimensional image. Then $((F(N \times N, Y), S_\alpha, S_\beta), a, (0,0))$ is a distinguishable 2-Commutative Linear Representation System which realizes a.

Remark: Note that the linear output map $a : K[z_\alpha, z_\beta] \to Y$ is introduced by the fact $F(N \times N, Y) = L(K[z_\alpha, z_\beta], Y)$. See Proposition (3-A.17).

Theorem 3.3. The following 2-Commutative Linear Representation Systems are canonical realizations of any two-dimensional image $a \in F(N \times N, Y)$.

1) $((K[z_\alpha, z_\beta]/_{=a}, \dot{z_\alpha}, \dot{z_\beta}), [\mathbf{1}], \dot{a})$,

where $K[z_\alpha, z_\beta]/_{=a}$ is a quotient space obtained by the following equivalence relation:

$$\sum_{i,j} \lambda_1(i,j) z_\alpha^i z_\beta^j = \sum_{i,j} \lambda_2(i,j) z_\alpha^i z_\beta^j \Longleftrightarrow \sum_{i,j} \lambda_1(i,j) a(i,j) = \sum_{i,j} \lambda_2(i,j) a(i,j).$$

$\dot{z_\alpha}$ is given by a map $\dot{z_\alpha} : K[z_\alpha, z_\beta]/_{=a} \to K[z_\alpha, z_\beta]/_{=a}; [\lambda] \mapsto [z_\alpha\lambda]$,
$\dot{z_\beta}$ is given by a map $\dot{z_\beta} : K[z_\alpha, z_\beta]/_{=a} \to K[z_\alpha, z_\beta]/_{=a}; [\lambda] \mapsto [z_\beta\lambda]$ and
$\dot{a}$ is given by $\dot{a} : K[z_\alpha, z_\beta]/_{=a} \to Y$; $[\lambda] \mapsto \dot{a}([\lambda]) = \sum_{i,j} \lambda(i,j) a(i,j)$,

where $\lambda = \sum_{i,j} \lambda(i,j) z_\alpha^i z_\beta^j \in K[z_\alpha, z_\beta]$.

(2) $((\ll S_\alpha^N S_\beta^N a \gg), S_\alpha S_\beta), a, (0,0))$,

where $\ll S_\alpha^N S_\beta^N a \gg$ is the smallest linear space which contains $S_\alpha^N S_\beta^N a := \{S_\alpha^i S_\beta^j a; (i,j) \in N \times N\}$.

Proof. See Remark 2 of Proposition (3-A.22) or (3-A.27). Also see Propositions (3-A.23), (3-A.28), (3-A.31) and Corollary (3-A.33).

Definition 3.4. Let $\sigma_1 = ((X_1, F_{\alpha_1} F_{\beta_1}), x_1^0, h_1)$ and $\sigma_2 = ((X_2, F_{\alpha_2} F_{\beta_2}), x_2^0, h_2)$ be 2-Commutative Linear Representation Systems. Then a linear operator $T : X_1 \to X_2$ is said to be a 2-Commutative Linear Representation System morphism $T : \sigma_1 \to \sigma_2$ if T satisfies $TF_{\alpha_1} = F_{\alpha_2} T$, $TF_{\beta_1} = F_{\beta_2} T$, $Tx_1^0 = x_2^0$ and $h_1 = h_2 T$.
If $T : X_1 \to X_2$ is bijective, then $T : \sigma_1 \to \sigma_2$ is said to be an isomorphism.

Theorem 3.5. Realization Theorem of 2-Commutative Linear Representation Systems

1) Existence: For any two-dimensional image $a \in F(N \times N, Y)$, there exist at least two canonical 2-Commutative Linear Representation Systems which realize a.
2) Uniqueness: Let σ_1 and σ_2 be any two canonical 2-Commutative Linear Representation Systems that realize $a \in F(N \times N, Y)$. Then there exists an isomorphism $T : \sigma_1 \to \sigma_2$.

Proof. The existence part is the same as Theorem (3.3). The uniqueness part is obtained by Remark of Lemma (3-A.37) in Appendix 3-A.5.

3.6 Relation to two-dimensional systems
The following system is said to be a two-dimensional system.

$$\begin{bmatrix} x^h(i+1,j) \\ x^v(i,j+1) \end{bmatrix} = \begin{bmatrix} A_1 & 0 \\ A_3 & A_4 \end{bmatrix} \begin{bmatrix} x^h(i,j) \\ x^v(i,j) \end{bmatrix} + \begin{bmatrix} b_1 \\ b_2 \end{bmatrix} u(i,j)$$

$$y(i,j) = [c_1, c_2] \begin{bmatrix} x^h(i,j) \\ x^v(i,j), \end{bmatrix}, \quad i,j \in N,$$

where $x^h(i,j) \in R^n$, $x^v(i,j) \in R^m$, $u(i,j) \in R^{n+m}$, $A_1 \in R^{n \times n}$, $A_3 \in R^{m \times n}$, $A_4 \in R^{m \times m}$, $b_1 \in R^n$, $b_2 \in R^m$, $c_1 \in R^{1 \times n}$, $c_2 \in R^{1 \times m}$ and R is the field of real numbers.

The two-dimensional system expressed by the above equation is written as $\Sigma = (R^n, R^m, A_1, A_3, A_4, b_1, b_2, c_1, c_2)$.
The input/output map, that is weight factor, of the two-dimensional system is defined by $\varphi_{i0} = c_1 A_1^{i-1} b_1$, $\varphi_{0j} = c_2 A_4^{j-1} b_2$ and $\varphi_{ij} = c_2 A_4^{j-1} A_3 A_1^{i-1} b_1$, $i,j \in N - \{0\}$.

The realization problem is considered to be a problem of determining the two-dimensional system from the input/output map, that is weight factor, satisfying the above relations.

We shall now explain a relation between our 2-Commutative Linear Representation Systems and two-dimensional systems. In particular, we will clearly connect 2-Commutative Linear Representation Systems displaying the behavior of two-dimensional images and two-dimensional systems with the input/output map. It should be noted that the canonical (minimum) property is not preserved in the procedure.

Proposition 3.7. The behavior of any 2-Commutative Linear Representation System $\sigma = ((X, F_\alpha, F_\beta), x^0, h)$ is the same as the input/output map, that is weight factor, of a two-dimensional system $\Sigma = (X, X, F_\alpha, F_\beta, F_\beta, F_\alpha x^0, F_\beta x^0, h, h)$.

Proof. A direct calculation based on the definition of the behavior of 2-Commutative Linear Representation Systems and of the input/output map of two-dimensional systems shows the validity of this proposition.

Proposition 3.8. The input/output map of any two-dimensional system $\Sigma = (R^n, R^m, A_1, A_3, A_4, b_1, b_2, c_1, c_2)$ has the same behavior of the 2-Commutative Linear Representation System $\sigma = ((R^{(1+4m)\times(1+4n)}, F_\alpha, F_\beta), x^0, h)$ defined as follows:

$$F_\alpha : R^{(1+4m)\times(1+4n)} \to R^{(1+4m)\times(1+4n)}; x \mapsto x\bar{F}_\alpha.$$
$$F_\beta : R^{(1+4m)\times(1+4n)} \to R^{(1+4m)\times(1+4n)}; x \mapsto \bar{F}_\beta x.$$
$$\bar{F}_\alpha = \begin{bmatrix} 0 & 0 & 0 & 0 & 0 \\ 0 & A_1 & 0 & 0 & 0 \\ 0 & 0 & 0 & 0 & 0 \\ 0 & 0 & 0 & A_1 & 0 \\ 0 & 0 & 0 & 0 & A_1 \end{bmatrix} \in R^{(1+4n)\times(1+4n)},$$
$$\bar{F}_\beta = \begin{bmatrix} 0 & 0 & 0 & 0 & 0 \\ 0 & 0 & 0 & 0 & 0 \\ 0 & 0 & A_4 & 0 & 0 \\ 0 & 0 & 0 & 0 & 0 \\ 0 & 0 & 0 & 0 & A_4 \end{bmatrix} \in R^{(1+4m)\times(1+4m)},$$
$$x^0 = \begin{bmatrix} 0 & c_1 & 0 & 0 & 0 \\ 0 & 0 & 0 & 0 & 0 \\ b_2 & 0 & A_3 & 0 & 0 \\ 0 & 0 & 0 & A_3 & 0 \\ 0 & 0 & 0 & 0 & A_3 \end{bmatrix} \in R^{(1+4m)\times(1+4n)}.$$
$$h : R^{(1+4m)\times(1+4n)} \to R; x \mapsto \begin{bmatrix} 1 & 0 & c_2 & c_2 & c_2 \end{bmatrix} x \begin{bmatrix} 1 & b_1 & -b_1 & -b_1 & b_1 \end{bmatrix}^T.$$

Proof. A direct calculation based on the definition of the input/output map and the behavior shows the validity of this proposition.

3.2 Definite Examples of Two-Dimensional Images Generated by Finite-Dimensional 2-Commutative Linear Representation Systems

In order to show some examples of two-dimensional images, we briefly introduce the finite-dimensional 2-Commutative Linear Representation System which can be treated by computer or non-linear circuits.

2-Commutative Linear Representation System $\sigma = ((X, F_\alpha, F_\beta), x^0, h)$ is called a finite (or n)-dimensional 2-Commutative Linear Representation System if the state space X is a finite (or n)-dimensional linear space.

Lemma 3.9. For any image $a \in F(N^2, Y)$, the following three conditions are equivalent to each other.

(1) a has the behavior of a finite-dimensional canonical 2-Commutative Linear Representation System.
(2) The quotient space $K[z_\alpha, z_\beta]/_{=a}$ is finite-dimensional.
(3) The linear space generated by $\{S_\alpha^i S_\beta^j a : i, j \in N\}$ is finite-dimensional, where $K[z_\alpha, z_\beta]/_{=a}$ is a quotient space given by the following equivalence relations: $a_1 = a_2 \iff a_1(i,j) = a_2(i,j)$ for any $i, j \in N$. Moreover $S_\alpha, S_\beta \in L(F(N^2, Y))$ are given by $S_\alpha a : N^2 \to Y; (i,j) \mapsto a(i+1, j)$ and $S_\beta a : N^2 \to Y; (i,j) \mapsto a(i, j+1)$.

Proof. This lemma is obtained by the direct consequence of Theorem (3.3).

Example 3.10. A three-dimensional 2-Commutative Linear Representation System

Let K be $N/3N$, which is the quotient field modulo the prime number 3, and let the set Y of output values be K.

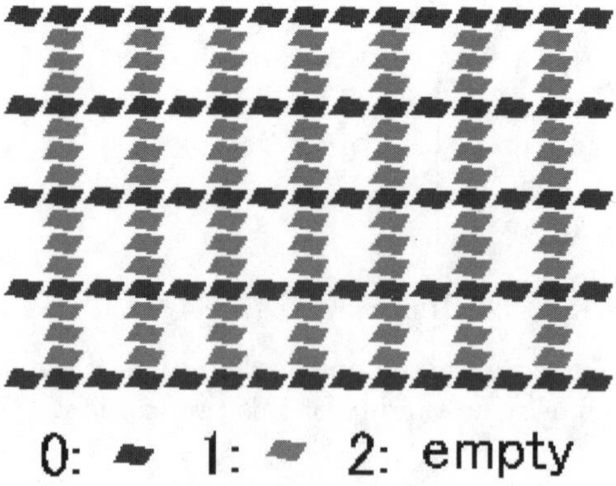

Fig. 3.1. The two-dimensional image generated by the three-dimensional 2-Commutative Linear Representation System σ

Let the state space be K^3 and let $F_\alpha, F_\beta \in K^{3\times3}, x^0 \in K^3$ and $h \in K^{1\times3}$ be as follows:

$$F_\alpha = \begin{bmatrix} 2\,0\,0 \\ 0\,2\,0 \\ 0\,0\,2 \end{bmatrix}, F_\beta = \begin{bmatrix} 2\,0\,0 \\ 0\,0\,2 \\ 0\,1\,0 \end{bmatrix}, x^0 = \begin{bmatrix} 1 \\ 1 \\ 0 \end{bmatrix}, h = \begin{bmatrix} 1\,2\,0 \end{bmatrix}.$$

Then $\sigma = ((X, F_\alpha, F_\beta), x^0, h)$ is the three-dimensional 2-Commutative Linear Representation System. See Figure 3.1.

Example 3.11. An eight-dimensional 2-Commutative Linear Representation System.

Let K be $N/3N$, which is the quotient field modulo the prime number 3, and let the set Y of output values be K.

Let the state space be K^8 and let $F_\alpha, F_\beta \in K^{8\times8}, x^0 \in K^8$ and $h \in K^{1\times8}$ be as follows:

$$F_\alpha = \begin{bmatrix} 1\,0\,0\,0\,0\,0\,0\,0 \\ 0\,1\,0\,0\,0\,0\,0\,0 \\ 0\,0\,1\,0\,0\,0\,0\,0 \\ 0\,0\,0\,1\,0\,0\,0\,0 \\ 0\,0\,0\,0\,2\,0\,0\,0 \\ 0\,0\,0\,0\,0\,2\,0\,0 \\ 0\,0\,0\,0\,0\,0\,2\,0 \\ 0\,0\,0\,0\,0\,0\,0\,2 \end{bmatrix}, \quad F_\beta = \begin{bmatrix} 1\,0\,0\,0\,0\,0\,0\,0 \\ 0\,2\,0\,0\,0\,0\,0\,0 \\ 0\,0\,0\,2\,0\,0\,0\,0 \\ 0\,0\,1\,0\,0\,0\,0\,0 \\ 0\,0\,0\,0\,1\,0\,0\,0 \\ 0\,0\,0\,0\,0\,2\,0\,0 \\ 0\,0\,0\,0\,0\,0\,0\,2 \\ 0\,0\,0\,0\,0\,0\,1\,0 \end{bmatrix},$$

$$x^0 = \begin{bmatrix} 1 \\ 1 \\ 1 \\ 0 \\ 1 \\ 1 \\ 1 \\ 0 \end{bmatrix}, h = \begin{bmatrix} 2\,1\,2\,2\,1\,1\,2\,0 \end{bmatrix}.$$

Then $\sigma = ((X, F_\alpha, F_\beta), x^0, h)$ is the eight-dimensional 2-Commutative Linear Representation System. See Figure 3.2.

3.3 Finite-Dimensional 2-Commutative Linear Representation Systems

This section deals with the fundamental structures of finite-dimensional 2-Commutative Linear Representation Systems based on the realization Theorem (3.5).

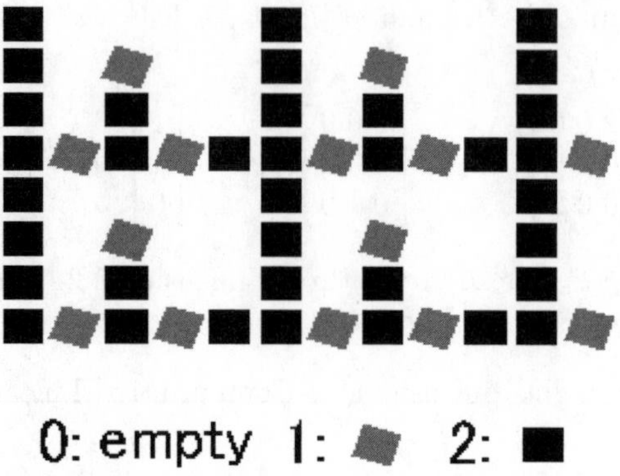

Fig. 3.2. The two-dimensional image generated by the eight-dimensional canonical 2-Commutative Linear Representation System σ

First, the conditions under which a finite-dimensional 2-Commutative Linear Representation System is canonical are given.

Second, the representation theorem for finite-dimensional canonical 2-Commutative Linear Representation Systems is obtained. This involves showing two standard systems are representatives in their equivalence class. One is the Quasi-reachable Standard System and the other is the Distinguishable Standard System.

Third, two criteria for the behavior of the finite-dimensional 2-Commutative Linear Representation Systems are given. One is the rank condition of the infinite Hankel matrix, and the other is the application of Kleene's theorem from automata theory.

Finally, a procedure for obtaining the Quasi-reachable Standard System which realizes a given two-dimensional image is presented.

Proofs for these matters are provided in Appendix 3-B.

Corollary 3.12. Let T be a 2-Commutative Linear Representation System morphism $T : \sigma_1 \to \sigma_2$. Then $a_{\sigma_1} = a_{\sigma_2}$ holds.

Proof. A direct calculation using the definition of the behavior and 2-Commutative Linear Representation System morphism shows the validity of this corollary.

There is a fact on finite-dimensional linear space that an n-dimensional linear space over the field K is isomorphic to K^n. Furthermore, $L(K^n, K^m)$ is isomorphic to $K^{m \times n}$. See Halmos [1958]. Therefore, without loss of generality, we can consider an n-dimensional 2-Commutative Linear Representation System as $\sigma = ((K^n, F_\alpha, F_\beta), x^0, h)$, where F_α, $F_\beta \in K^{n \times n}$, $x^0 \in K^n$ and $h \in K^{p \times n}$.

Theorem 3.13. A 2-Commutative Linear Representation System $\sigma = ((K^n, F_\alpha, F_\beta), x^0, h)$ is canonical if and only if the following conditions 1) and 2) hold:

(1) rank $[x^0, F_\alpha x^0, F_\alpha^2 x^0, \cdots, F_\alpha^{n-1} x^0, F_\beta x^0, F_\alpha F_\beta x^0, F_\alpha^2 F_\beta x^0, \cdots, F_\alpha^{n-2} F_\beta x^0, F_\beta^2 x^0, F_\alpha F_\beta^2 x^0, F_\alpha^2 F_\beta^2 x^0, \cdots, F_\alpha^{n-3} F_\beta^2 x^0, F_\beta^3 x^0, F_\alpha F_\beta^3 x^0, F_\alpha^2 F_\beta^3 x^0, \cdots, F_\alpha^{n-4} F_\beta^3 x^0, \cdots, \cdots, F_\beta^{n-3} x^0, F_\alpha F_\beta^{n-3} x^0, F_\alpha^2 F_\beta^{n-3} x^0, F_\beta^{n-2} x^0, F_\alpha F_\beta^{n-2} x^0, F_\beta^{n-1} x^0] = n.$

(2) rank $[h^T, (hF_\alpha)^T, (hF_\alpha^2)^T, \cdots, (hF_\alpha^{n-1})^T, (hF_\beta)^T, (hF_\alpha F_\beta)^T, (hF_\alpha^2 F_\beta)^T, \cdots, (hF_\alpha^{n-2} F_\beta)^T, \cdots, (hF_\beta^{n-3})^T, (hF_\alpha F_\beta^{n-3})^T, (hF_\alpha^2 F_\beta^{n-3})^T, (hF_\beta^{n-2})^T, (hF_\alpha F_\beta^{n-2})^T, (hF_\beta^{n-1})^T] = n.$

Proof. See Propositions (3-B.8) and (3-B.17) in Appendix 3-B.

Definition 3.14. A canonical 2-Commutative Linear Representation System $\sigma_s = ((K^n, F_{\alpha_s}, F_{\beta_s}), \mathbf{e}_1, h_s)$ is said to be a Quasi-reachable Standard System with a vector index $\nu = (\nu_1, \nu_2, \cdots, \nu_k)$ if the following conditions hold:

(1) An integer ν_j $(1 \le j \le k)$ satisfies $n = \sum_{j=1}^k \nu_j$ and $0 \le \nu_k \le \nu_{k-1} \le \cdots \le \nu_2 \le \nu_1$.

(2) For any $i,\ j$ $(1 \le j \le k,\ 1 \le i \le \nu_j)$, $F_\beta^{j-1} F_\alpha^{i-1} \mathbf{e}_1 = \mathbf{e}_{\nu_1+\nu_2+\cdots+\nu_{j-1}+i}$, $F_\beta^{j-1} F_\alpha^{\nu_j} \mathbf{e}_1 = \sum_{m=1}^{j} \sum_{l=1}^{\nu_m} c_{ml}^j F_\beta^{m-1} F_\alpha^{l-1} \mathbf{e}_1$, where $c_{ml}^j \in K$, $F_\beta^k \mathbf{e}_1 = \sum_{m=1}^{k} \sum_{l=1}^{\nu_m} c_{ml}^{k+1} F_\beta^{m-1} F_\alpha^{l-1} \mathbf{e}_1$, where $c_{ml}^{k+1} \in K$, where $\mathbf{e}_i = [0, \cdots, 0, \overset{i}{1}, 0, \cdots, 0]^T \in K^n$, and T denotes the transposition of matrices or vectors.

The $F_{\alpha s}$ and $F_{\beta s}$ of the Quasi-reachable Standard System with a vector index $\nu = (\nu_1, \nu_2, \cdots, \nu_k)$ are characterized by Figure 3.3 and Figure 3.4. See also Proposition (3-B.11).

Each $\mathbf{c}^i$ in the figures is given by $\mathbf{c}^i := [c_{11}^i, \cdots, c_{1\nu_1}^i, \cdots, c_{i1}^i, \cdots, c_{i\nu_i}^i, \mathbf{0}]^T$ for $1 \le i \le k,\ 1 \le j \le i$, and $\mathbf{c}^{k+1} := [c_{11}^{k+1}, \cdots, c_{1\nu_1}^{k+1}, \cdots, c_{k1}^{k+1}, \cdots, c_{k\nu_k}^{k+1}]^T$.

Theorem 3.15. Representation Theorem for equivalence classes.
For any finite-dimensional canonical 2-Commutative Linear Representation System, there exists a uniquely determined isomorphic Quasi-reachable Standard System.

Proof. See (3-B.21) in Appendix 3-B.

Definition 3.16. Let Y be a field K for convenience. A canonical 2-Commutative Linear Representation System $\sigma_d = ((K^n, F_{\alpha_d}, F_{\beta_d}), x_d^0, h_d)$ is said to be a Distinguishable Standard System with a vector index $\mu = (\mu_1, \mu_2, \cdots, \mu_k)$ if the following conditions hold:

(1) An integer μ_j $(1 \le j \le k)$ satisfies $n = \sum_{j=1}^k \mu_j$ and $0 \le \mu_k \le \mu_{k-1} \le \cdots \le \mu_2 \le \mu_1$.

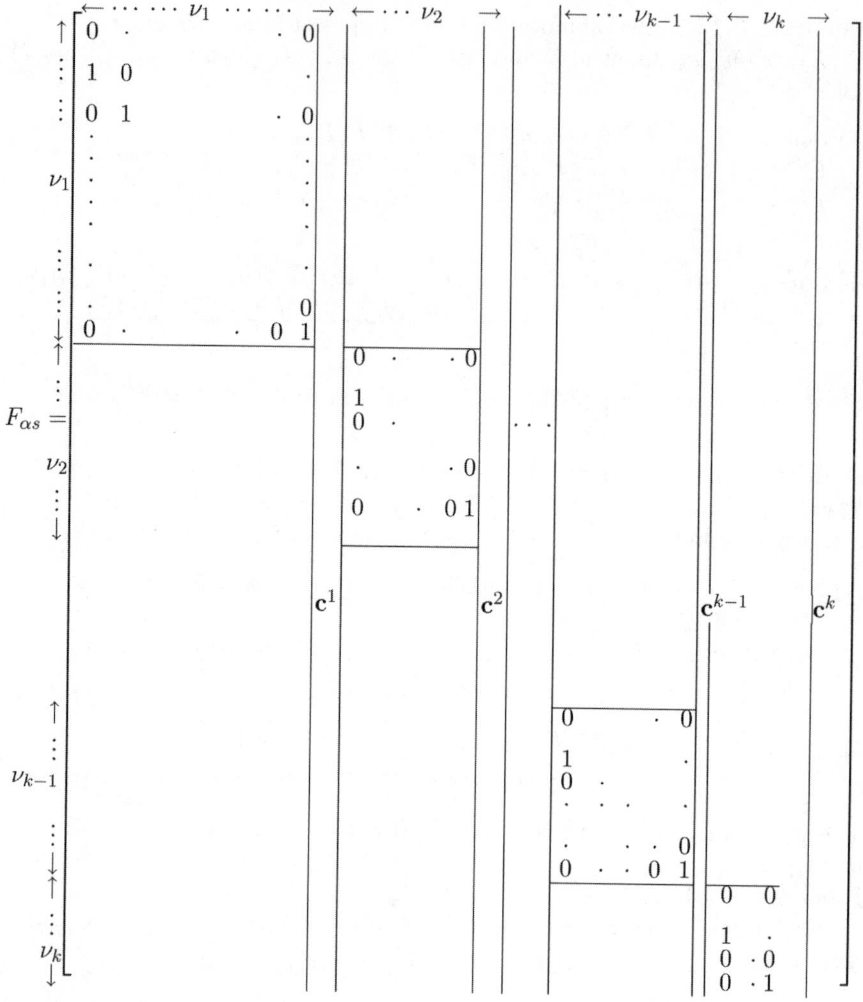

Fig. 3.3. $F_{\alpha s}$ of the Quasi-reachable Standard System σ_s defined in Definition (3.14)

(2) For any i, j $(1 \le j \le k,\ 1 \le i \le \mu_j)$, $(h_d F_\beta^{j-1} F_\alpha^{i-1})^T = (\mathbf{e}_{\mu_1+\mu_2+\cdots+\mu_{j-1}+i})^T$, $(h_d F_\beta^{j-1} F_\alpha^{\mu_j})^T = \sum_{m=1}^{j} \sum_{l=1}^{\mu_m} c_{ml}^{j} (h_d F_\beta^{m-1} F_\alpha^{l-1})^T$, where $c_{ml}^{j} \in K$, $(h_d F_\beta^{k})^T = \sum_{m=1}^{k} \sum_{l=1}^{\mu_m} c_{ml}^{k+1} (h_d F_\beta^{m-1} F_\alpha^{l-1})^T$, where $c_{ml}^{k+1} \in K$, $\mathbf{e}_i = [0, \cdots, 0, \overset{i}{1}, 0, \cdots, 0]^T$, and T denotes the transposition of matrices or vectors.

The $F_{\alpha d}$ and $F_{\beta d}$ of the Distinguishable Standard System with a vector index $\mu = (\mu_1, \mu_2, \cdots, \mu_k)$ are characterized by Figure 3.5 and Figure 3.6.

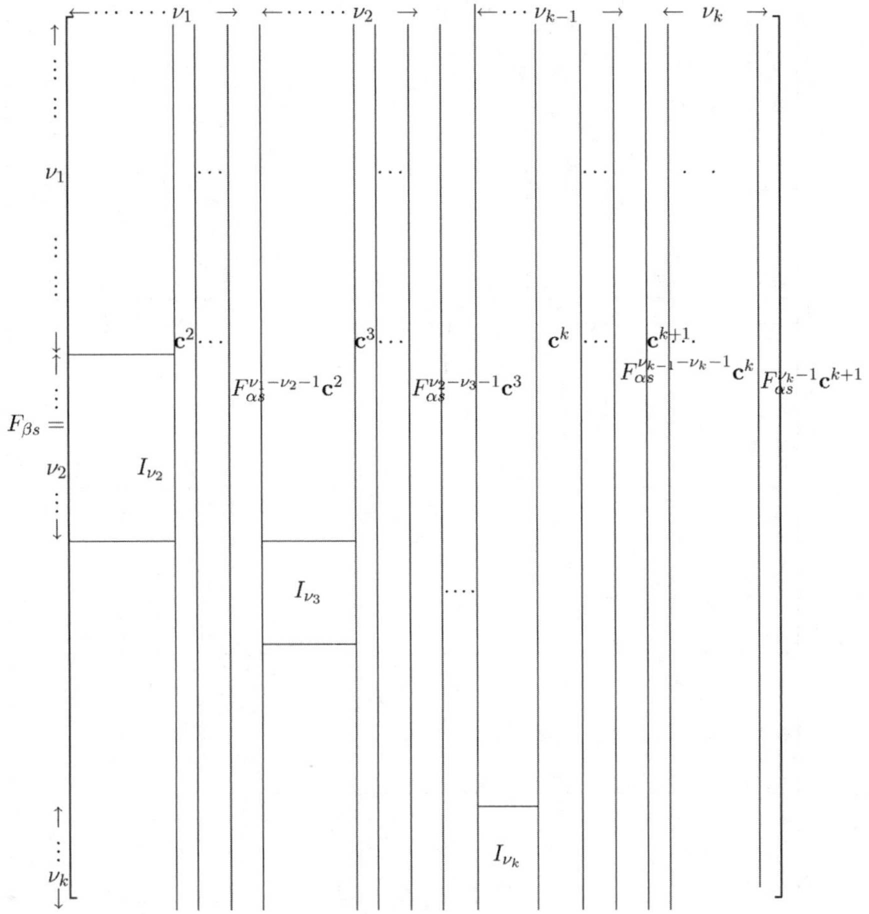

Fig. 3.4. $F_{\beta s}$ of the Quasi-reachable Standard System σ_s defined in Definition (3.14)

Each $\mathbf{c}^i$ in the figures is given by $\mathbf{c}^i := [c^i_{11}, \cdots, c^i_{1\nu_1}, \cdots, c^i_{i1}, \cdots, c^i_{i\nu_i}, \mathbf{0}]^T$ for $1 \le i \le k,\ 1 \le j \le i$, and $\mathbf{c}^{k+1} := [c^{k+1}_{11}, \cdots, c^{k+1}_{1\nu_1}, \cdots, c^{k+1}_{k1}, \cdots, c^{k+1}_{k\nu_k}]^T$.

Theorem 3.17. Representation Theorem for equivalence classes.
Let the set of output values Y be K. Then for any finite-dimensional canonical 2-Commutative Linear Representation System, there exists a uniquely determined isomorphic Distinguishable Standard System.

Proof. See (3-B.22) in Appendix 3-B.

Definition 3.18. For any two-dimensional image $a \in F(N \times N, Y)$, the corresponding linear input/output map $A : (K[z_\alpha, z_\beta], z_\alpha, z_\beta) \to (F(N \times N, Y), S_\alpha, S_\beta)$ satisfies $A(z^i_\alpha z^j_\beta) = S^i_\alpha S^j_\beta a$ for $i,\ j \in N$.

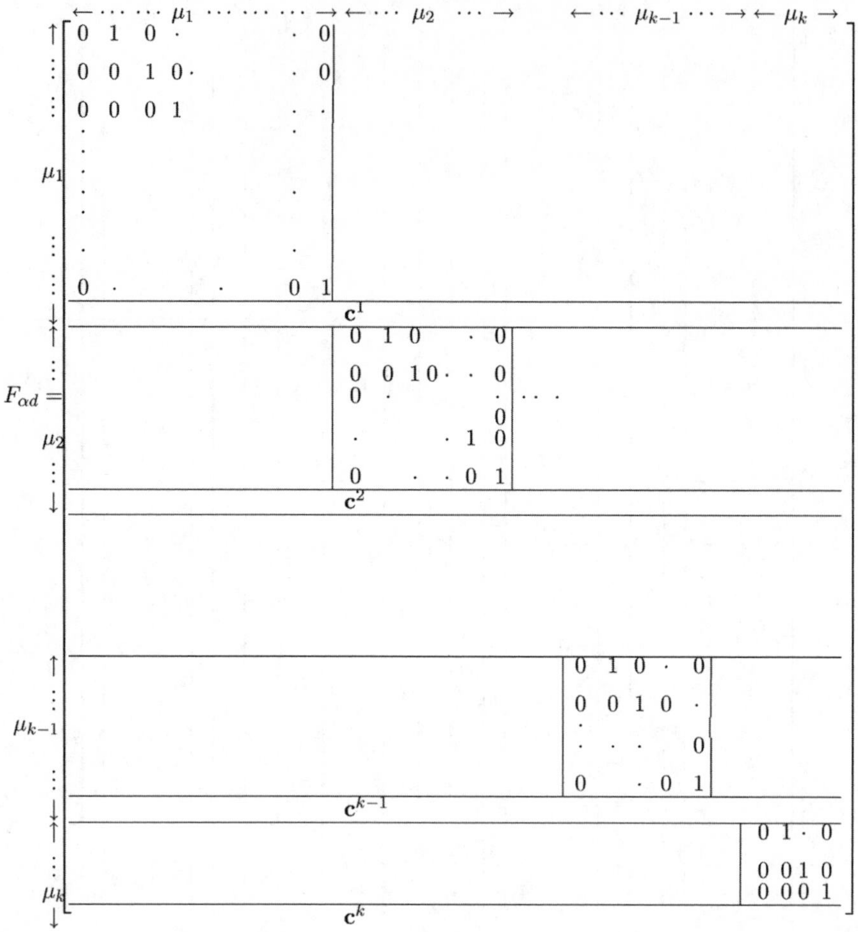

Fig. 3.5. $F_{\alpha d}$ of the Distinguishable Standard System σ_d defined in Definition (3.16)

Hence, A can be represented by the following infinite matrix H_a. This H_a is said to be a Hankel matrix of a:

$$H_a = \begin{array}{c} \\ \\ \\ (\tilde{i},\tilde{j}) \end{array} \left(\begin{array}{ccc} & & (i,j) \\ & & \vdots \\ & & \vdots \\ \cdots & \cdots & a(\tilde{i}+i,\tilde{j}+j) \end{array} \right)$$

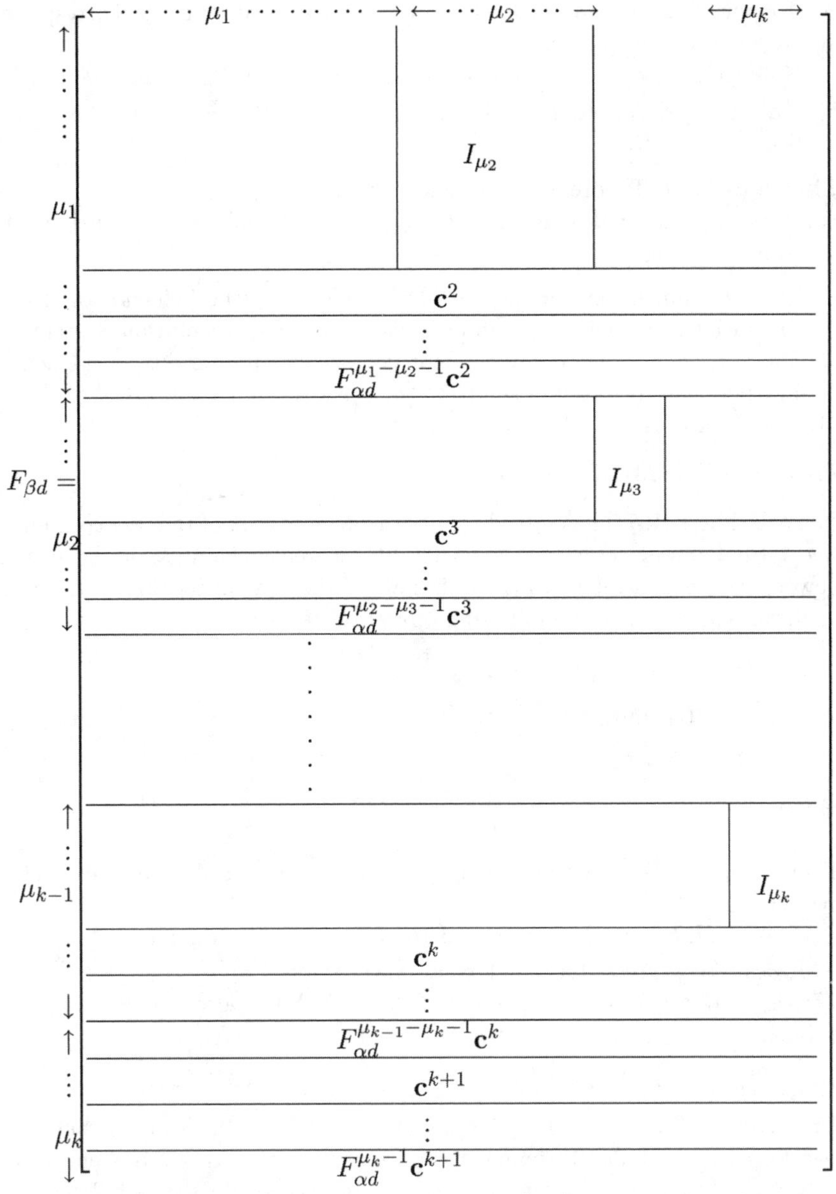

Fig. 3.6. $F_{\beta d}$ of the Distinguishable Standard System σ_d defined in Definition (3.16)

See Remark 2 concerning Proposition (3-A.27) in relation to the corresponding linear input/output map A.

Note that the column vectors of the Hankel matrix H_a are represented by $S_\alpha^i S_\beta^j a$ for $i,\ j \in N$. Moreover, $a(\tilde{i}+i,\tilde{j}+j)$=$(0,0)S_\alpha^{\tilde{i}} S_\beta^{\tilde{j}} S_\alpha^i S_\beta^j a$ holds.

See Example (3.2).

Theorem 3.19. Theorem for existence criteria
For a two-dimensional image $a \in F(N \times N, Y)$, the following conditions are equivalent:

(1) The two-dimensional image $a \in F(N \times N, Y)$ is the behavior of the n-dimensional canonical 2-Commutative Linear Representation System.
(2) There exist n linearly independent vectors and no more than n linearly independent vectors in a set $\{S_\alpha^i S_\beta^j a;\ i+j \le n-1 \text{ for } i,j \in N\}$.
(3) The rank of the Hankel matrix H_a is n.

Proof. See (3-B.23) in Appendix 3-B.

Remark: Fliess [1974] has introduced the Hankel matrix of the non-commutative formal power series and shown that the recognizability of the formal power series is equivalent to the finiteness of the rank of its Hankel matrix.

Let $K[z_\alpha, z_\beta]$ have the following operation $\times$.

$$\times : K[z_\alpha,z_\beta] \times K[z_\alpha,z_\beta] \to K[z_\alpha,z_\beta];$$
$$\Big(\sum_{i_1,j_1} \lambda_1(i_1,j_1) z_\alpha^{i_1} z_\beta^{j_1}, \sum_{i_2,j_2} \lambda_2(i_2,j_2) z_\alpha^{i_2} z_\beta^{j_2}\Big) \mapsto \sum_{i=i_1+i_2, j=j_1+j_2} \lambda(i,j) z_\alpha^i z_\beta^j$$
$$= \Big(\sum_{i_1,j_1} \lambda_1(i_1,j_1) z_\alpha^{i_1} z_\beta^{j_1}\Big) \times \Big(\sum_{i_2,j_2} \lambda(i_2,j_2) z_\alpha^{i_2} z_\beta^{j_2}\Big).$$

Then $K[z_\alpha, z_\beta]$ is an algebra over K. Moreover $K[z_\alpha, z_\beta]$ is a free algebra over K.

$F(N \times N, K)$ can be considered to contain $K[z_\alpha, z_\beta]$, since $a \in F(N \times N, K)$ can be expressed as the formal power series $\bar{a} = \sum_{i,j} a(i,j) z_\alpha^i z_\beta^j$.

Any two-dimensional image $a \in F(N \times N, Y)$ is also expressed by the following formal power series, where $Y = K^p$:

$\bar{a} = \sum_{i=0}^{\infty} \sum_{j=0}^{\infty} a(i,j) z_\alpha^{-i} z_\beta^{-j}$.

Theorem 3.20. A two-dimensional image $a \in F(N \times N, K^p)$ is the behavior of a finite-dimensional 2-Commutative Linear Representation System if and only if the formal power series $\bar{a} = [\bar{a}_1 \bar{a}_2 \cdots \bar{a}_p]^T$ is expressed as follows:

$$\bar{a_k} = \frac{z_\alpha z_\beta \left(\sum_{j=0}^{m} \sum_{i=0}^{n} \lambda_k(i,j) z_\alpha^i z_\beta^j\right)}{q_\alpha(z_\alpha) q_\beta(z_\beta)},$$

where k $(1 \le k \le p)$ is an integer, $q_\alpha(z_\alpha)$ and $q_\beta(z_\beta)$ are a monic polynomial of z_α with order n, a monic polynomial of z_β with order m, respectively. Furthermore, $\lambda_k(i,j) \in K$.

Proof. See (3-B.24) in Appendix 3-B.

Remark 1: The equivalent condition for the commutative formal power series in one variable to be rational has already been established. See Gantmacher [1959] and Kalman, Falb and Arbib [1969]. Also see Matsuo and Hasegawa [1981] for the case of two variables.

Remark 2: Fliess [1970],[1974] has pointed out that the recognizability of the three variable commutative formal power series can be characterized by the same form as the rational function in Theorem (3.20).

Theorem 3.21. Theorem for a realization procedure
Let a two-dimensional image $a \in F(N \times N, Y)$ satisfy the condition of Theorem (3.19). Then the Quasi-reachable Standard System $\sigma = ((K^n, F_{\alpha_s}, F_{\beta_s}), \mathbf{e}_1, h_s)$ which realizes the two-dimensional image a is obtained by the following procedure:

1) Find an integer ν_1 and coefficients $\{c_{1l}^1; 1 \leq l \leq \nu_1\}$ such that the vectors $\{S_\alpha^i a; 1 \leq i \leq \nu_1 - 1\}$ of the set $\{S_\alpha^i a; i \leq n-1, i \in N\}$ are linearly independent and $S_\alpha^{\nu_1} a = \sum_{l=1}^{\nu_1} c_{1l}^1 S_\alpha^{l-1} a$.
2) Find an integer ν_2 and coefficients $\{c_{ml}^2; 1 \leq l \leq \nu_m, 1 \leq m \leq 2\}$ such that the vectors $\{S_\beta^{j-1} S_\alpha^{i-1} a; 1 \leq i \leq \nu_j - 1, 1 \leq j \leq 2\}$ of the set $\{S_\beta^j S_\alpha^i a; i \leq n-1, j \leq n-2 \in N\}$ are linearly independent and $S_\beta S_\alpha^{\nu_2} a = \sum_{m=1}^{2} \sum_{l=1}^{\nu_m} c_{ml}^2 S_\beta^{m-1} S_\alpha^{l-1} a$.

$\vdots$

k) Find an integer ν_k and coefficients $\{c_{ml}^k; 1 \leq l \leq \nu_l, 1 \leq m \leq k\}$ such that the vectors $\{S_\beta^{j-1} S_\alpha^{i-1} a; 1 \leq i \leq \nu_j, 1 \leq j \leq k\}$ of the set $\{S_\beta^j S_\alpha^i a; i \leq n-1, j \leq k-1 \in N\}$ are linearly independent, $S_\beta^{k-1} S_\alpha^{\nu_k} a = \sum_{m=1}^{k} \sum_{l=1}^{\nu_m} c_{ml}^k S_\beta^{m-1} S_\alpha^{l-1} a$ and $S_\beta^k a = \sum_{m=1}^{k} \sum_{l=1}^{\nu_m} c_{ml}^{k+1} S_\beta^{m-1} S_\alpha^{l-1} a$.
k+1) Let the state space be K^n, and let the initial state be $\mathbf{e}_1$, where $n = \sum_{i=1}^{k} \nu_i$.
k+2) Let matrices F_{α_s} and F_{β_s} be those given in the proof of Theorem (3.15).
k+3) Let the output map h_s be $h_s = [a(0,0), a(1,0), \cdots, a(\nu_1 - 1, 0), a(0,1), \cdots, a(\nu_2 - 1, 1), \cdots, a(0, k-1), \cdots, a(\nu_k - 1, k-1)]$.

Proof. See (3-B.25) in Appendix 3-B.

3.4 Partial Realization Theory of Two-Dimensional Images

Here we consider a partial realization problem for two-dimensional images, namely, we will obtain a 2-Commutative Linear Representation System which describes given finite-sized two-dimensional images. Let $\underline{a}$ be an

$(L+1)\times(M+1)$ sized two-dimensional image, that is $\underline{a} \in F(\mathbf{L}\times\mathbf{M}, Y)$, where $L, M \in N$, $\mathbf{L} := \{0, 1, \cdots, L-1, L\}$ and $\mathbf{M} := \{0, 1, \cdots, M-1, M\}$. The $\underline{a}$ is said to be a finite-sized two-dimensional image. A finite-dimensional 2-Commutative Linear Representation System $\sigma = ((X, F_\alpha, F_\beta), x^0, h)$ is called a partial realization of $\underline{a}$ if $hF_\alpha^i F_\beta^j x^0 = \underline{a}(i, j)$ holds for any $(i, j) \in \mathbf{L}\times\mathbf{M}$.

A partial realization problem of 2-Commutative Linear Representation Systems can be stated as follows:

< For any given $\underline{a} \in F(\mathbf{L}\times\mathbf{M}, Y)$, find a partial realization σ of $\underline{a}$ such that the dimension of state space X of σ is minimum. This σ is said to be a minimal partial realization of $\underline{a}$. Moreover, show the minimal partial realizations are unique modulo isomorphism.>

Proposition 3.22. For any given $\underline{a} \in F(\mathbf{L}\times\mathbf{M}, Y)$, there always exists a minimal partial realization of $\underline{a}$.

Proof. Set $\underline{a}(i, j) = 0$ for any $(i, j) \notin \mathbf{L}\times\mathbf{M}$. Then $\underline{a} \in F(N\times N, Y)$, and Theorem (3.19) implies that there exists a finite-dimensional partial realization of $\underline{a}$. Therefore, there exists a minimal partial realization.

Minimal partial realizations are, in general, not unique modulo isomorphisms. Therefore, a natural partial realization will be introduced, and it will be shown that natural partial realizations exist if and only if they are isomorphic.

Definition 3.23. For a 2-Commutative Linear Representation System $\sigma = ((X, F_\alpha, F_\beta), x^0, h)$ and some $l_1, m_1 \in N$, if $X = \ll \{F_\alpha^i F_\beta^j x^0; i \leq l_1,\ j \leq m_1\} \gg$, then σ is said to be (l_1, m_1)-quasi-reachable, where $\ll S \gg$ denotes the smallest linear space which contains a set S.

Let l_2, m_2 be some integer. If $hF_\alpha^i F_\beta^j x = 0$ implies $x = 0$ for any $i \leq l_2$ and $j \leq m_2$, then σ is said to be (l_2, m_2)-distinguishable.

For a given $\underline{a} \in F(\mathbf{L}\times\mathbf{M}, Y)$, if there exist l_1, m_1 and $l_2, m_2 \in N$ such that $l_1 + l_2 < L$, $m_1 + m_2 < M$ and σ of its partial realization is (l_1, m_1)-quasi-reachable and (l_2, m_2)-distinguishable, then σ is said to be a natural partial realization of $\underline{a}$.

For a partial finite-sized image $\underline{a} \in F(\mathbf{L}\times\mathbf{M}, Y)$, the following matrix, $H_{\underline{a}\,(l_1,m_1,L-l_1,M-m_1)}$, is said to be a finite-sized Hankel matrix of $\underline{a}$.

$$H_{\underline{a}\,(l_1,m_1,L-l_1,M-m_1)} = \begin{array}{r} \\ \\ \\ \\ (l,m) \end{array} \left(\begin{array}{ccc} & & (i,j) \\ & & \vdots \\ & & \vdots \\ & & \vdots \\ \cdots & \cdots & \underline{a}(i+l, j+m) \end{array} \right),$$

where $0 \leq i \leq l_1,\ 0 \leq j \leq m_1,\ 0 \leq l \leq L - l_1,\ 0 \leq m \leq M - m_1$.

Theorem 3.24. Let $H_{\underline{a}\ (l_1,m_1,L-l_1,M-m_1)}$ be the finite Hankel matrix of $\underline{a} \in F(\mathbf{L}\times\mathbf{M},Y)$. Then there exists a natural partial realization of $\underline{a}$ if and only if the following conditions hold:

$$\begin{aligned}&\text{rank } H_{\underline{a}\ (l_1,m_1,L-l_1,M-m_1)}\\ &= \text{rank } H_{\underline{a}\ (l_1+1,m_1,L-l_1-1,M-m_1-1)}\\ &= \text{rank } H_{\underline{a}\ (l_1,m_1+1,L-l_1,M-m_1-1)}\\ &= \text{rank } H_{\underline{a}\ (l_1,m_1+1,L-l_1-1,M-m_1-1)}\\ &= \text{rank } H_{\underline{a}\ (l_1,m_1,L-l_1-1,M-m_1-1)} \qquad \text{for some } l_1 \in \mathbf{L}, m_1 \in \mathbf{M}.\end{aligned}$$

Proof. See (3-C.9) in Appendix 3-C.

Theorem 3.25. There exists a natural partial realization of a given partial finite-sized image $\underline{a} \in F(\mathbf{L}\times\mathbf{M},Y)$ if and only if the minimal partial realizations of $\underline{a}$ are unique modulo isomorphisms.

Proof. See (3-C.11) in Appendix 3-C.

In order to discuss the partial realization problem for finite-sized two-dimensional images, define the following operators $\underline{S_\alpha}$ and $\underline{S_\beta}$ as:

$$\begin{aligned}&\underline{S_\alpha} : F(\mathbf{L}\times\mathbf{M},Y) \to F((\mathbf{L}-\mathbf{1})\times\mathbf{M},Y); \underline{a} \mapsto \underline{S_\alpha}\underline{a}\ [;(i,j)\mapsto \underline{a}(i+1,j)],\\ &\underline{S_\beta} : F(\mathbf{L}\times\mathbf{M},Y) \to F(\mathbf{L}\times(\mathbf{M}-\mathbf{1}),Y); \underline{a} \mapsto \underline{S_\beta}\underline{a}\ [;(i,j)\mapsto \underline{a}(i,j+1)].\end{aligned}$$

Then the column vectors in $H_{\underline{a}\ (l_1,m_1,L-l_1,M-m_1)}$ are expressed as $\underline{S_\alpha^i}\underline{S_\beta^j}\underline{a}$ for $0 \le i \le l_1$ and $0 \le j \le m_1$.

Theorem 3.26. Let a partial finite-sized image be $\underline{a} \in F(\mathbf{L}\times\mathbf{M},Y)$.
There exists a natural partial realization of $\underline{a}$ if and only if the Quasi-reachable Standard System $\sigma_s = ((K^n, F_{\alpha s}, F_{\beta s}), \mathbf{e}_1, h_s)$ which realizes $\underline{a}$ can be obtained by the following algorithm.
Here, n is given by $n := \sum_{i=1}^{k} \nu_i$.

1) Check the independences of column vectors of the finite-sized Hankel matrices $H_{\underline{a}\ (0,0,L,M)}$, $H_{\underline{a}\ (1,0,L-1,M)}$, $H_{\underline{a}\ (2,0,L-2,M)}, \cdots$, in turn.
   Find a smallest integer ν_1 such that all column vectors $\{\underline{S_\alpha}^{i-1}\underline{a}; 1 \le i \le \nu_1\}$
   in $F((\mathbf{L}-\nu_1+\mathbf{1})\times\mathbf{M},Y)$ are linearly independent and column vectors $\{\underline{S_\alpha}^{i}\underline{a}; 0 \le i \le \nu_1\}$ in $F((\mathbf{L}-\nu_1)\times\mathbf{M},Y)$ are linearly dependent.
   Determine a set of coefficients $\{c_{1l}^1; 1 \le l \le \nu_1\}$ such that
   $\underline{S_\alpha}^{\nu_1}\underline{a} = \sum_{l=1}^{\nu_1} c_{1l}^1 \underline{S_\alpha}^{l-1}\underline{a}$ holds in the sense of $F((\mathbf{L}-\nu_1)\times\mathbf{M},Y)$.

2) If column vectors $\{\underline{S_\alpha}^{i-1}\underline{a}; 1 \leq i \leq \nu_1\}$ in $F((\mathbf{L}-\nu_1+\mathbf{1})\times(\mathbf{M}-\mathbf{1}), Y)$ in finite-sized Hankel matrix $H_{\underline{a}\ (\nu_1-1,1,L-\nu_1+1,M-1)}$ are linearly dependent, then stop this algorithm.
Otherwise, find the smallest integer ν_2 such that column vectors
$\{\underline{S_\beta}^{m-1}\underline{S_\alpha}^{l-1}\underline{a} \in F((\mathbf{L}-\nu_1+\mathbf{1})\times(\mathbf{M}-\mathbf{1}), Y); 1 \leq m \leq 2,\ 1 \leq l \leq \nu_m\}$
in finite-sized Hankel matrix $H_{\underline{a}\ (\nu_1-1,1,L-\nu_1+1,M-1)}$ are linearly independent and column vectors
$\{\underline{S_\beta}^{m-1}\underline{S_\alpha}^{l-1}\underline{a},\ \underline{S_\beta}\underline{S_\alpha}^{\nu_2}\underline{a} \in F((\mathbf{L}-\nu_1)\times(\mathbf{M}-\mathbf{1}), Y); 1 \leq m \leq 2,\ 1 \leq l \leq \nu_m\}$ in finite-sized Hankel matrix $H_{\underline{a}\ (\nu_1,1,L-\nu_1,M-1)}$ are linearly dependent.
Determine a set of coefficients $\{c^2_{ml}; 1 \leq m \leq 2,\ 1 \leq l \leq \nu_m\}$ such that $\underline{S_\beta}\underline{S_\alpha}^{\nu_2}\underline{a} = \sum_{m=1}^{2}\sum_{l=1}^{\nu_m} c^2_{ml}\underline{S_\beta}^{m-1}\underline{S_\alpha}^{l-1}\underline{a}$ holds in the sense of
$F((\mathbf{L}-\nu_1)\times(\mathbf{M}-\mathbf{1}), Y)$.

⋮

k) If column vectors $\{\underline{S_\beta}^{m-1}\underline{S_\alpha}^{l-1}\underline{a} \in F((\mathbf{L}-\nu_1+\mathbf{1})\times(\mathbf{M}-\mathbf{k}+\mathbf{1}), Y); 1 \leq m \leq k,\ 1 \leq l \leq \nu_m\}$ in finite-sized Hankel matrix $H_{\underline{a}\ (\nu_1-1,k-1,L-\nu_1+1,M-k+1)}$ are linearly dependent, then stop this algorithm.
Otherwise find the smallest integer ν_k such that column vectors
$\{\underline{S_\beta}^{m-1}\underline{S_\alpha}^{l-1}\underline{a} \in F((\mathbf{L}-\nu_1+\mathbf{1})\times(\mathbf{M}-\mathbf{k}+\mathbf{1}), Y); 1 \leq m \leq k,\ 1 \leq l \leq \nu_m\}$
in finite-sized Hankel matrix $H_{\underline{a}\ (\nu_1-1,k-1,L-\nu_1+1,M-k+1)}$ are linearly independent and column vectors
$\{\underline{S_\beta}^{m-1}\underline{S_\alpha}^{l-1}\underline{a},\ \underline{S_\beta}^{k-1}\underline{S_\alpha}^{\nu_k}\underline{a} \in F((\mathbf{L}-\nu_1)\times(\mathbf{M}-\mathbf{k}+\mathbf{1}), Y); 1 \leq m \leq k,\ 1 \leq l \leq \nu_m\}$ in finite-sized Hankel matrix $H_{\underline{a}\ (\nu_1-1,k-1,L-\nu_1,M-k+1)}$ are linearly dependent.
Determine a set of coefficients $\{c^k_{ml}; 1 \leq m \leq k,\ 1 \leq l \leq \nu_m\}$ such that $\underline{S_\beta}^{k-1}\underline{S_\alpha}^{\nu_k}\underline{a} = \sum_{m=1}^{k}\sum_{l=l}^{\nu_m} c^k_{ml}\underline{S_\beta}^{m-1}\underline{S_\alpha}^{l-1}\underline{a}$ holds in the sense of
$F((\mathbf{L}-\nu_1)\times(\mathbf{M}-\mathbf{k}+\mathbf{1}), Y)$.

k+1) If column vectors $\{\underline{S_\beta}^{m-1}\underline{S_\alpha}^{l-1}\underline{a} \in F((\mathbf{L}-\nu_1+\mathbf{1})\times(\mathbf{M}-\mathbf{k}), Y); 1 \leq m \leq k,\ 1 \leq l \leq \nu_m\}$ in finite-sized Hankel matrix $H_{\underline{a}\ (\nu_1-1,k,L-\nu_1+1,M-k)}$ are linearly dependent, then stop this algorithm.
Otherwise find the smallest integer k such that column vectors
$\{\underline{S_\beta}^{k}\underline{a},\ \underline{S_\beta}^{m-1}\underline{S_\alpha}^{l-1}\underline{a} \in F((\mathbf{L}-\nu_1+\mathbf{1})\times(\mathbf{M}-\mathbf{k}), Y); 1 \leq m \leq k,\ 1 \leq l \leq \nu_m\}$ in finite-sized Hankel matrix $H_{\underline{a}\ (\nu_1-1,k-1,L-\nu_1+1,M-k)}$ are linearly dependent.
Determine a set of coefficients $\{c^{k+1}_{ml}; 1 \leq m \leq k,\ 1 \leq l \leq \nu_m\}$ such that $\underline{S_\beta}^{k}\underline{a} = \sum_{m=1}^{k}\sum_{l=1}^{\nu_m} c^{k+1}_{ml}\underline{S_\beta}^{m-1}\underline{S_\alpha}^{l-1}\underline{a}$ holds in the sense of
$F((\mathbf{L}-\nu_1+\mathbf{1})\times(\mathbf{M}-\mathbf{k}), Y)$.

k+2) For a set of obtained coefficients $\{c_{ml}^{j} \in K; 1 \le m \le k,\ 1 \le l \le \nu_m\}$ for $1 \le j \le k+1$, set $\mathbf{c}_i^j = [c_{i1}^j, c_{i2}^j, \cdots, c_{i\nu_i}^j]^T \in K^{\nu_i}$ $(1 \le i \le k)$, $\mathbf{c}^j = [\mathbf{c}_1^j, \mathbf{c}_2^j, \cdots, \mathbf{c}_k^j, 0, 0, \cdots, 0]^T$ and $\mathbf{c}^{k+1} = [\mathbf{c}_1^{k+1}, \mathbf{c}_2^{k+1}, \cdots, \mathbf{c}_k^{k+1}]^T$.

k+3) Insert the vectors $\mathbf{c}^j$ $(1 \le j \le k+1)$ obtained in the k+2) step into the $F_{\alpha s}$ and $F_{\beta s}$ of Figure 3.3 and Figure 3.4.

k+4) Let $h_s = [\underline{a}(0,0), \underline{a}(1,0), \cdots, \underline{a}(\nu_1 - 1, 0), \underline{a}(0,1), \cdots, \underline{a}(\nu_2 - 1, 1), \cdots, \underline{a}(0, k-1), \cdots, \underline{a}(\nu_k - 1, k-1)]$.

Proof. When this algorithm can be executed, note that
$n :=$ rank $H_{\underline{a}\ (\nu_1-1,k-1,L-\nu_1+1,M-k+1)} =$ rank $H_{\underline{a}\ (\nu_1,k-1,L-\nu_1,M-k+1)}$
$=$ rank $H_{\underline{a}\ (\nu_1-1,k,L-\nu_1+1,M-k)} =$ rank $H_{\underline{a}\ (\nu_1,k-1,L-\nu_1,M-k)}$ hold.

For the details of this proof, see (3-C.12) in Appendix 3-C.

Example 3.27. A two-dimensional canonical 2-Commutative Linear Representation System.

Following is the derivation of the minimal 2-Commutative Linear Representation System which partially realizes the geometrical image depicted in Figure 3.7 by using the proposed algorithm. First, we encode the geometrical pattern depicted in Figure 3.7 using Figure 3.8.

Then let K be $N/2N$ which is the quotient field modulo the prime number 2, and let the set Y of output values be K. Figure 3.7 is encoded in the following two dimensional array:

$$
\begin{matrix}
0&1&0&1&0&1&0&1&0&1\\
1&0&1&0&1&0&1&0&1&0\\
0&1&0&1&0&1&0&1&0&1\\
1&0&1&0&1&0&1&0&1&0\\
0&1&0&1&0&1&0&1&0&1\\
1&0&1&0&1&0&1&0&1&0\\
0&1&0&1&0&1&0&1&0&1\\
1&0&1&0&1&0&1&0&1&0\\
0&1&0&1&0&1&0&1&0&1\\
1&0&1&0&1&0&1&0&1&0
\end{matrix}
$$

$\nu_1 = 2$ and the coefficients $\{c_{11}^1 = 1, c_{12}^1 = 0\}$ can be obtained by applying the proposed algorithm to the finite-sized Hankel matrix $H_{\underline{a}\ (2,0,3,5)}$. Similarly, $\nu_2 = 0$ and the coefficients $\{c_{11}^2 = 0, c_{12}^2 = 1\}$ can be obtained from the finite-sized Hankel matrix $H_{\underline{a}\ (2,1,3,4)}$. These Hankel matrices are shown just below:

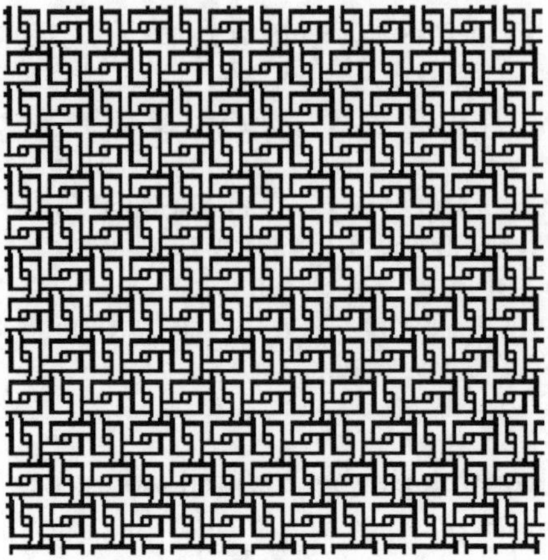

Fig. 3.7. The geometrical image for Example (3.27)

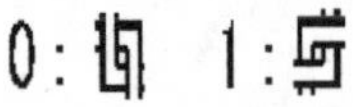

Fig. 3.8. The encoding chart for Example (3.27)

$$H_{\underline{a}\ (2,0,3,5)} = \begin{bmatrix} 0\,1\,0 \\ 1\,0\,1 \\ 1\,0\,1 \\ 0\,1\,0 \\ 0\,1\,0 \\ 0\,1\,0 \\ 1\,0\,1 \\ 1\,0\,1 \\ 1\,0\,1 \\ 1\,0\,1 \\ 0\,1\,0 \\ 0\,1\,0 \\ 0\,1\,0 \\ 0\,1\,0 \\ 0\,1\,0 \\ 1\,0\,1 \\ 1\,0\,1 \\ 1\,0\,1 \\ 1\,0\,1 \\ 1\,0\,1 \\ 1\,0\,1 \end{bmatrix}, \quad H_{\underline{a}\ (2,1,3,4)} = \begin{bmatrix} 0\,1\,1 \\ 1\,0\,0 \\ 1\,0\,0 \\ 0\,1\,1 \\ 0\,1\,1 \\ 0\,1\,1 \\ 1\,0\,0 \\ 1\,0\,0 \\ 1\,0\,0 \\ 1\,0\,0 \\ 0\,1\,1 \\ 0\,1\,1 \\ 0\,1\,1 \\ 0\,1\,1 \\ 0\,1\,1 \\ 1\,0\,0 \\ 1\,0\,0 \\ 1\,0\,0 \\ 1\,0\,0 \\ 1\,0\,0 \\ 1\,0\,0 \end{bmatrix}.$$

This leads to the 2-Commutative Linear Representaion System $\sigma = ((K^2, F_{\alpha s}, F_{\beta s}), \mathbf{e}_1, h_s)$ as follows:

$$F_{\alpha s} = \begin{bmatrix} 0 & 1 \\ 1 & 0 \end{bmatrix}, F_{\beta s} = \begin{bmatrix} 0 & 1 \\ 1 & 0 \end{bmatrix}, \mathbf{e}_1 = \begin{bmatrix} 1 \\ 0 \end{bmatrix}, h_s = \begin{bmatrix} 0 & 1 \end{bmatrix}.$$

Example 3.28. A five-dimensional canonical 2-Commutative Linear Representation System.

Next, we observe the derivation of the minimal 2-Commutative Linear Representation System which partially realizes the geometrical image depicted in Figure 3.9 by means of the proposed algorithm. First, the geometrical pattern depicted in Figure 3.9 is encoded as we see in Figure 3.10

Fig. 3.9. The geometrical image for Example (3.28)

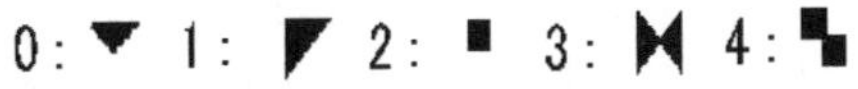

Fig. 3.10. The encoding chart for Example (3.28)

Then let K be $N/5N$ which is the quotient field modulo the prime number 5, and let the set Y of output values be K. Figure 3.9 is encoded in the following two dimensional array:

$$
\begin{array}{c}
3\,2\,1\,4\,0\,3\,2\,1\,4\,0 \\
1\,4\,0\,3\,2\,1\,4\,0\,3\,2 \\
0\,3\,2\,1\,4\,0\,3\,2\,1\,4 \\
2\,1\,4\,0\,3\,2\,1\,4\,0\,3 \\
4\,0\,3\,2\,1\,4\,0\,3\,2\,1 \\
3\,2\,1\,4\,0\,3\,2\,1\,4\,0 \\
1\,4\,0\,3\,2\,1\,4\,0\,3\,2 \\
0\,3\,2\,1\,4\,0\,3\,2\,1\,4 \\
2\,1\,4\,0\,3\,2\,1\,4\,0\,3 \\
4\,0\,3\,2\,1\,4\,0\,3\,2\,1
\end{array}
$$

$\nu_1 = 5$ and the coefficients $\{c^1_{11} = 1, c^1_{12} = 0, c^1_{13} = 0, c^1_{14} = 0, c^1_{15} = 0\}$ can be found by applying the proposed algorithm to the finite-sized Hankel matrix $H_{\underline{a}\ (5,0,0,5)}$. Similarly, $\nu_2 = 0$, $k = 1$ and the coefficients $\{c^2_{11} = 0, c^2_{12} = 0, c^2_{13} = 0, c^2_{14} = 1, c^2_{15} = 0\}$ can be obtained from the finite-sized Hankel matrix $H_{\underline{a}\ (5,1,0,4)}$. Both Hankel matrices are shown below.

$$
H_{\underline{a}\ (5,0,0,5)} = \begin{bmatrix}
3\,1\,0\,2\,4\,3 \\
1\,0\,2\,4\,3\,1 \\
2\,4\,3\,1\,0\,2 \\
0\,2\,4\,3\,1\,0 \\
4\,3\,1\,0\,2\,4 \\
1\,0\,2\,4\,3\,1 \\
2\,4\,3\,1\,0\,2 \\
3\,1\,0\,2\,4\,3 \\
0\,2\,4\,3\,1\,0 \\
4\,3\,1\,0\,2\,4 \\
4\,3\,1\,0\,2\,4 \\
1\,0\,2\,4\,3\,1 \\
2\,4\,3\,1\,0\,2 \\
3\,1\,0\,2\,4\,3 \\
0\,2\,4\,3\,1\,0 \\
3\,1\,0\,2\,4\,3 \\
0\,2\,4\,3\,1\,0 \\
4\,3\,1\,0\,2\,4 \\
1\,0\,2\,4\,3\,1 \\
2\,4\,3\,1\,0\,2 \\
3\,1\,0\,2\,4\,3
\end{bmatrix}, \quad
H_{\underline{a}\ (5,1,0,4)} = \begin{bmatrix}
3\,1\,0\,2\,4\,2 \\
1\,0\,2\,4\,3\,4 \\
2\,4\,3\,1\,0\,1 \\
0\,2\,4\,3\,1\,3 \\
4\,3\,1\,0\,2\,0 \\
1\,0\,2\,4\,3\,4 \\
2\,4\,3\,1\,0\,1 \\
3\,1\,0\,2\,4\,2 \\
0\,2\,4\,3\,1\,3 \\
4\,3\,1\,0\,2\,0 \\
4\,3\,1\,0\,2\,0 \\
1\,0\,2\,4\,3\,4 \\
2\,4\,3\,1\,0\,1 \\
3\,1\,0\,2\,4\,2 \\
0\,2\,4\,3\,1\,3 \\
3\,1\,0\,2\,4\,2 \\
0\,2\,4\,3\,1\,3 \\
4\,3\,1\,0\,2\,0 \\
1\,0\,2\,4\,3\,4 \\
2\,4\,3\,1\,0\,1 \\
3\,1\,0\,2\,4\,2
\end{bmatrix}
$$

This leads to the 2-Commutative Linear Representaion System $\sigma = ((K^5, F_{\alpha_s}, F_{\beta s}), \mathbf{e}_1, h_s)$ as follows:

$$F_{\alpha s} = \begin{bmatrix} 0\,0\,0\,0\,1 \\ 1\,0\,0\,0\,0 \\ 0\,1\,0\,0\,0 \\ 0\,0\,1\,0\,0 \\ 0\,0\,0\,1\,0 \end{bmatrix}, F_{\beta s} = \begin{bmatrix} 0\,0\,1\,0\,0 \\ 0\,0\,0\,1\,0 \\ 0\,0\,0\,0\,1 \\ 1\,0\,0\,1\,0 \\ 0\,1\,0\,0\,0 \end{bmatrix}, \mathbf{e}_1 = \begin{bmatrix} 1 \\ 0 \\ 0 \\ 0 \\ 0 \end{bmatrix}, h_s = \begin{bmatrix} 3\,1\,0\,2\,4 \end{bmatrix}.$$

Examples (3.27) and (3.28) demonstrate that the dimension of the state space of the 2-Commutative Linear Representation System is affected by the period of the image, that is, if the period of the image is small, the dimension of the state space is small. We emphasize that reploducing the given image implies the calculation of the behavior of the 2-Commutative Linear Representation System.

Example 3.29. A four-dimensional canonical 2-Commutative Linear Representation System.

This example shows the derivation of the minimal order 2-Commutative Linear Representation System which partially realizes the geometrical image depicted in Figure by means of the proposed algorithm. First, Figure 3.11 is encoded as we see in Figure .

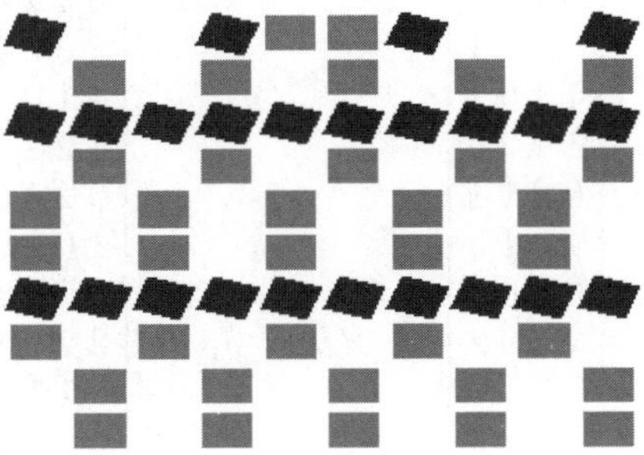

Fig. 3.11. The geometrical image for Example (3.29)

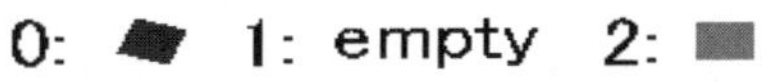

Fig. 3.12. The encoding chart for Example (3.29)

Then let K be $N/3N$ which is the quotient field modulo the prime number 3, and let the set Y of output values be K. This yields the following two dimensional array for Figure 3.11:

$$
\begin{array}{c}
0\,1\,1\,0\,2\,2\,0\,1\,1\,0\\
1\,2\,1\,2\,1\,2\,1\,2\,1\,2\\
0\,0\,0\,0\,0\,0\,0\,0\,0\,0\\
1\,2\,1\,2\,1\,2\,1\,2\,1\,2\\
2\,1\,2\,1\,2\,1\,2\,1\,2\,1\\
2\,1\,2\,1\,2\,1\,2\,1\,2\,1\\
0\,0\,0\,0\,0\,0\,0\,0\,0\,0\\
2\,1\,2\,1\,2\,1\,2\,1\,2\,1\\
1\,2\,1\,2\,1\,2\,1\,2\,1\,2\\
1\,2\,1\,2\,1\,2\,1\,2\,1\,2
\end{array}
$$

$\nu_1 = 3$ and the coefficients $\{c_{11}^1 = 0, c_{12}^1 = 1, c_{13}^1 = 2\}$ can be found by applying the proposed algorithm to the finite-sized Hankel matrix $H_{\underline{a}\ (3,0,2,5)}$. Similarly, $\nu_2 = 1$ and the coefficients $\{c_{11}^2 = 0, c_{12}^2 = 2, c_{13}^2 = 0, c_{14}^2 = 0\}$ can be found from the finite-sized Hankel matrix $H_{\underline{a}\ (3,1,2,4)}$. Moreover, $k = 2$ and the coefficients $\{c_{11}^3 = 2, c_{12}^3 = 0, c_{13}^3 = 0, c_{14}^3 = 1\}$ can be obtained from the finite-sized Hankel matrix $H_{\underline{a}\ (2,1,3,4)}$. These Hankel matrices appear below:

$$
H_{\underline{a}\ (3,0,2,5)} = \begin{bmatrix}
0\,1\,0\,1\\
1\,0\,1\,2\\
1\,2\,0\,2\\
0\,1\,2\,2\\
2\,0\,2\,1\\
1\,1\,0\,1\\
1\,2\,2\,0\\
0\,2\,1\,1\\
1\,0\,1\,2\\
0\,2\,0\,2\\
2\,2\,0\,2\\
2\,1\,1\,0\\
0\,1\,2\,2\\
2\,0\,2\,1\\
2\,1\,0\,1\\
2\,0\,2\,1\\
1\,1\,0\,1\\
1\,2\,2\,0\\
0\,2\,1\,1\\
1\,0\,1\,2\\
2\,2\,0\,2
\end{bmatrix}, \quad
H_{\underline{a}\ (3,1,2,4)} = \begin{bmatrix}
0\,1\,0\,1\,2\\
1\,0\,1\,2\,0\\
1\,2\,0\,1\,1\\
0\,1\,2\,0\,2\\
2\,0\,2\,1\,0\\
1\,1\,0\,0\,2\\
1\,2\,2\,2\,1\\
0\,2\,1\,0\,1\\
1\,0\,1\,2\,0\\
0\,2\,0\,2\,1\\
2\,2\,0\,1\,1\\
2\,1\,1\,1\,2\\
0\,1\,2\,0\,2\\
2\,0\,2\,1\,0\\
2\,1\,0\,2\,2\\
2\,0\,2\,1\,0\\
1\,1\,0\,2\,2\\
1\,2\,2\,2\,1\\
0\,2\,1\,0\,1\\
1\,0\,1\,2\,0\\
2\,2\,0\,0\,1
\end{bmatrix}
$$

$$H_{\underline{a}\ (2,1,3,4)} = \begin{bmatrix} 0\,1\,0\,1\,1 \\ 1\,0\,1\,2\,1 \\ 1\,2\,0\,1\,0 \\ 0\,1\,2\,0\,0 \\ 2\,0\,2\,1\,2 \\ 1\,1\,0\,0\,2 \\ 1\,2\,2\,2\,1 \\ 0\,2\,1\,0\,0 \\ 1\,0\,1\,2\,1 \\ 0\,2\,0\,2\,2 \\ 2\,2\,0\,1\,2 \\ 2\,1\,1\,1\,2 \\ 0\,1\,2\,0\,0 \\ 2\,0\,2\,1\,2 \\ 2\,1\,0\,2\,0 \\ 2\,0\,2\,1\,2 \\ 1\,1\,0\,2\,1 \\ 1\,2\,2\,2\,1 \\ 0\,2\,1\,0\,0 \\ 1\,0\,1\,2\,1 \\ 2\,2\,0\,0\,1 \end{bmatrix}$$

The following 2-Commutative Linear Representation System $\sigma = ((K^4, F_{\alpha_s}, F_{\beta s}), \mathbf{e}_1, h_s)$ can be obtained:

$$F_{\alpha s} = \begin{bmatrix} 0\,0\,0\,0 \\ 1\,0\,1\,2 \\ 0\,1\,2\,0 \\ 0\,0\,0\,0 \end{bmatrix}, F_{\beta s} = \begin{bmatrix} 0\,0\,0\,2 \\ 0\,2\,0\,0 \\ 0\,0\,2\,0 \\ 1\,0\,0\,1 \end{bmatrix}, \mathbf{e}_1 = \begin{bmatrix} 1 \\ 0 \\ 0 \\ 0 \end{bmatrix}, h_s = \begin{bmatrix} 0\,1\,0\,1 \end{bmatrix}.$$

3.5 Historical Notes and Concluding Remarks

Clearly, there may already be notions of 2-Commutative Linear Representation Systems in Linear Representation Systems, which came from the ideas of Sussmann [1976, 1977]. we also remember that homogeneous bilinear systems and $K-U$ automaton are a sort of Linear Representation Systems. See Tarn and Nonoyama [1976] and Fliess [1978] for discrete-time homogeneous bilinear systems. See also Brockett [1976] for continuous-time systems and Paz [1966] for probabilistic automaton. Schuzenberger [1961] considered generalized automata, which are called the $K-U$ automata by Eilenberg [1974]. See also Matsuo and Hasegawa [2003].
The set $A(\Omega)$ in Linear Representation Systems may be equivalent to the algebra of polynomial in non-commutative variable introduced by Fliess [1970] and [1974]. $K[z_\alpha, z_\beta]$ in Example (3.2) is an algebra of polynomial in commutative variables which is a special case of $A(\Omega)$.

Nerode equivalence for $K[z_\alpha, z_\beta]$ in Theorem (3.3) is a new result. Note that Nerode [1958] proposed the so-called Nerode equivalence for automata with linear input/output maps.
Kalman, et al. [1969] treated the state space of discrete-time linear systems as $K[z]$-modules. Our state space of 2-Commutative Linear Representation Systems as $K[z_\alpha, z_\beta]$ are a natural extension of Kalman's idea. Fornasini and Marchesini [1976] intended to treat the state space of 2-D systems as $K[z_\alpha, z_\beta]$-modules, however their treatment is unnatural and incomplete.

It is shown that the uniqueness Theorem (3.5) holds in the sense of 2-Commutative Linear Representation Systems, namely, the theorem is stronger than in the sense of Linear Representation Systems including commutativity.

Theorem (3.13) states the condition for canonicality of finite-dimensional 2-Commutative Linear Representation Systems. It can be easily understood that this theorem is an extension of the theorem establishing condition for canonicality of finite-dimensional linear systems.

We gave the Quasi-reachable Standard System and Distinguishable Standard System that correspond to companion forms of linear systems.

We gave the two criteria for the behavior of finite-dimensional Linear Representation Systems. See the Remark in Theorems (3.19) and (3.20) regarding this.

We also described a realization procedure to obtain the Quasi-reachable Standard System from a given image.

In discrete-time case, Isidori [1973] gave only a sufficient condition for uniqueness and an algorithm for inhomogeneous bilinear systems.

In regard to other kinds of nonlinear systems, Sontag [1979] gave a procedure to obtain realization from a given input/output map, but the procedure is less clear than ours. The partial Realization Theorems (3.24) and (3.25) which we obtained in this recent study have not been obtained before, and the partial realization algorithm shown in Theorem (3.26) is completely new.

Thus far, the linear system theories based on the state space method have been well established. Speaking bravely, we can say that these developments in view of approximations depend on the following three fields.

(1) Modeling and control
    [Davison, 1966], [Moore, 1984], [Glover,1984],
    [Doyle, Glover, Kargonekar and Francis, 1989],

(2) Signal processing
    [Millis and Roberts, 1976], [Hassibi, Sayed and Kailath, 1996], and

(3) Related mathematics
    [Prony, 1795], [Pade, 1892], [Eckart and Young, 1936], [Mirsky, 1960], [Caratheodory and Fejer, 1911], [Adamjan, Arov and Krein,1971,1978].

To the best of our knowledge, the treatments including approximation or compression of two-dimensional image have not been established yet because mathematical models have not existed.

From this point of view, we think that our new treatment for two-dimensional images in this chapter are highly significant.

Appendix to Chapter 3

This appendix furnishes the detailed proofs for assertions made in Chapter 3 concerning 2-Commutative Linear Representation Systems.

3-A Realization Theorem

Proof of the Realization Theorem (3.5) for 2-Commutative Linear Representation Systems is provided in this section. To prove the theorem, we equivalently convert 2-Commutative Linear Representation Systems to sophisticated 2-Commutative Linear Representation Systems by virtue of the result of this appendix. Proving the realization theorem in the sophisticated 2-Commutative Linear Representation Systems implies proving Theorem (3.5).

3-A.1 Linear State Structure: $\{\alpha, \beta\}$-Actions

Definition 3-A.1. A system given by the following equations is written as a pair (X, F_α, F_β) and it is said to be a $\{\alpha, \beta\}$-action.

$$\begin{cases} x(i+1, j) = F_\alpha x(i, j) \\ x(i, j+1) = F_\beta x(i, j), \end{cases}$$

where X is a linear space over the field K. F_α and F_β are linear operators on X which satisfy $F_\alpha F_\beta = F_\beta F_\alpha$.

Let $(X_1, F_{\alpha_1}, F_{\beta_1})$ and $(X_2, F_{\alpha_2}, F_{\beta_2})$ be $\{\alpha, \beta\}$-actions. Then a linear map $T : X_1 \rightarrow X_2$ is said to be a $\{\alpha, \beta\}$-morphism $T : (X_1, F_{\alpha_1}, F_{\beta_1}) \rightarrow (X_2, F_{\alpha_2}, F_{\beta_2})$ if T satisfies $TF_{\alpha_1} = F_{\alpha_2}T$ and $TF_{\beta_1} = F_{\beta_2}T$.

Example 3-A.2. Let $K[z_\alpha, z_\beta]$, z_α and z_β be the same as in Example (3.2). Then $(K[z_\alpha, z_\beta], z_\alpha, z_\beta)$ is a $\{\alpha, \beta\}$-action.

Example 3-A.3. In the set $F(N \times N, Y)$ of any two-dimensional images, let S_α and S_β be the same as in Example (3.2). Then $(F(N \times N, Y), S_\alpha, S_\beta)$ is a $\{\alpha, \beta\}$-action.

Definition 3-A.4. For the $\{\alpha, \beta\}$-actions $(K[z_\alpha, z_\beta], z_\alpha, z_\beta)$ and $(F(N \times N, Y), S_\alpha, S_\beta)$ considered in Examples (3-A.2) and (3-A.3), a $\{\alpha, \beta\}$-morphism $A : (K[z_\alpha, z_\beta], z_\alpha, z_\beta) \to (F(N \times N, Y), S_\alpha, S_\beta)$ is called a linear input/output map.
For a $\{\alpha, \beta\}$-action (X, F_α, F_β), a $\{\alpha, \beta\}$-morphism $G : (K[z_\alpha, z_\beta], z_\alpha, z_\beta) \to (X, F_\alpha, F_\beta)$ is called a linear input map, and a $\{\alpha, \beta\}$-morphism $H : (X, F_\alpha, F_\beta) \to (F(N \times N, Y), S_\alpha, S_\beta)$ is called a linear observation map.

Remark: A linear input/output map $A : (K[z_\alpha, z_\beta], z_\alpha, z_\beta) \to (F(N \times N, Y), S_\alpha, S_\beta)$ is different from the map discussed in Kalman [1968]. He discussed the bi-linear operator $A : K[z_\alpha] \times K[z_\beta] \to K[z^{-1}]$.

Next, we introduce $K[z_\alpha, z_\beta]$-modules. Then we will show the connection between $\{\alpha, \beta\}$-actions and $K[z_\alpha, z_\beta]$-modules.

Recall that $N \times N$ is a commutative monoid with a unit (0, 0) and the operation

$$+ : (N \times N) \times (N \times N) \to N \times N; ((i, j), (l, m)) \mapsto (i + l, j + m).$$

If a map $\phi : N \times N \to L(X)$ is a monoid morphism, then a pair (X, ϕ) is called a $N \times N$-module, where $L(X)$ is a monoid with a composition of operation in $L(X)$. Let (X_1, ϕ_1) and (X_2, ϕ_2) be $N \times N$-modules. If a linear operator $T : X_1 \to X_2$ satisfies $T\phi_1(i, j) = \phi_2(i, j)T$, then T is called an $N \times N$-morphism : $(X_1, \phi_1) \to (X_2, \phi_2)$. Note that $N \times N$ is a free monoid over the set $\{\alpha, \beta\}$.

Definition 3-A.5. Let X be a linear space over the field K and let $\tilde{\phi} : K[z_\alpha, z_\beta] \to L(X)$ be an algebra morphism, i.e., $\tilde{\phi}(\lambda_1 \cdot \lambda_2) = \tilde{\phi}(\lambda_1) \cdot \tilde{\phi}(\lambda_2)$ and $\tilde{\phi}(\mathbf{1}) = I$ hold for any $\lambda_1,\ \lambda_2 \in K[z_\alpha, z_\beta]$. Then the pair $(X, \tilde{\phi})$ is called a $K[z_\alpha, z_\beta]$-module.

Let $(X_1, \tilde{\phi}_1)$ and $(X_2, \tilde{\phi}_2)$ be $K[z_\alpha, z_\beta]$-modules.

If a linear operator $T : X_1 \to X_2$ satisfies $T\tilde{\phi}_1(\lambda) = \tilde{\phi}_2(\lambda)T$ for any $\lambda \in K[z_\alpha, z_\beta]$, T is called a $K[z_\alpha, z_\beta]$-morphism $T : (X_1, \tilde{\phi}_1) \to (X_2, \tilde{\phi}_2)$.

Let's introduce the map

$$\begin{aligned} \times : K[z_\alpha, z_\beta] \times K[z_\alpha, z_\beta] &\to K[z_\alpha, z_\beta]; \\ \left(\textstyle\sum_{i_1,j_1} \lambda_1(i_1, j_1) z_\alpha^{i_1} z_\beta^{j_1}, \sum_{i_2,j_2} \lambda_2(i_2, j_2) z_\alpha^{i_2} z_\beta^{j_2}\right) \\ \mapsto \left(\textstyle\sum_{i_1,j_1} \lambda_1(i_1, j_1) z_\alpha^{i_1} z_\beta^{j_1}\right) &\times \left(\textstyle\sum_{i_2,j_2} \lambda_2(i_2, j_2) z_\alpha^{i_2} z_\beta^{j_2}\right) \\ &= \textstyle\sum_{i=i_1+i_2,\ j=j_1+j_2} \lambda(i, j) z_\alpha^i z_\beta^j \end{aligned}$$

Then $K[z_\alpha, z_\beta]$ is an algebra and the free algebra over the $N \times N$ which is a free monoid over $\{\alpha, \beta\}$.

The following lemma is important for the free algebra.

Lemma 3-A.6. Let A be any algebra, and let a map $e : N \times N \to K[z_\alpha, z_\beta]$ be ; $(i, j) \mapsto z_\alpha^i z_\beta^j$. For any monoid morphism $f : N \times N \to A$, there uniquely exists $\tilde{f} : K[z_\alpha, z_\beta] \to A$ such that $f = \tilde{f} \cdot e$. Moreover, $\tilde{f}(\lambda) = \tilde{f}(\sum_{i,j} \lambda(i,j) z_\alpha^i z_\beta^j)$ holds for any $\lambda = \sum_{i,j} \lambda(i,j) z_\alpha^i z_\beta^j \in K[z_\alpha, z_\beta]$. Conversely, for any algebra morphism $\tilde{f} : K[z_\alpha, z_\beta] \to A$, $f = \tilde{f} \cdot e$ is a monoid morphism : $N \times N \to A$.

Furthermore, there is an injection from the set $\{\alpha, \beta\}$ to the monoid $N \times N$ by a map $i : \{\alpha, \beta\} \to N \times N; \alpha \mapsto (1, 0)$ (or $\beta \mapsto (0, 1)$).

Proposition 3-A.7. Let X be a linear space, let Mor $(\{\alpha, \beta\}, L(X))$ be a set of any monoid morphisms : $\{\alpha, \beta\} \to L(X)$ and let AMor $(K[z_\alpha, z_\beta], L(X))$ be a set of any algebra morphisms : $K[z_\alpha, z_\beta] \to L(X)$.
Then the map $\tilde{}$: Mor $(\{\alpha, \beta\}, L(X)) \to$ AMor $(K[z_\alpha, z_\beta], L(X))$;
$\phi \mapsto \tilde{\phi}$ [; $\lambda \mapsto (\sum_{i,j} \lambda(i,j) \phi(z_\alpha^i z_\beta^j))$] is bijective, where $\{\alpha, \beta\}$ is the set of alphabets α and β.

Proof. $L(X)$ is a linear space and an algebra by the composition of operation. Letting A in Lemma (3-A.6) be $L(X)$ yields this proposition.

Lemma 3-A.8. The $K[z_\alpha, z_\beta]$-module corresponding to $(K[z_\alpha, z_\beta], z_\alpha, z_\beta)$ is $(K[z_\alpha, z_\beta], \times)$, where $(K[z_\alpha, z_\beta], z_\alpha, z_\beta)$ has been considered in Example (3-A.2) and the operation $\times$ was introduced in (3-A.5).

Proof. Clearly, $K[z_\alpha, z_\beta]$ is an algebra. By Lemma (3-A.6), an algebra morphism is given by $\tilde{S}(\lambda) = \lambda \cdot e$ for $\lambda = \sum_{i,j} \lambda(i,j) z_\alpha^i z_\beta^j \in K[z_\alpha, z_\beta]$, where the map $e : N \times N \to L(K[z_\alpha, z_\beta])$ is given as follows:
$e : N \times N \to L(K[z_\alpha, z_\beta]); (i, j) \mapsto z_\alpha^i z_\beta^j$.

Example 3-A.9. For the $\{\alpha, \beta\}$-action $(F(N \times N, Y), S_\alpha, S_\beta)$ considered in (3-A.3), the corresponding $K[z_\alpha, z_\beta]$-module $(F(N \times N, Y), \tilde{S})$ is given by setting $\tilde{S}(\lambda) := \sum_{i,j} \lambda(i,j) S_\alpha^i S_\beta^j$ for $\lambda = \sum_{i,j} \lambda(i,j) z_\alpha^i z_\beta^j \in K[z_\alpha, z_\beta]$.

Definition 3-A.10. Let $(X_1, F_{\alpha_1}, F_{\beta_1})$ [$(X_1, \phi_1), (X_1, \tilde{\phi}_1)$] and $(X_2, F_{\alpha_2}, F_{\beta_2})$ [(X_2, ϕ_2) , $(X_2, \tilde{\phi}_2)$] be $\{\alpha, \beta\}$-actions [$N \times N$-module , $K[z_\alpha, z_\beta]$-module]. The linear map $T : X_1 \to X_2$ is called a $\{\alpha, \beta\}$-morphism $T : (X_1, F_{\alpha_1}, F_{\beta_1}) \to (X_2, F_{\alpha_2}, F_{\beta_2})$ [an $N \times N$-morphism $T : (X_1, \phi_1) \to (X_2, \phi_2)$, $K[z_\alpha, z_\beta]$-morphism $T : (X_1, \tilde{\phi}_1) \to (X_2, \tilde{\phi}_2)$] if T satisfies $TF_{\alpha_1} = F_{\alpha_2} T$ and $TF_{\beta_1} = F_{\beta_2} T$ [$T\phi_1 = \phi_2 T$, $T\tilde{\phi}_1 = \tilde{\phi}_2 T$], where terms in [ ] are corresponding to each other.

Proposition 3-A.11. Let $(X, \tilde{\phi})$ be an $K[z_\alpha, z_\beta]$-module. A map T is an $K[z_\alpha, z_\beta]$-morphism $T : (K[z_\alpha, z_\beta], \times) \to (X, \tilde{\phi})$ if and only if $T(\lambda) = \tilde{\phi}(\lambda)T(\mathbf{1})$ holds for any $\lambda \in K[z_\alpha, z_\beta]$.

Proof. Let T be an $K[z_\alpha, z_\beta]$-morphism, i.e., T satisfies $T(\lambda) = \tilde{\phi}(\lambda)T$. Then $T(\lambda) = T(\lambda \times \mathbf{1}) = \tilde{\phi}(\lambda)T(\mathbf{1})$ holds. Conversely, let $T(\lambda) = \tilde{\phi}(\lambda)T(\mathbf{1})$ holds. Since T is an algebra morphism, $T(\lambda \times \tilde{\lambda}) = \tilde{\phi}(\lambda) \times \tilde{\phi}(\tilde{\lambda})T(\mathbf{1}) = \tilde{\phi}(\lambda)T(\tilde{\lambda})$ holds for any $\lambda, \tilde{\lambda} \in K[z_\alpha, z_\beta]$; i.e., $T(\lambda \cdot e) = \tilde{\phi}(\lambda)T$ holds. Clearly T is a linear operator. Therefore, T is a $K[z_\alpha, z_\beta]$-morphism
$T : (K[z_\alpha, z_\beta], \times) \to (X, \tilde{\phi})$.

Let (X, F_α, F_β) be an $\{\alpha, \beta\}$-action, and let (X, ϕ) be a $N \times N$-module. Then there is a one to one correspondence between (X, F_α, F_β) and (X, ϕ) by setting $\phi(i, 0) = F_\alpha^i$, $\phi(0, j) = F_\beta^j$.

Moreover, let $(X, \tilde{\phi})$ be an $K[z_\alpha, z_\beta]$-module. Then there is a one to one correspondence between (X, F_α, F_β) and $(X, \tilde{\phi})$ by setting $\tilde{\phi}(z_\alpha^i) = F_\alpha^i$ and $\tilde{\phi}(z_\beta^j) = F_\beta^j$.

Proposition 3-A.12. Let $(X_i, F_{\alpha_i}, F_{\beta_i})$ be a $\{\alpha, \beta\}$-action, (X_i, ϕ_i) be a corresponding $N \times N$-module and $(X_i, \tilde{\phi}_i)$ be a corresponding $K[z_\alpha, z_\beta]$-module, where $i = 1,\ 2$. Then the following three conditions are equivalent.

1) A morphism T is a $\{\alpha, \beta\}$-morphism : $(X_1, F_{\alpha_1}, F_{\beta_1}) \to (X_2, F_{\alpha_2}, F_{\beta_2})$.
2) A morphism T is a $N \times N$-morphism : $(X_1, \phi_1) \to (X_2, \phi_2)$.
3) A morphism T is a $K[z_\alpha, z_\beta]$-morphism : $(X_1, \tilde{\phi}_1) \to (X_2, \tilde{\phi}_2)$.

Proof. It is trivial that 1) is equivalent to 2). Therefore, we will show that 1) is equivalent to 3). Let the condition 1) holds. Then clearly, equations $T(F_{\alpha 1}^i \cdot F_{\beta 1}^j) = (F_{\alpha 2}^i \cdot F_{\beta 2}^j)T$ and $TF_{\beta_1}(F_{\alpha 1}^i \cdot F_{\beta 1}^j) = F_{\beta_2}(F_{\alpha 2}^i \cdot F_{\beta 2}^j)T$ hold for any $i, j \in N$. For any $\lambda = \sum_{i,j} \lambda(i,j) z_\alpha^i z_\beta^j \in K[z_\alpha, z_\beta]$, we obtain $T(\sum_{i,j} \lambda(i,j) F_{\alpha_1}^i F_{\beta_1}^j) = (\sum_{i,j} \lambda(i,j) T F_{\alpha_1}^i F_{\beta_1}^j) = (\sum_{i,j} \lambda(i,j) F_{\alpha_2}^i F_{\beta_2}^j)T$.

Therefore, condition 3) holds. Conversely, let condition 3) hold, i.e., $T(F_{\alpha_1}^i F_{\beta_1}^j) = (F_{\alpha_2}^i F_{\beta_2}^j)T$ holds for any $i, j \in N$.

Then $TF_{\alpha_1} = F_{\alpha_2}T$ and $TF_{\beta_1} = F_{\beta_2}T$ hold. Therefore, T is a $\{\alpha, \beta\}$-morphism $T : (X_1, F_{\alpha_1}, F_{\beta_1}) \to (X_2, F_{\alpha_2}, F_{\beta_2})$.

In set theory, the concept of subset, quotient set and product set are discussed for any given set.

In the same manner, for $\{\alpha, \beta\}$-action, we introduce sub-$\{\alpha, \beta\}$-action, quotient $\{\alpha, \beta\}$-action and product $\{\alpha, \beta\}$-action.

3-A.13. Sub-$\{\alpha, \beta\}$-action.
Let (X, F_α, F_β) be a $\{\alpha, \beta\}$-action and $Y \subseteq X$ be invariant subspace under F_α and F_β, i.e., $F_\alpha y$ and $F_\beta y \in Y$ for any $y \in Y$. Let $F_{\alpha Y} := F_\alpha|_Y$ and $F_{\beta Y} := F_\beta|_Y$ (restriction of the maps F_α and F_β to Y). Then $(Y, F_{\alpha Y}, F_{\beta Y})$ is a $\{\alpha, \beta\}$-action, and it is said to be a sub-$\{\alpha, \beta\}$-action of (X, F_α, F_β).

3-A.14. Quotient $\{\alpha, \beta\}$-action.
Let (X, F_α, F_β) be a $\{\alpha, \beta\}$-action and let a linear equivalence relation R in X be consistent with F_α and F_β. In other words, an equivalence relation R is given by $x_1 \; R \; x_2 \iff x_1 - x_2 \in S$ for some linear subspace $S \subseteq X$ and $x_1 \; R \; x_2$ implies $F_\alpha x_1 \; R \; F_\alpha x_2$ and $F_\beta x_1 \; R \; F_\beta x_2$. Then we consider the quotient linear space $X/_R = X/S$. Therefore, we obtain a quotient $\{\alpha, \beta\}$-action $(X/S, \dot{F}_\alpha, \dot{F}_\beta)$, where $\dot{F}_\alpha : X/S \to X/S; [x] \mapsto [F_\alpha x]$ and $\dot{F}_\beta : X/S \to X/S; [x] \mapsto [F_\beta x]$.

Corollary 3-A.15. Any $\{\alpha, \beta\}$-morphism $T : (X_1, F_{\alpha_1}, F_{\beta_1}) \to (X_2, F_{\alpha_2}, F_{\beta_2})$ can be normally decomposed into $X_1 \xrightarrow{\pi} X_1/\ker T \xrightarrow{T^b} \mathrm{im}\, T \xrightarrow{j} X_2$, where $X_1/\ker T$ is the quotient set induced by the map T, π is the canonical surjection, T^b is the bijection associated with T and j is the canonical injection. In addition, π, T^b and j are also $\{\alpha, \beta\}$-morphisms.

3-A.16. Product $\{\alpha, \beta\}$-action
Let $(X_1, F_{\alpha_1}, F_{\beta_1})$ and $(X_2, F_{\alpha_2}, F_{\beta_2})$ be $\{\alpha, \beta\}$-actions and define $F_{\alpha_1} \times F_{\alpha_2} : X_1 \times X_2 \to X_1 \times X_2; (x_1, x_2) \mapsto (F_{\alpha_1} x_1, F_{\alpha_2} x_2)$ and
$F_{\beta_1} \times F_{\beta_2} : X_1 \times X_2 \to X_1 \times X_2; (x_1, x_2) \mapsto (F_{\beta_1} x_1, F_{\beta_2} x_2)$ for the product space $X_1 \times X_2$. Then $(X_1 \times X_2, (F_{\alpha_1}, F_{\beta_1}) \times (F_{\alpha_2}, F_{\beta_2}))$ becomes a $\{\alpha, \beta\}$-action; it is called a product $\{\alpha, \beta\}$-action of $(X_1, F_{\alpha_1}, F_{\beta_1})$ and $(X_2, F_{\alpha_2}, F_{\beta_2})$.

Proposition 3-A.17. $(K[z_\alpha, z_\beta])' = F(N \times N, Y)$, where $(K[z_\alpha, z_\beta])'$ is a set of any linear maps from $K[z_\alpha, z_\beta]$ to Y.

Proof. For any $a \in F(N \times N, Y)$, set $\tilde{} : a \mapsto \tilde{a} \; [; \sum_{i,j} \lambda(i,j) z_\alpha^i z_\beta^j \mapsto \sum_{i,j} \lambda(i,j) a(i,j)]$. Then $\tilde{a} \in (K[z_\alpha, z_\beta])'$ holds. For any $\tilde{a} \in (K[z_\alpha, z_\beta])'$, set $e* : \tilde{a} \mapsto \tilde{a} \cdot e[; z_\alpha^i z_\beta^j \mapsto z_\alpha^i z_\beta^j]$, and then $\tilde{a} \cdot e \in F(N \times N, Y)$ holds. Here, $e * \cdot \tilde{} = I$ and $\tilde{} \cdot e* = I$ hold.

Since $F(N \times N, Y)$ is a concrete expression of $(K[z_\alpha, z_\beta])'$, we obtain $(K[z_\alpha, z_\beta])' = F(N \times N, Y)$.

3-A.2 Pointed $\{\alpha, \beta\}$-Actions

In this section, we introduce pointed $\{\alpha, \beta\}$-actions and $\{\alpha, \beta\}$-actions with a linear input map and show that they are equivalent. Moreover, we discuss the quasi-reachability of pointed $\{\alpha, \beta\}$-actions.

Definition 3-A.18. For a $\{\alpha, \beta\}$-action (X, F_α, F_β) and an initial state $x^0 \in X$, a collection $((X, F_\alpha, F_\beta), x^0)$ is called a pointed $\{\alpha, \beta\}$-action. A pointed $\{\alpha, \beta\}$-action $((X, F_\alpha, F_\beta), x^0)$ represents the following equations:

$$\begin{cases} x(i+1,j) = F_\alpha x(i,j) \\ x(i,j+1) = F_\beta x(i,j) \\ x(0,0) \quad = x^0 , \end{cases}$$

for any $i, j \in N$ and $x(i,j) \in X$.

For the reachable set $R(x^0) = \{F_\alpha^i F_\beta^j x^0; i,j \in N\}$, if the smallest linear space which contains $R(x^0)$ equals X, then $((X, F_\alpha, F_\beta), x^0)$ is called quasi-reachable.

Example 3-A.19. For the $\{\alpha,\beta\}$-action $(K[z_\alpha, z_\beta], z_\alpha, z_\beta)$ considered in Example (3-A.2) and the unit element $\mathbf{1}$ of multiplication, $((K[z_\alpha, z_\beta], z_\alpha, z_\beta), \mathbf{1})$ is a pointed $\{\alpha,\beta\}$-action and quasi-reachable.

Example 3-A.20. For the $\{\alpha,\beta\}$-action $(F(N \times N, Y), S_\alpha, S_\beta)$ considered in Example (3-A.3) and a two-dimensional image $a \in F(N \times N, Y)$, $((F(N \times N, Y), S_\alpha, S_\beta), a)$ is a pointed $\{\alpha,\beta\}$-action.

Definition 3-A.21. For pointed $\{\alpha,\beta\}$-actions $((X_1, F_{\alpha 1}, F_{\beta 1}), x_1^0)$ and $((X_2, F_{\alpha 2}, F_{\beta 2}), x_2^0)$, a $\{\alpha,\beta\}$-morphism $T : (X_1, F_{\alpha 1}, F_{\beta 1}) \to (X_2, F_{\alpha 2}, F_{\beta 2})$ which satisfies $Tx_1^0 = x_2^0$ is said to be a pointed $\{\alpha,\beta\}$-morphism $T : ((X_1, F_{\alpha 1}, F_{\beta 1}), x_1^0) \to ((X_2, F_{\alpha 2}, F_{\beta 2}), x_2^0)$.

Proposition 3-A.22. For any pointed $\{\alpha,\beta\}$-action $((X, F_\alpha, F_\beta), x^0)$, there exists a unique pointed $\{\alpha,\beta\}$-morphism $G : ((K[z_\alpha, z_\beta], z_\alpha, z_\beta), \mathbf{1}) \to ((X, F_\alpha, F_\beta), x^0)$. The G is said to be a linear input map.

Proof. Set $G(\mathbf{1}) = x^0$ in Propositions (3-A.11) and (3-A.12). Since $\{z_\alpha^i z_\beta^j; i,j \in N\}$ is the basis in $K[z_\alpha, z_\beta]$, G is unique.

Remark 1: According to Propositions (3-A.12) and (3-A.22), a linear input map $G : (K[z_\alpha, z_\beta], z_\alpha, z_\beta) \to (X, F_\alpha, F_\beta)$ corresponds to an initial state $x^0 \in X$ uniquely, and this correspondence is isomorphic.

Remark 2: If a pointed $\{\alpha,\beta\}$-action $((X, F_\alpha, F_\beta), x^0)$ in Proposition (3-A.22) is replaced with the $((F(N \times N, Y), S_\alpha, S_\beta), a)$ considered in Example (3-A.20), then the linear input/output map $A : (K[z_\alpha, z_\beta], z_\alpha, z_\beta) \to (F(N \times N, Y), S_\alpha, S_\beta)$ corresponds to a two-dimensional image $a \in F(N \times N, Y)$ uniquely, and this correspondence is isomorphic.

A linear input map $G : (K[z_\alpha, z_\beta], z_\alpha, z_\beta) \to (X, F_\alpha, F_\beta)$ satisfies an equation $G(\sum_{i,j} \lambda(i,j) z_\alpha^i z_\beta^j) = \sum_{i,j} \lambda(i,j) F_\alpha^i F_\beta^j G(\mathbf{1})$ for any $\sum_{i,j} \lambda(i,j) z_\alpha^i z_\beta^j \in K[z_\alpha, z_\beta]$. By using this equation and the definition of quasi-reachability, the following proposition can be obtained easily:

Proposition 3-A.23 A pointed $\{\alpha,\beta\}$-action $((X,F_\alpha,F_\beta),x^0)$ is quasi-reachable if and only if the corresponding linear input map $G:(K[z_\alpha,z_\beta],z_\alpha, z_\beta)\to(X,F_\alpha,F_\beta)$ is surjective.

3-A.3 $\{\alpha,\beta\}$-Actions with a Readout Map

In this section, we introduce $\{\alpha,\beta\}$-actions with a readout map and $\{\alpha,\beta\}$-actions with a linear output map, and we show that they are equivalent. Moreover, we discuss the distinguishablity of $\{\alpha,\beta\}$-actions with a readout map.

Definition 3-A.24. For a $\{\alpha,\beta\}$-action (X,F_α,F_β) and a linear map $h: X\to Y$, a collection $((X,F_\alpha,F_\beta),h)$ is called a $\{\alpha,\beta\}$-action with a readout map. A $\{\alpha,\beta\}$-action with a readout map $((X,F_\alpha,F_\beta),h)$ represents the following equations:

$$\begin{cases} x(i+1,j)=F_\alpha x(i,j)\\ x(i,j+1)=F_\beta x(i,j)\\ \gamma(i,j) \quad\;\; =hx(i,j)\end{cases}$$

for any $i,j\in N$, where $x(i,j)\in X$ and $\gamma(i,j)\in Y$.

For any $i,j\in N$, if $hF_\alpha^iF_\beta^jx_1 = hF_\alpha^iF_\beta^jx_2$ implies $x_1=x_2$, then $((X,F_\alpha,F_\beta),h)$ is said to be distinguishable.

Let $((X_1,F_{\alpha_1},F_{\beta_1}),h_1)$ and $((X_2,F_{\alpha_2},F_{\beta_2}),h_2)$ be $\{\alpha,\beta\}$-actions with a readout map. Then the $\{\alpha,\beta\}$-morphism $T:(X_1,F_{\alpha_1},F_{\beta_1})\to(X_2,F_{\alpha_2},F_{\beta_2})$ which satisfies $h_1=h_2T$ is called a $\{\alpha,\beta\}$-morphism with a readout map $T:((X_1,F_{\alpha_1},F_{\beta_1}),h_1)\to((X_2,F_{\alpha_2},F_{\beta_2}),h_2)$.

Example 3-A.25. For the $\{\alpha,\beta\}$-action $(K[z_\alpha,z_\beta],z_\alpha,z_\beta)$ considered in (3-A.2) and any image $a\in F(N\times N,Y)$, $((K[z_\alpha,z_\beta],z_\alpha,z_\beta),a)$ is a $\{\alpha,\beta\}$-action with a readout map. See Proposition (3-A.17).

Example 3-A.26. Considering the $\{\alpha,\beta\}$-action $(F(N\times N,Y),S_\alpha,S_\beta)$ in Example (3-A.3), by defining a linear map $(0,0):F(N\times N,Y)\to Y; a\mapsto a(0,0)$, $((F(N\times N,Y),S_\alpha,S_\beta),(0,0))$ is a $\{\alpha,\beta\}$-action with a readout map and it is distinguishable.

Proposition 3-A.27. For any $\{\alpha,\beta\}$-action with a readout map $((X,F_\alpha, F_\beta),h)$, there exists a unique $\{\alpha,\beta\}$-morphism $H:(X,F_\alpha,F_\beta)\to(F(N\times N,Y),S_\alpha,S_\beta)$ which satisfies $h=(0,0)H$, where $(Hx)(i,j)=hF_\alpha^iF_\beta^jx$ holds for any $x(i,j)\in X$, $i,j\in N$. This H is called a linear observation map.

Proof. Let $((X,F_\alpha,F_\beta),h)$ be any $\{\alpha,\beta\}$-action with a readout map. Defining $(Hx)(i,j)=hF_{\alpha}^i,F_\beta^jx$ for any $x\in X$, $i,j\in N$, we can obtain a linear observation map $H:(X,F_\alpha,F_\beta)\to(F(N\times N,Y),S_\alpha,S_\beta)$ and H satisfies $h=(0,0)H$. Next, we show the uniqueness of H. Let H:$(X,F_\alpha,F_\beta)\to(F(N\times$

$N, Y), S_\alpha, S_\beta)$ be a linear observation map which satisfies $h = (0,0)H$. Then $(Hx)(i,j) = S_\alpha^i S_\beta^j Hx(0,0) = (0,0)S_\alpha^i S_\beta^j Hx = (0,0)(HF_\alpha^i F_\beta^j x) = hF_\alpha^i F_\beta^j x$ holds for any $x \in X$, $i, j \in N$.

Therefore, H is unique.

Remark 1: According to Proposition (3-A.27), a linear observation map $H : (X, F_\alpha, F_\beta) \to (F(N \times N, Y), S_\alpha, S_\beta)$ corresponds to a linear map $h : X \to Y$ uniquely and this correspondence is isomorphic.
Remark 2: If $((X, F_\alpha, F_\beta), h)$ in Proposition (3-A.27) is replaced with the $((K[z_\alpha, z_\beta], z_\alpha, z_\beta), a)$ considered in Example (3-A.25), a linear observation map A: $(K[z_\alpha, z_\beta], z_\alpha, z_\beta) \to (F(N \times N, Y), S_\alpha, S_\beta)$ is a linear input/output map.

The definition of distinguishability and Proposition (3-A.27) lead to the following proposition.

Proposition 3-A.28. A $\{\alpha, \beta\}$-action with a readout map $((X, F_\alpha, F_\beta), h)$ is distinguishable if and only if the corresponding linear observation map $H : (X, F_\alpha, F_\beta) \to (F(N \times N, Y), S_\alpha, S_\beta)$ is injective.

3-A.4 2-Commutative Linear Representation Systems

In this section, we introduce sophisticated 2-Commutative Linear Representation Systems. For 2-Commutative Linear Representation Systems which is also called a naive 2-Commutative Linear Representation Systems dealt in Section 3.1 and sophisticated 2-Commutative Linear Representation Systems, we show that they are the same.

Definition 3-A.29. A collection $\Sigma = ((X, F_\alpha, F_\beta), G, H)$ is called a sophisticated 2-Commutative Linear Representation System if G is the linear input map $G : (K[z_\alpha, z_\beta], z_\alpha, z_\beta) \to (X, F_\alpha, F_\beta)$ and H is the linear observation map $H : (X, F_\alpha, F_\beta) \to (F(N \times N, Y), S_\alpha, S_\beta)$.

The linear input/output map $A_\Sigma := H \cdot G : (K[z_\alpha, z_\beta], z_\alpha, z_\beta) \to (F(N \times N, Y), S_\alpha, S_\beta)$ is called the behavior of Σ.

For a linear input/output map A, if $A_\Sigma = A$, then the sophisticated 2-Commutative Linear Representation System Σ is called a realization of A.

A sophisticated 2-Commutative Linear Representation System $\Sigma = ((X, F_\alpha, F_\beta), G, H)$ is called canonical if G is surjective and H is injective.

For $\Sigma_1 = ((X, F_{\alpha_1}, F_{\beta_1}), G_1, H_1)$ and $\Sigma_2 = ((X, F_{\alpha_2}, F_{\beta_2}), G_2, H_2)$, the $\{\alpha, \beta\}$-morphism $T : (X, F_{\alpha_1}, F_{\beta_1}) \to (X, F_{\alpha_2}, F_{\beta_2})$ which satisfies $TG_1 = G_2$ and $H_1 = H_2 T$ is called the sophisticated 2-Commutative Linear Representation System morphism $T : \Sigma_1 \to \Sigma_2$. If T is surjective and injective, then $T : (X, F_{\alpha_1}, F_{\beta_1}) \to (X, F_{\alpha_2}, F_{\beta_2})$ is said to be an isomorphism.

Example 3-A.30. For the $\{\alpha,\beta\}$-action $(K[z_\alpha,z_\beta],z_\alpha,z_\beta)$ in Example (3-A.2), identity map I on $K[z_\alpha,z_\beta]$ and a linear input/output map A : $(K[z_\alpha,z_\beta],z_\alpha,z_\beta) \to (F(N\times N,Y),S_\alpha,S_\beta)$,
the collection $\Sigma = ((K[z_\alpha,z_\beta],z_\alpha,z_\beta),I,A)$ is the sophisticated 2-Commutative Linear Representation System with the behavior A.

For the $\{\alpha,\beta\}$-action $(F(N\times N,Y),S_\alpha,S_\beta)$ in Example (3-A.3), a linear input/output map A and identity map I on $F(N\times N,Y)$, a collection $\Sigma_s = ((F(N\times N,Y),S_\alpha,S_\beta),A,I)$ is the sophisticated 2-Commutative Linear Representation System with the behavior A.

On this basis, we now investigate the relation between sophisticated 2-Commutative Linear Representation Systems and naive ones.

Proposition 3-A.31. For any sophisticated 2-Commutative Linear Representation System $\Sigma = ((X,F_\alpha,F_\beta),G,H)$, there exists a unique naive 2-Commutative Linear Representation System $\sigma = ((X,F_\alpha,F_\beta),x^0,h)$ corresponding to the sophisticated 2-Commutative Linear Representation System given by the following two equations:

$$\textstyle\sum_{i,j}\lambda(i,j)F_\alpha^iF_\beta^jx^0 = G(\sum_{i,j}\lambda(i,j)z_\alpha^iz_\beta^j) \quad \text{(a.1)}$$

$$hF_\alpha^iF_\beta^jx = (Hx)(i,j) \text{ for any } x\in X,\ i,j\in N \quad \text{(a.2)}$$

This correspondence is isomorphic in the category's sense (Pareigis [1970]).

Proof. Remark 1 of Proposition (3-A.22) and Remark 1 of Proposition (3-A.27) leads to this proposition.

3-A.5 Sophisticated 2-Commutative Linear Representation System

In this section, we will prove Realization Theorem (3.5) for sophisticated 2-commutative Linear Representation System. According to Remark 2 in Proposition (3-A.22) (or Remark 2 in Proposition (3-A.27)) and Proposition (3-A.31), the realization theorem can be replaced with the Theorem (3-A.32) just below. Therefore, the proof of Theorem (3-A.32) implies the proof of Realization Theorem (3.5).

Theorem 3-A.32. Sophisticated Realization Theorem.

(1) Existence: For any linear input/output map
$A : (K[z_\alpha,z_\beta],z_\alpha,z_\beta) \to (F(N\times N,Y),S_\alpha,S_\beta)$, there exist at least two sophisticated canonical 2-Commutative Linear Representation Systems that realize A.

(2) Uniqueness: Let $\Sigma_1 = ((X_1, F_{\alpha_1}, F_{\beta_1}), G_1, H_1)$ and $\Sigma_2 = ((X_2, F_{\alpha_2}, F_{\beta_2}), G_2, H_2)$ be sophisticated canonical 2-Commutative Linear Representation Systems that have the same behavior. Then there exists an isomorphism $T : \Sigma_1 \to \Sigma_2$.

Proof. The following Corollary (3-A.33) indicates the proof of existence, while the Remark in Corollary (3-A.37) indicates the proof of uniqueness.

Corollary 3-A.33. For any linear input/output map $A : (K[z_\alpha, z_\beta], z_\alpha, z_\beta) \to (F(N \times N, Y), S_\alpha, S_\beta)$, the following sophisticated 2-Commutative Linear Representation Systems
indexcanonical (naive) are both canonical realizations of A:

(1) $\Sigma_q = ((K[z_\alpha, z_\beta]/\text{ker } A, \dot{z}_\alpha, \dot{z}_\beta), \pi, A^i)$, where π is the canonical surjection : $K[z_\alpha, z_\beta] \to K[z_\alpha, z_\beta]/\text{ker } A$ and A^i is given by $A^i = jA^b$ for $A^b : K[z_\alpha, z_\beta]/\text{ker } A \to \text{im } A$ being isomorphic with A and j being the canonical injection : im $A \to F(N \times N, Y)$.
(2) $\Sigma_s = ((\text{im } A, S_\alpha, S_\beta), A^s, j)$, where $A^s = A^b \cdot \pi$.

Σ_q and Σ_s are called the quotient canonical realization and the subspace canonical realization, respectively.

Proof. This can be obtained easily by Corollary (3-A.15), Example (3-A.30), the definition of canonicality and the definition of behavior.

To prove the uniqueness part of Theorem (3-A.32), we introduce the following set of morphisms Mor (Σ_1, Σ_2) from a sophisticated 2-Commutative Linear Representation System Σ_1 to another sophisticated 2-Commutative Linear Representation System Σ_2, where Σ_1 and Σ_2 are given by $\Sigma_1 = ((X_1, F_{\alpha_1}, F_{\beta_1}), G_1, H_1)$ and $\Sigma_2 = ((X_2, F_{\alpha_2}, F_{\beta_2}), G_2, H_2)$, respectively.
Mor $(\Sigma_1, \Sigma_2) :=$ {a relation $T_{12} : X_1 \to X_2; GrT_{12}^{min} \subseteq GrT_{12} \subseteq GrT_{12}^{max}$}, where GrT_{12}^{min}, GrT_{12} and GrT_{12}^{max} denote the graph of $T_{12}^{min} := G_2 \cdot G_1^{-1}$, T_{12} and $T_{12}^{max} = H_2^{-1} \cdot H_1$, respectively. The reason for introducing this morphism depends on the following lemma:

Lemma 3-A.34. $A_{\Sigma_1} = A_{\Sigma_2}$ if and only if Mor $(\Sigma_1, \Sigma_2) \neq \emptyset$.

Proof. This can be proved the same as in Matsuo [1977] and [1981].

Lemma 3-A.35. Let $A_{\Sigma_1} = A_{\Sigma_2}$ hold.

(1) If G_1 of Σ_1 is surjective, then dom $T_{12}^{min} = X_1$ holds, where dom T_{12}^{min} denotes the domain of T_{12}^{min}.
(2) If H_2 of Σ_2 is injective, then T_{12}^{max} is a partial function : $X_1 \to X_2$.

Proof. This can be proved the same as in Matsuo [1977] and [1981].

Lemma 3-A.36. Let $A_{\Sigma_1} = A_{\Sigma_2}$ hold. Then GrT_{12}^{max} is an invariant subproduct $\{\alpha, \beta\}$-action of $(X_1, F_{\alpha_1}, F_{\beta_1})$ and $(X_2, F_{\alpha_2}, F_{\beta_2})$.

Proof. By the definition of T_{12}^{max}, $GrT_{12}^{max} = \{(x_1, x_2) \in X_1 \times X_2; H_1x_1 = H_2x_2\}$ holds. Let (x_1, x_2) and $(x_{1'}, x_{2'}) \in GrT_{12}^{max}$, i.e., $H_1x_1 = H_2x_2$ and $H_1x_{1'} = H_2x_{2'}$ hold. Then $H_1(x_1 + x_{1'}) = H_1x_1 + H_1x_{1'} = H_2x_2 + H_2x_{2'} = H_2(x_2 + x_{2'})$ holds. This implies $(x_1 + x_{1'}, x_2 + x_{2'}) \in GrT_{12}^{max}$. Then $(kx_1, kx_2) \in GrT_{12}^{max}$ holds for any scalar $k \in K$ and $(x_1, x_2) \in GrT_{12}^{max}$. Moreover, for $(x_1, x_2) \in GrT_{12}^{max}$, $H_1F_{\alpha_1}x_1 = S_\alpha H_1x_1 = S_\alpha H_2x_2 = H_2F_{\alpha_2}x_2$ holds and $H_1F_{\beta_1}x_1 = S_\beta H_1x_1 = S_\beta H_2x_2 = H_2F_{\beta_2}x_2$ holds. Hence, we obtain $(F_{\alpha_1}x_1, F_{\alpha_2}x_2) \in GrT_{12}^{max}$ and $(F_{\beta_1}x_1, F_{\beta_2}x_2) \in GrT_{12}^{max}$. This implies that $GrT_{12}^{max} \subseteq X_1 \times X_2$ is invariant under $F_{\alpha_1} \times F_{\alpha_2}$ and $F_{\beta_1} \times F_{\beta_2}$. Therefore, $(GrT_{12}^{max}, F_{\alpha_1} \times F_{\alpha_2}, F_{\beta_1} \times F_{\beta_2})$ is a $\{\alpha, \beta\}$-action.

Lemma 3-A.37. Let $A_{\Sigma_1} = A_{\Sigma_2}$ hold, G_1 be surjective and H_2 be injective. Then $T_{12}^{min} = T_{12}^{max}$ holds and T_{12} is a 2-Commutative Linear Representation System morphism T_{12}: $\Sigma_1 \to \Sigma_2$ by setting $T_{12} := T_{12}^{min}$.

Proof. If G_1 is surjective and H_2 is injective, then Lemma (3-A.35) implies that $T_{12} \in$ Mor (Σ_1, Σ_2) is unique, i.e., $T_{12}^{min} = T_{12} = T_{12}^{max}$. Therefore, $T_{12}G_1 = G_2$ and $H_2T_{12} = H_1$ hold. By virtue of Lemma (3-A.36), $(x_1, x_2) \in GrT_{12}$ implies that $(F_{\alpha 1}x_1, F_{\alpha 2}x_2)$ and $(F_{\beta 1}x_1, F_{\beta 2}x_2) \in GrT_{12}$. Hence T_{12} is a $\{\alpha, \beta\}$-morphism $T_{12} : (X_1, F_{\alpha_1}, F_{\beta_1}) \to (X_2, F_{\alpha_2}, F_{\beta_2})$.
Remark: The uniqueness of Sophisticated Realization Theorem (3-A.32) for two-dimensional images is proved by the canonicality of sophisticated 2-Commutative Linear Representation Systems and Lemma (3-A.37).

3-B Finite-Dimensional 2-Commutative Linear Representation Systems

In this section, we will supply proofs of the theorems, propositions and corollaries for Finite- dimensional 2-Commutative Linear Representation Systems set forth in Section 3.2.

3-B.1 Finite-Dimensional $\{\alpha, \beta\}$-Actions and Pointed $\{\alpha, \beta\}$-Actions

Appendix 3-A examined $\{\alpha, \beta\}$-actions. Here, we investigate cases whose state spaces are finite-dimensional. We will find that finite-dimensional $\{\alpha, \beta\}$-actions can be represented by matrix expressions.

Definition 3-B.1. A $\{\alpha, \beta\}$-action (X, F_α, F_β) of which X is finite (n)-dimensional is called a finite-dimensional (n-dimensional) $\{\alpha, \beta\}$-action.

Let (X, F_α, F_β) be an n-dimensional $\{\alpha, \beta\}$-action. Then according to the fact discussed in [Halmos, 1958], X can be viewed as $X = K^n$ and F_α,$F_\beta \in K^{n \times n}$. See the comment just above Theorem (3.13).

Proposition 3-B.2. Let (K^n, F_α, F_β) be an n-dimensional $\{\alpha, \beta\}$-action and (K^n, ϕ) be the $N \times N$-module corresponding to the $\{\alpha, \beta\}$-action (K^n, F_α, F_β).
Then a formal power series $\bar{\phi}$ of ϕ can be represented as the following matrix-valued rational function:

$$\bar{\phi} = z_\alpha z_\beta [z_\beta I - F_\beta]^{-1}[z_\alpha I - F_\alpha]^{-1} \in (K(z_\alpha, z_\beta))^{n\times n}.$$

The (l, m) element $\bar{\phi}_{l,m}$ of matrix $\bar{\phi}$ can be expressed thus:

$$\bar{\phi}_{l,m} = z_\alpha z_\beta (\textstyle\sum_{i=0}^{n-1} \sum_{j=0}^{n-1} \lambda_{l,m}(i,j) z_\alpha^i z_\beta^j)/q_\alpha(z_\alpha) q_\beta(z_\beta),$$

where $q_\alpha(z_\alpha) = z_\alpha^n + \alpha_{n-1} z_\alpha^{n-1} + \alpha_{n-2} z_\alpha^{n-2} + \cdots + \alpha_0$ and $q_\beta(z_\beta) = z_\beta^n + \beta_{n-1} z_\beta^{n-1} + \beta_{n-2} z_\beta^{n-2} + \cdots + \beta_0$.

Proof. ϕ is expressed by the following matrix-valued formal power series: $\bar{\phi} = \sum_{i=0}^\infty \sum_{j=0}^\infty \phi(i,j) z_\alpha^{-i} z_\beta^{-j} = \sum_{i=0}^\infty \sum_{j=0}^\infty F_\alpha^i F_\beta^j z_\alpha^{-i} z_\beta^{-j} \in K^{n\times n}[[z_\alpha^{-1}, z_\beta^{-1}]]$.

Since the equation $K^{n\times n}[[z_\alpha^{-1}, z_\beta^{-1}]] = (K[[z_\alpha^{-1}, z_\beta^{-1}]])^{n\times n}$ holds and also $(K[[z_\alpha^{-1}, z_\beta^{-1}]])^{n\times n} \subseteq (K((z_\alpha^{-1}, z_\beta^{-1})))^{n\times n} = (K((z_\alpha, z_\beta)))^{n\times n}$ holds, the following equation holds in the sense of $(K((z_\alpha, z_\beta)))^{n\times n}$: $[z_\alpha I - F_\alpha][z_\beta I - F_\beta]\bar{\phi} = z_\alpha z_\beta I$.

Since $[z_\alpha I - F_\alpha]$ and $[z_\beta I - F_\beta]$ are bijective in $(K((z_\alpha^{-1}, z_\beta^{-1})))^{n\times n}$, we obtain $\bar{\phi} = z_\alpha z_\beta [z_\beta I - F_\beta]^{-1}[z_\alpha I - F_\alpha]^{-1}$.
The (l, m) element $\bar{\phi}_{l,m}$ of matrix $\bar{\phi}$ is obvious from the matrix calculation.

This proposition implies that any n-dimensional $\{\alpha, \beta\}$-action can be expressed by a rational function matrix.

Proposition 3-B.3. Let $((K^n, F_\alpha, F_\beta), x^0)$ be a pointed n-dimensional $\{\alpha, \beta\}$-action. Then a formal power series $X(z_\alpha^{-1}, z_\beta^{-1})$ of this pointed $\{\alpha, \beta\}$-action can be expressed by the following vector-valued rational function: $X(z_\alpha^{-1}, z_\beta^{-1}) = z_\alpha z_\beta [z_\beta I - F_\beta]^{-1}[z_\alpha I - F_\alpha]^{-1} x^0 \in (K((z_\alpha, z_\beta)))^{n\times n}$.

Proof. Let (K^n, ϕ) be the $N\times N$-module corresponding to the $\{\alpha, \beta\}$-action (K^n, F_α, F_β). Then by Proposition (3-A.11), ϕ is expressed by the following matrix-valued formal power series $\bar{\phi}$:
$\bar{\phi} = \sum_{i=0}^\infty \sum_{j=0}^\infty F_\alpha^i F_\beta^j z_\alpha^{-i} z_\beta^{-j} = z_\alpha z_\beta [z_\beta I - F_\beta]^{-1}[z_\alpha I - F_\alpha]^{-1}$.
Therefore, $((K^n, F_\alpha, F_\beta), x^0)$ can be expressed as follows:
$\bar{\phi} x^0 = z_\alpha z_\beta [z_\beta I - F_\beta]^{-1}[z_\alpha I - F_\alpha]^{-1} x^0$.

In Appendix 3-A, we presented an $\{\alpha, \beta\}$-morphism $G : (K[z_\alpha, z_\beta], z_\alpha, z_\beta) \to (X, F_\alpha, F_\beta)$ and we showed that an initial object of any pointed $\{\alpha, \beta\}$-action $((X, F_\alpha, F_\beta), x^0)$ is $((K[z_\alpha, z_\beta], z_\alpha, z_\beta), \mathbf{1})$ by using the $\{\alpha, \beta\}$-morphism $G : (K[z_\alpha, z_\beta], z_\alpha, z_\beta) \to (X, F_\alpha, F_\beta)$ and $G(\mathbf{1}) = x^0$.

Moreover, we demonstrated that the quasi-reachability of $((X, F_\alpha, F_\beta), x^0)$ implies a surjectivity of the corresponding $\{\alpha, \beta\}$-morphism G which is also called linear input map.

In this section, we provide a criterion for the quasi-reachability of pointed finite-dimensional $\{\alpha, \beta\}$-actions. Arriving at the quasi-reachable standard form, we then show that it is a representative of pointed $\{\alpha, \beta\}$-actions.

Let (X, F_α, F_β) be a pointed $\{\alpha, \beta\}$-action and G be the linear input map corresponding to an initial state x^0, namely, a $\{\alpha, \beta\}$-morphism $G : (K[z_\alpha, z_\beta], z_\alpha, z_\beta) \to (X, F_\alpha, F_\beta)$ which satisfies $G(\mathbf{1}) = x^0$.

Let $QR(r)$ be the linear hull of the reachable set obtained by $F_\alpha^i F_\beta^j x^0$ with a condition $i+j \leq r$, i.e., $QR(r) := \{\sum \lambda_j x_j; x_j = F_\alpha^i F_\beta^j x^0, \lambda_j \in K,\ i+j \leq r\}$. Then the following formula holds:

$$QR(r+1) = QR(r) + \ll \{F_\alpha x, F_\beta x; x \in QR(r)\} \gg.$$

Therefore, the following sequence can be obtained:

$$QR(0) \subseteq QR(1) \subseteq \cdots QR(i) \subseteq \cdots \subseteq QR(\infty).$$

Here, let $P(\leq m_1, \leq n_1)$ be $P(\leq m_1, \leq n_1) := \sum_{i=0}^{m_1} \sum_{j=0}^{n_1} \lambda(i,j) z_\alpha^i z_\beta^j$ and especially, let $P(\leq l) := \ll \{z_\alpha^i z_\beta^j; i+j \leq l, i, j \in N\} \gg$.
Moreover, let $G_l = G \cdot J_l$, where J_l is the canonical injection $J_l : P(\leq l) \to K[z_\alpha, z_\beta]$.
Then the above sequence of $QR(r)$ can be rewritten thus:

$$\text{im } G_0 \subseteq \text{im } G_1 \subseteq \cdots \subseteq \text{im } G_i \subseteq \cdots \subseteq \text{im } G_\infty.$$

Then we can obtain the following lemma easily:

Lemma 3-B.4. If im G_{j-1} = im G_j for any integer $j \in N$ then im G_j = im G_{j+1}.

Proof. By formula, im G_j = im $G_{j-1} + \{F_\alpha x, F_\beta x;\ x \in \text{im } G_{j-1}\}$ holds. By assumption, im G_{j-1} = im G_j holds. Then im G_{j+1} = im $G_j + \{F_\alpha x, F_\beta x;\ x \in \text{im } G_j\}$ = im $G_{j-1} + \{F_\alpha x, F_\beta x;\ x \in \text{im } G_{j-1}\}$ = im G_j holds.

Lemma 3-B.5. For any pointed $\{\alpha, \beta\}$-action $((K^n, F_\alpha, F_\beta), x^0)$, im G_{n-1} = im G always holds. Therefore, $((\text{im } G_{n-1}, F_\alpha, F_\beta), x^0)$ is a quasi-reachable pointed $\{\alpha, \beta\}$-action.

Proof. This is a direct consequence of Lemma (3-B.4) and the definition of quasi-reachability.

Proposition 3-B.6. Let $((K^n, F_\alpha, F_\beta), x^0)$ be a pointed $\{\alpha, \beta\}$-action. Then $((K^n, F_\alpha, F_\beta), x^0)$ is quasi-reachable if and only if im $G_{n-1} = K^n$ holds.

Proof. The necessary and sufficient condition that $((K^n, F_\alpha, F_\beta), x^0)$ is quasi-reachable is im $G = K^n$. By Lemma (3-B.5), this is equivalent to im $G_{n-1} = K^n$. Consequently, the proposition holds.

Proposition 3-B.7. Let $((K^n, F_\alpha, F_\beta), x^0)$ be a quasi-reachable pointed $\{\alpha, \beta\}$-action. Then the dimension of im G_{j-1} is more than j for any $1 \le j \le n$.

Proof. For any integer j, assume that there do not exist j linearly independent vectors in im G_{j-1}. If im $G_{j-2} \subseteq$ im G_{j-1} holds, then this contradicts the nonexistence of j independent vectors. Hence, im $G_{j-2} =$ im $G_{j-1} = \cdots =$ im G_∞ holds and im G_∞ has no more than j vectors. This contradicts the quasi-reachability of $((K^n, F_\alpha, F_\beta), x^0)$.

Proposition 3-B.8. Let $((K^n, F_\alpha, F_\beta), x^0)$ be a pointed $\{\alpha, \beta\}$-action. $((K^n, F_\alpha, F_\beta), x^0)$ is quasi-reachable if and only if the following equation holds:

$$\text{rank}\,[x^0, F_\alpha x^0, F_\beta x^0, F_\alpha^2 x^0, F_\alpha F_\beta x^0, F_\beta^2 x^0, \cdots,$$
$$F_\alpha^{n-1} x^0, F_\alpha^{n-2} F_\beta x^0, \cdots, F_\beta^{n-1} x^0] = n$$

Proof. This can be obtained by Proposition (3-B.6).

Definition 3-B.9. Let $((K^n, F_\alpha, F_\beta), x^0)$ be a quasi-reachable pointed $\{\alpha, \beta\}$-action. If $((K^n, F_\alpha, F_\beta), x^0)$ satisfies the following conditions, then it is called the quasi-reachable standard form:

1) An integer ν_j $(1 \le j \le k)$ satisfies $n := \sum_{j=1}^{k} \nu_j$ and $0 \le \nu_k \le \nu_{k-1} \le \cdots \le \nu_2 \le \nu_1$.
2) For any $1 \le j \le k$ and $1 \le i \le \nu_j$, $F_\beta^{j-1} F_\alpha^{i-1} \mathbf{e}_1 = \mathbf{e}_{\nu_1+\nu_2+\cdots+\nu_{j-1}+i}$.
3) For any $1 \le j \le k$,
$F_\beta^{j-1} F_\alpha^{\nu_j} \mathbf{e}_1 = \sum_{m=1}^{j} \sum_{l=1}^{\nu_m} c_{ml}^j F_\beta^{m-1} F_\alpha^{l-1} \mathbf{e}_1$, where $c_{ml}^j \in K$,
$F_\beta^k \mathbf{e}_1 = \sum_{m=1}^{k} \sum_{l=1}^{\nu_m} c_{ml}^{k+1} F_\beta^{m-1} F_\alpha^{l-1} \mathbf{e}_1$, where $c_{ml}^{k+1} \in K$,
$\mathbf{e}_i = [0, \cdots, 0, \overset{i}{1}, 0, \cdots, 0]^T$.

Remark: If $((K^n, F_\alpha, F_\beta), x^0)$ is the quasi-reachable standard form, note that $x^0 = \mathbf{e}_1$.

Proposition 3-B.10. $((K^n, F_\alpha, F_\beta), \mathbf{e}_1)$ is the quasi-reachable standard form if and only if F_α and F_β are given by Figure 3.3 and Figure 3.4 respectively, and $F_{\alpha s}^{\nu_{j-1}-\nu_j} \mathbf{c}^j = F_{\beta s} \mathbf{c}^{j-1}$ holds for $2 \le j \le k+1$.

Proof. First, we prove necessity.
For any i , j $(1 \le i \le \nu_j,\ 1 \le j \le k)$, $F_\beta^{j-1} F_\alpha^{i-1} \mathbf{e}_1 = \mathbf{e}_{\nu_1+\nu_2+\cdots+\nu_{j-1}+i}$ and $F_\beta^{j-1} F_\alpha^{\nu_j} \mathbf{e}_1 = \sum_{m=1}^{j} \sum_{l=1}^{\nu_m} c_{ml}^j F_\beta^{m-1} F_\alpha^{l-1} \mathbf{e}_1$ imply $F_\alpha = F_{\alpha s}$.

The commutativity $F_{\alpha s}F_\beta = F_\beta F_{\alpha s}$ and

$$F_\beta^k \mathbf{e}_1 = \sum_{m=1}^{k}\sum_{l=1}^{\nu_m} c_{ml}^{k+1} F_\beta^{m-1} F_\alpha^{l-1}\mathbf{e}_1$$

imply that $F_\beta = F_{\beta s}$ and

$$F_{\alpha s}^{\nu_{j-1}-\nu_j}\mathbf{c}^j = F_{\beta s}\mathbf{c}^{j-1} \ (2 \le j \le k+1).$$

Next, we show sufficiency. Let $F_\alpha = F_{\alpha s}$, $F_\beta = F_{\beta s}$ and

$$F_{\alpha s}^{\nu_{j-1}-\nu_j}\mathbf{c}^j = F_{\beta s}\mathbf{c}^{j-1} \ (2 \le j \le k+1)$$

hold. Then we can easily check by direct calculation that $((K^n, F_\alpha, F_\beta), \mathbf{e}_1)$ is the quasi-reachable standard form.

Proposition 3-B.11. For any quasi-reachable pointed $\{\alpha,\beta\}$-action $((K^n, F_\alpha, F_\beta), x^0)$, there exists the unique quasi-reachable standard form $((K^n, F_{\alpha s}, F_{\beta s}), \mathbf{e}_1)$ which is isomorphic to it.

Proof. Find an integer ν_1 such that vectors $\{F_\alpha^i x^0; 0 \le i \le \nu_1 - 1\}$ are linearly independent and vectors $\{F_\alpha^i x^0; 0 \le i \le \nu_1\}$ are linearly dependent. Set $F_\alpha^{\nu_1} x^0 = \sum_{i=1}^{\nu_1} c_{1i}^1 F_\alpha^{i-1} x^0$ and $\mathbf{c}^1 = [c_{11}^1, c_{12}^1, c_{13}^1, \cdots, c_{1\nu_1}^1, 0, \cdots, 0]^T$.

Next, find an integer ν_2 such that vectors
$B_2 := \{F_\beta^j F_\alpha^i x^0; 0 \le j \le 1, 0 \le i \le \nu_{j+1} - 1\}$ are linearly independent and vectors $\{B_2, F_\beta F_\alpha^{\nu_2} x^0\}$ are linearly dependent.
Set $F_\beta F_\alpha^{\nu_2} x^0 = \sum_{m=1}^{2}\sum_{l=1}^{\nu_m} c_{ml}^2 F_\beta^{m-1} F_\alpha^{l-1} x^0$ and
$\mathbf{c}^2 = [c_{11}^2, c_{12}^2, c_{13}^2, \cdots, c_{1\nu_1}^2, c_{21}^2, c_{22}^2, \cdots, c_{2\nu_2}^2, 0, \cdots, 0]^T$.
These procedures are continued till the next procedure.
Find an integer ν_k such that vectors
$B_k := \{F_\beta^j F_\alpha^i x^0; 0 \le j \le k-1, 0 \le i \le \nu_{j+1} - 1\}$ are linearly independent and vectors $\{B_k, F_\beta^{k-1} F_\alpha^{\nu_k} x^0\}$ and $\{B_k, F_\beta^k x^0\}$ are linearly dependent.
Set $\mathbf{c}^k = [c_{11}^k, c_{12}^k, \cdots, c_{1\nu_1}^k, c_{21}^k, c_{22}^k, \cdots, c_{2\nu_2}^k, \cdots, c_{k\nu_k}^k]^T$ and
$\mathbf{c}^{k+1} = [c_{11}^{k+1}, c_{12}^{k+1}, \cdots, c_{1\nu_1}^{k+1}, c_{21}^{k+1}, c_{22}^{k+1}, \cdots, c_{2\nu_2}^{k+1}, \cdots, c_{k\nu_k}^{k+1}]^T$ in terms of
$F_\beta^{k-1} F_\alpha^{\nu_k} x^0 = \sum_{m=1}^{k}\sum_{l=1}^{\nu_m} c_{ml}^k F_\beta^{m-1} F_\alpha^{l-1} x^0$ and
$F_\beta^k x^0 = \sum_{m=1}^{k}\sum_{l=1}^{\nu_m} c_{ml}^{k+1} F_\beta^{m-1} F_\alpha^{l-1} x^0$.

By the selection of vectors and indices $\{\nu_i : 1 \le i \le k\}$, $\sum_{i=1}^{k}\nu_i = n$ holds. At this point, let us bring in a linear operator $K^n \to K^n$ by setting $TF_\beta^i F_\alpha^j x^0 = \mathbf{e}_{\nu_1+\nu_2+\cdots+\nu_i+j+1}$. Then T becomes a regular matrix.
Set $F_{\alpha s} = TF_\alpha T^{-1}$ and $F_{\beta s} = TF_\beta T^{-1}$. Then $F_{\alpha s}, F_{\beta s} \in K^{n\times n}$ and $F_{\alpha s}F_{\beta s} = F_{\beta s}F_{\alpha s}$ hold.

Finally, we conclude that $((K^n, F_{\alpha s}, F_{\beta s}), \mathbf{e}_1)$ is a pointed $\{\alpha,\beta\}$-action. Since the linear operator T preserves the linear independence and linear dependence of each vector, $((K^n, F_{\alpha s}, F_{\beta s}), \mathbf{e}_1)$ is a $\nu = (\nu_1, \nu_2, \cdots, \nu_k)$-quasi-reachable. Hence, T is a pointed $\{\alpha,\beta\}$-morphism $T : ((K^n, F_\alpha, F_\beta), x^0) \to$

$((K^n, F_{\alpha s}, F_{\beta s}), \mathbf{e}_1)$. Its uniqueness is proved by the definition of the quasi-reachable standard form.

Remark: There are many equivalence classes in the category of pointed $\{\alpha, \beta\}$-actions; this proposition states that the equivalence classes can be represented as quasi-reachable standard forms.

3-B.2 Finite-Dimensional $\{\alpha, \beta\}$-Actions with a Readout Map

Proposition 3-B.12. Let $((K^n, F_\alpha, F_\beta), h)$ be a n-dimensional $\{\alpha, \beta\}$-action with a readout map. Then a formal power series $Y(z_\alpha^{-1}, z_\beta^{-1})$ can be expressed by the following vector-valued rational function:
$Y(z_\alpha^{-1}, z_\beta^{-1}) = z_\alpha z_\beta h[z_\alpha I - F_\alpha]^{-1}[z_\beta I - F_\beta]^{-1}$.

Proof. This proposition can be obtained in the same way as Proposition (3-B.3).

In Appendix 3-A, we established that a final object of any $\{\alpha, \beta\}$-action with a readout map $((X, F_\alpha, F_\beta), h)$ is $((F(N \times N, Y), S_\alpha, S_\beta), (0, 0))$, and the distinguishability of $((X, F_\alpha, F_\beta), h)$ implies the injectivity of the corresponding linear observation map H.

Now, we will set forth a criterion for distinguishability of finite-dimensional $\{\alpha, \beta\}$-actions with a readout map. Introducing the distinguishable standard form, we show that it is a representative of $\{\alpha, \beta\}$-actions with a readout map.

Let $((X, F_\alpha, F_\beta), h)$ be a $\{\alpha, \beta\}$-action with a readout map and H be the linear observation map corresponding to a readout map h, namely, a $\{\alpha, \beta\}$-morphism $H : (X, F_\alpha, F_\beta) \to (F(N \times N, Y), S_\alpha, S_\beta)$ which satisfies $(0, 0)H = h$. Let $LO(i)$ be the linear hull of the reachable set by the readout map $\{hF_\alpha^{\tilde{i}} F_\beta^{\tilde{j}}; \tilde{i} + \tilde{j} \leq i\}$. Namely, $LO(i) := \{\sum \lambda_j x_j^*; x_j^* = hF_\alpha^{\tilde{i}} F_\beta^{\tilde{j}}; \tilde{i} + \tilde{j} \leq i,\ \lambda_j \in K\}$.

Then the following sequence holds.

$$LO(0) \subseteq LO(1) \subseteq \cdots \subseteq LO(i) \subseteq \cdots \subseteq LO(\infty).$$

Let $H_l = P_l \cdot H$, where P_l is the canonical surjection
$P_l : F(N \times N, Y) \to F((N \times N)_l, Y)$,
where $F((N \times N)_l, Y) := \{\underline{a} \in F((N \times N)_l, Y); \underline{a} : (N \times N)_l \to Y\}$ and $(N \times N)_l := \{(\tilde{i}, \tilde{j}); \tilde{i} + \tilde{j} \leq l\}$.
Then $\ker H_l = LO(l)^0$ holds, where
$LO(l)^0 := \ker H_l = \{x \in X; hx = 0 \text{ for } h \in LO(l)\}$.
Moreover, $\ker H = LO(\infty)^0$ holds.

Lemma 3-B.13. For any $\{\alpha, \beta\}$-action with a readout map $((K^n, F_\alpha, F_\beta), h)$, $LO(n-1) = \ll hF_\alpha^i F_\beta^j; i, j \in N \gg$ holds.

Proof. This lemma can be proved the same way as in Lemma (3-B.6).

Proposition 3-B.14. For any $\{\alpha, \beta\}$-action with a readout map $((K^n, F_\alpha, F_\beta), h)$, $(\ker H_{n-1}, F_\alpha, F_\beta)$ is a sub-$\{\alpha, \beta\}$-action of (K^n, F_α, F_β) and $((K^n/\ker H_{n-1}, \dot{F}_\alpha, \dot{F}_\beta), \dot{h})$ is a distinguishable $\{\alpha, \beta\}$-action with a readout map.

Proof. Let H be the corresponding linear observation map to h. By Lemma (3-B.13), $LO(n-1) = \ll hF_\alpha^i F_\beta^j; i, j \in N \gg$ holds. Therefore, $\ker H_{n-1} = \ker H$ holds. Because H is a $\{\alpha, \beta\}$-morphism, $H : (K^n, F_\alpha, F_\beta) \to (F(N \times N, Y), S_\alpha, S_\beta)$, $(\ker H_{n-1}, F_\alpha, F_\beta)$ is a sub $\{\alpha, \beta\}$-action of (K^n, F_α, F_β). Therefore, $((K^n/\ker H_{n-1}, \dot{F}_\alpha, \dot{F}_\beta), \dot{h})$ can be defined. It is a distinguishable $\{\alpha, \beta\}$-action with a readout map.

Proposition 3-B.15. Let $((K^n, F_\alpha, F_\beta), h)$ be a $\{\alpha, \beta\}$-action with a readout map. $((K^n, F_\alpha, F_\beta), h)$ is distinguishable if and only if $LO(n-1) = K^n$.

Proof. This proposition is handled in the same way as Proposition (3-B.6).

Proposition 3-B.16. If $((K^n, F_\alpha, F_\beta), h)$ is distinguishable, then the dimension of $LO(j-1)$ is more than j for any j $(1 \le j \le n)$.

Proof. This proposition is obtained in the same way as Proposition (3-B.7).

Proposition 3-B.17. Let $((K^n, F_\alpha, F_\beta), h)$ be a $\{\alpha, \beta\}$-action with a readout map. $((K^n, F_\alpha, F_\beta), h)$ is distinguishable if and only if rank $[h^T, (hF_\alpha)^T, (hF_\beta)^T \cdots, (hF_\alpha^2)^T, (hF_\alpha F_\beta)^T, (hF_\beta^2)^T, \cdots, (hF_\alpha^{n-1})^T, \cdots, (hF_\beta^{n-1})^T] = n$.

Proof. This proposition is proved in the same way as Proposition (3-B.8).

Definition 3-B.18. Let $((K^n, F_{\alpha_d}, F_{\beta_d}), h_d)$ be a distinguishable $\{\alpha, \beta\}$-action with a readout map, and let the set of output values Y be K. If $((K^n, F_{\alpha_d}, F_{\beta_d}), h_d)$ satisfies the following conditions, then it is called the distinguishable standard form with a vector index $\mu = (\mu_1, \mu_2, \cdots, \mu_k)$:

1) An integer μ_j $(1 \le j \le k)$ satisfies $n := \sum_{j=1}^k \mu_j$ and $0 \le \mu_k \le \mu_{k-1} \le \cdots \le \mu_2 \le \mu_1$.
2) For any i, j $(1 \le j \le k, 1 \le i \le \mu_j)$,
$(h_d F_\beta^{j-1} F_\alpha^{i-1})^T = (\mathbf{e}_{\mu_1+\mu_2+\cdots+\mu_{j-1}+i})^T$,
$(h_d F_\beta^{j-1} F_\alpha^{\mu_j})^T = \sum_{m=1}^j \sum_{l=1}^{\mu_m} c_{ml}^j (h_d F_\beta^{m-1} F_\alpha^{l-1})^T$ and
$(h_d F_\beta^k)^T = \sum_{m=1}^k \sum_{l=1}^{\mu_m} c_{ml}^{k+1} (h_d F_\beta^{m-1} F_\alpha^{l-1})^T$,

where $\mathbf{e}_i = [0, \cdots, 0, \overset{i}{1}, 0, \cdots, 0]^T$ and T denotes transposition of matrices and vectors.

Remark: If $((K^n, F_{\alpha_d}, F_{\beta_d}), h_d)$ is the distinguishable standard form, note that $h_d = \mathbf{e}_1^T$.

Proposition 3-B.19. $((K^n, F_{\alpha_d}, F_{\beta_d}), h_d)$ is the distinguishable standard form if and only if F_{α_d} and F_{β_d} are given by Figure 3.5 and Figure 3.6 respectively, and $F_{\alpha d}^{\mu_{j-1}-\mu_j}\mathbf{c}^j = F_{\beta d}\mathbf{c}^{j-1}$ holds for j $(2 \le j \le k+1)$.

Proof. First, we show the necessity. For any i, j $(1 \le j \le k,\ 1 \le i \le \mu_j)$, $(h_d F_\beta^{j-1} F_\alpha^{i-1})^T = \mathbf{e}_{\mu_1+\mu_2+\cdots+\mu_{j-1}+i}$ and $(h_d F_\beta^{j-1} F_\alpha^{\mu_j})^T = \sum_{m=1}^{j} \sum_{l=1}^{\nu_m} c_{ml}^j (h_d F_\beta^{m-1} F_\alpha^{l-1})^T$ imply $F_\alpha = F_{\alpha d}$. The commutativity $F_{\alpha d} F_\beta = F_\beta F_{\alpha d}$ and $(h_d F_\beta^k)^T = \sum_{m=1}^{k} \sum_{l=1}^{\mu_m} c_{ml}^{k+1} (h_d F_\beta^{m-1} F_\alpha^{l-1})^T$ imply that $F_\beta = F_{\beta d}$ and $F_{\alpha d}^{\mu_{j-1}-\mu_j}\mathbf{c}^j = F_{\beta d}\mathbf{c}^{j-1}$ $(2 \le j \le k+1)$.

Second, we show the sufficiency. Let $F_\alpha = F_{\alpha d}$, $F_\beta = F_{\beta d}$ and $F_{\alpha d}^{\mu_{j-1}-\mu_j}\mathbf{c}^j = F_{\beta d}\mathbf{c}^{j-1}$ $(2 \le j \le k+1)$ hold. Then we can easily ascertain by a direct calculation that $((K^n, F_{\alpha d}, F_{\beta d}), \mathbf{e}_1^T)$ is the distinguishable standard form.

Proposition 3-B.20. Let the set of output values Y be K. For any distinguishable $\{\alpha, \beta\}$-action with a readout map $((K^n, F_\alpha, F_\beta), h)$, there uniquely exists the distinguishable standard form $((K^n, F_{\alpha d}, F_{\beta d}), \mathbf{e}_1^T)$ which is isomorphic to it.

Proof. Find an integer μ_1 such that vectors $\{(\mathbf{e}_1^T F_\alpha^i)^T; 0 \le i \le \mu_1 - 1\}$ are linearly independent and vectors $\{(\mathbf{e}_1^T F_\alpha^i)^T; 0 \le i \le \mu_1\}$ are linearly dependent.

Set$(\mathbf{e}_1^T F_\alpha^{\mu_1})^T = \sum_{i=1}^{\mu_1} c_{1i}^1 (\mathbf{e}_1^T F_\alpha^{i-1})^T$ and $\mathbf{c}^1 = [c_{11}^1, c_{12}^1, c_{13}^1, \cdots, c_{1\mu_1}^1, 0, \cdots, 0]^T$.

Next, find an integer μ_2 such that vectors $B_2 := \{(\mathbf{e}_1^T F_\beta^j F_\alpha^i)^T; 0 \le j \le 1, 0 \le i \le \mu_{j+1} - 1\}$ are linearly independent and vectors $\{B_2, (\mathbf{e}_1^T F_\beta F_\alpha^{\mu_2})^T\}$ are linearly dependent.

Set $(\mathbf{e}_1^T F_\beta F_\alpha^{\mu_2})^T = \sum_{m=1}^{2} \sum_{l=1}^{\mu_m} c_{ml}^2 (\mathbf{e}_1^T F_\beta^{m-1} F_\alpha^{l-1})^T$ and $\mathbf{c}^2 = [c_{11}^2, c_{12}^2, c_{13}^2, \cdots, c_{1\mu_1}^2, c_{21}^2, c_{22}^2, \cdots, c_{2\mu_2}^2, 0, \cdots, 0]^T$.
These procedures are followed till the next procedure.

Find an integer μ_k such that vectors $B_k := \{(\mathbf{e}_1^T F_\beta^j F_\alpha^i)^T; 0 \le j \le k-1, 0 \le i \le \mu_{j+1} - 1\}$ are linearly independent and vectors $\{B_k, (\mathbf{e}_1^T F_\beta^{k-1} F_\alpha^{\mu_k})^T\}$ and $\{B_k, (\mathbf{e}_1^T F_\beta^k)^T\}$ are linearly dependent. Set $\mathbf{c}^k = [c_{11}^k, c_{12}^k, \cdots, c_{1\mu_1}^k, c_{21}^k, c_{22}^k, \cdots, c_{2\mu_2}^k, \cdots, c_{k1}^k, \cdots, c_{k\mu_k}^k]^T$ and $\mathbf{c}^{k+1} = [c_{11}^{k+1}, c_{12}^{k+1}, \cdots, c_{1\mu_1}^{k+1}, c_{21}^{k+1}, c_{22}^{k+1}, \cdots, c_{2\mu_2}^{k+1}, \cdots, c_{k1}^{k+1}, \cdots, c_{k\mu_k}^{k+1}]^T$ in terms of $(\mathbf{e}_1^T F_\beta^{k-1} F_\alpha^{\mu_k})^T = \sum_{m=1}^{k} \sum_{l=1}^{\mu_m} c_{ml}^k (\mathbf{e}_1^T F_\beta^{m-1} F_\alpha^{l-1})^T$ and $(\mathbf{e}_1^T F_\beta^k)^T = \sum_{m=1}^{k} \sum_{l=1}^{\mu_m} c_{ml}^{k+1} (\mathbf{e}_1^T F_\beta^{m-1} F_\alpha^{l-1})^T$.

By the selection of vectors and indices $\{\mu_i : 1 \le i \le k\}$, $\sum_{i=1}^{k} \mu_i = n$ holds. Let us supply a linear operator $K^n \to K^n$ by setting $T(\mathbf{e}_1^T F_\beta^i F_\alpha^j)^T$

$= \mathbf{e}_{\mu_1+\mu_2+\cdots+\mu_i+j+1}$. Then T is a regular matrix. Set $F_{\alpha d} = TF_\alpha T^{-1}$ and $F_{\beta d} = TF_\beta T^{-1}$. Then $F_{\alpha d}, F_{\beta d} \in K^{n\times n}$, $F_{\alpha d}F_{\beta d} = F_{\beta d}F_{\alpha d}$ hold.

Finally, we can conclude that $((K^n, F_{\alpha d}, F_{\beta d}), \mathbf{e}_1^T)$ is an $\{\alpha,\beta\}$-action with a readout map. Since the linear operator T preserves the linear independence and linear dependence of each vector, $((K^n, F_{\alpha d}, F_{\beta d}), \mathbf{e}_1^T)$ is a $\mu = (\mu_1, \mu_2, \cdots, \mu_k)$-distinguishable. T is an $\{\alpha,\beta\}$-morphism with a readout map $T : ((K^n, F_\alpha, F_\beta), h) \to ((K^n, F_{\alpha d}, F_{\beta d}), \mathbf{e}_1^T)$. Its uniqueness can be demonstrated by means of the definition of the distinguishable standard form.

Remark: There are many equivalence classes in the category of $\{\alpha,\beta\}$-actions with a readout map; this proposition asserts that the equivalence classes can be represented as distinguishable standard forms.

3-B.3 Finite-Dimensional 2-Commutative Linear Representation Systems

This section offers the proofs of Representation Theorems (3.15) and (3.17) for finite-dimensional canonical 2-Commutative Linear Representation Systems.

3-B.21. Proof of Representation Theorem (3.15)
Note that a pointed $\{\alpha,\beta\}$-action in the Quasi-reachable Standard System is the quasi-reachable standard form. Let $\sigma = ((K^n, F_\alpha, F_\beta), x^0, h)$ be any finite-dimensional canonical 2-Commutative Linear Representation System. For the quasi-reachable standard form $((K^n, F_{\alpha s}, F_{\beta s}), \mathbf{e}_1)$ and a pointed $\{\alpha,\beta\}$-morphism $T : ((K^n, F_\alpha, F_\beta), x^0) \to ((K^n, F_{\alpha s}, F_{\beta s}), \mathbf{e}_1)$ introduced in the proof of Theorem (3-B.10), let $h_s := h \cdot T^{-1}$. Then T is a 2-Commutative Linear Representation System morphism $T : ((K^n, F_\alpha, F_\beta), x^0, h) \to ((K^n, F_{\alpha s}, F_{\beta s}), \mathbf{e}_1, h_s)$. T is bijective and $((K^n, F_{\alpha s}, F_{\beta s}), \mathbf{e}_1, h_s)$ is the unique Quasi-reachable Standard System. By Corollary (3.12), the behaviors of $((K^n, F_\alpha, F_\beta), x^0, h)$ and $((K^n, F_{\alpha s}, F_{\beta s}), \mathbf{e}_1, h_s)$ are the same.

3-B.22. Proof of Representation Theorem (3.17)
Note that a $\{\alpha,\beta\}$-action with a readout map in the Distinguishable Standard System is the distinguishable standard form.

Let $\sigma = ((K^n, F_\alpha, F_\beta), x^0, h)$ be any finite-dimensional canonical 2-Commutative Linear Representation System. For the distinguishable standard form $((K^n, F_{\alpha d}, F_{\beta d}), \mathbf{e}_1^T)$ and a $\{\alpha,\beta\}$-morphism with readout map $T : ((K^n, F_\alpha, F_\beta), h) \to ((K^n, F_{\alpha d}, F_{\beta d}), \mathbf{e}_1^T)$ introduced in the proof of Theorem (3-B.20), let $x_d^0 := Tx^0$. Then T is a 2-Commutative Linear Representation System morphism $T : ((K^n, F_\alpha, F_\beta), x^0, h) \to ((K^n, F_{\alpha d}, F_{\beta d}), x_d^0, \mathbf{e}_1^T)$. T is bijective and $\sigma_d = ((K^n, F_{\alpha d}, F_{\beta d}), x_d^0, \mathbf{e}_1^T)$ is the unique Distinguishable Standard System. By Corollary (3.12), the behaviors of σ and σ_d are the same.

3-B.4 Existence Criterion for 2-Commutative Linear Representation Systems

This section furnishes the proofs of the existence criterion Theorem (3.19). Let $G_l = G \cdot J_l$, where J_l is the canonical injection : $P(\le l) \to K[z_\alpha, z_\beta]$. Let $H_m = P_m \cdot H$, where P_m is the canonical surjection $P_m : F(N \times N, Y) \to F((N \times N)_m, Y)$, where $P(\le l) := \{\lambda(i,j) z_\alpha^i z_\beta^j;\ \lambda(i,j) \in K,\ i+j \le l$ for any $i,\ j,\ l \in N\}$ and $F((N \times N)_m, Y) := \{a(i,j) \in Y;\ a \in F(N \times N, Y), i+j \le m$ for $i,\ j,\ m \in N\}$.

3-B.23. Proof of Theorem (3.19)
Let A be the linear input/output map corresponding to a two-dimensional image $a \in F(N \times N, Y)$. Obviously, im $A = \ll \{S_\alpha^i S_\beta^j a;\ i,j \in N\} \gg$. Let $A_l := A \cdot J_l$, and let a linear operator $A_{(l,m)} : P(\le l) \to F((N \times N)_m, Y)$ be defined by setting $A_{(l,m)} := P_m \cdot A \cdot J_l$. Then $A_{(l,m)}$ can be represented by a partial Hankel matrix $H_{a\ (l,m)}$ of the Hankel matrix H_a, where $H_{a\ (l,m)} = [a(\bar{i}+i, \bar{j}+j)]$ for $\bar{i}+\bar{j} \le l$ and $i+j \le m$.

First, we show 1) $\Longrightarrow$ 2). By Theorem (3.3) and Corollary (3-A.33), im A is n-dimensional. If im $A_{n-1} \ne$ im A_n, then the dimension of im A_n is greater than or equal to $n+1$ by Lemma (3-B.7).

Therefore, im $A_{n-1} =$ im $A_n = \cdots =$ im A. Consequently, there exist n linearly independent vectors in $\{S_\alpha^i S_\beta^j a;\ i+j \le n-1\}$, but not including greater than or equal to $n+1$ linearly independent vectors in it.

Second, we show 2) $\Longrightarrow$ 3). Since im $A_{n-1} =$ im A_n, im $A_{n-1} =$ im $A_n = \cdots =$ im A holds. Therefore, the dimension of im A_r is n for $r \le n-1$. On the other hand, by Corollary (3-A.33) and Lemma (3-B.7), ker $P_s = 0$ for $s \le n-1$. Consequently, the dimension of im $P_s \cdot A \cdot J_r$ is n. Therefore, the rank of partial Hankel matrix $H_{a\ (r,s)}$ corresponding to $P_s \cdot A \cdot J_r$ is n.

Finally, we show 3) $\Longrightarrow$ 1). Since the rank of the Hankel matrix H_a is n, im A of the linear input/output map A corresponding to H_a is n dimensional. By im $A = \ll \{S_\alpha^i S_\beta^j a;\ i,j \in N\} \gg$ and Corollary (3-A.33), 1) is obtained.

3-B.24. Proof of Theorem (3.20) The necessary condition is clear from Proposition (3-B.2) and Proposition (3-B.12). The sufficient condition will be proved only for the case of p=1. It is similar for all other p. Assume that $\bar{a}$ is given in Theorem (3.20), where $q_\alpha(z_\alpha) = z_\alpha^n + \alpha_{n-1} z_\alpha^{n-1} + \alpha_{n-2} z_\alpha^{n-2} + \cdots + \alpha_0$, and $q_\beta(z_\beta) = z_\beta^m + \beta_{m-1} z_\beta^{m-1} + \beta_{m-2} z_\beta^{m-2} + \cdots + \beta_0$. Consider a 2-Commutative Linear Representation System $\sigma = ((K^{m \times n}, F_\alpha, F_\beta), X^0, h)$, where $F_\alpha : K^{m \times n} \to K^{m \times n}; X \mapsto XA$, $F_\beta : K^{m \times n} \to K^{m \times n}; X \mapsto BX$ and X^0 and h are given as follows:

$$A = \begin{bmatrix} 0 & 1 & 0 \cdots & 0 & 0 \\ 0 & 0 & 1 & 0 \cdots & \vdots \\ \vdots & \vdots & & & 0 \\ \vdots & \vdots & & \ddots & 0 \\ 0 & 0 & \cdots\cdots & 0 & 1 \\ -\alpha_0 & -\alpha_1 & \cdots & & -\alpha_{n-1} \end{bmatrix} \in K^{n\times n},$$

$$B = \begin{bmatrix} 0 \cdots & \cdots & \cdots 0 & -\beta_0 \\ 1 \; 0 & \cdots & \cdots 0 & -\beta_1 \\ 0 \; \vdots & & & \vdots \\ 0 \; \vdots & \ddots & \vdots & \vdots \\ 0 \; \vdots & \cdots & \ddots \; 0 & \vdots \\ 0 \cdots & \cdots & 0 \; 1 & -\beta_{m-1} \end{bmatrix} \in K^{m\times m}.$$

$$X^0 = \begin{bmatrix} \lambda(0,0) & \cdots & \cdots & \lambda(0,n-1) \\ \vdots & & & \lambda(1,n-1) \\ \vdots & & & \vdots \\ \vdots & & & \vdots \\ \vdots & & & \vdots \\ \lambda(m-1,0) & \cdots & \cdots & \lambda(m-1,n-1) \end{bmatrix} \in K^{m\times n}.$$

$$h : K^{m\times n} \to K; X = \begin{bmatrix} x_{0,0} & \cdots & \cdots & x_{0,n-1} \\ x_{1,1} & & & x_{1,n-1} \\ \vdots & & & \vdots \\ \vdots & & & \\ x_{m-1,0} & \cdots & \cdots & x_{m-1,n-1} \end{bmatrix} \mapsto x_{m-1,n-1}.$$

Then, $\mathbf{e}_m^T[z_\beta I - F_\beta]^{-1} = (1/(q_\beta(z_\beta)))[1, z_\beta, \cdots, z_\beta^{m-1}]$ and $[z_\alpha I - F_\alpha]^{-1}\mathbf{e}_n = (1/(q_\alpha(z_\alpha)))[1, z_\alpha, \cdots, z_\alpha^{n-1}]^T$ hold. Consequently, the formal power series $\bar{a}_\sigma = \sum_i \sum_j a_\sigma(i,j) z_\alpha^{-i} z_\beta^{-j}$ which corresponds to the behavior a_σ of the 2-Commutative Linear representation System σ is:
$\bar{a}_\sigma = \mathbf{e}_m^T[z_\beta I - F_\beta]^{-1} X^0 [z_\alpha I - F_\alpha]^{-1}\mathbf{e}_n$
$= \bar{a} = z_\alpha z_\beta (\sum_{i=0}^{n-1} \sum_{j=0}^{m-1} \lambda(i,j) z_\alpha^i z_\beta^j)/(q_\alpha(z_\alpha) q_\beta(z_\beta))$.

Therefore, there exists a finite-dimensional 2-Commutative Linear Representation System σ which expresses $\bar{a}$.

Remark: Fliess [1970] has pointed out that the recognizability of two-variable formal power series can be characterized by a rational function as in Theorem (3.20).

3-B.5 Realization Procedure for 2-Commutative Linear Representation Systems

This section supplies the proof of the realization procedure Theorem (3.21).

3-B.25 Proof of Theorem (3.21)
Let $R(a) := \{S_\alpha^i S_\beta^j a;\ i,j \in N\}$. By Theorem (3.3), $((\ll R(a) \gg, S_\alpha, S_\beta), a, (0,0))$ is a canonical 2-Commutative Linear Representation System that realizes $a \in F(N \times N, Y)$. The linearly independent vectors $\{S_\alpha^i S_\beta^j a;\ 0 \leq i \leq \nu_{j+1} - 1,\ 0 \leq j \leq k-1,\}$ satisfy
$\ll \{S_\alpha^i S_\beta^j a;\ 0 \leq i \leq \nu_{j+1} - 1,\ 0 \leq j \leq k-1\} \gg = \ll R(a) \gg$.
Let $T : \ll R(a) \gg \rightarrow K^n$ be a linear map. Then
$T \cdot S_\beta^j S_\alpha^i \mathbf{e}_1 = \mathbf{e}_{\nu_1+\nu_2+\cdots+\nu_{j-1}+i+1}$ holds for $0 \leq i \leq \nu_{j+1}-1$ and $0 \leq j \leq k-1$. The procedure from the step 1) through the step k+3) ascerts $F_{\alpha s}T = TS_\alpha$ and $F_{\beta s}T = TS_\beta$. $h_s \cdot T = (0,0)$ follows from the step k+4).

Consequently, T is bijective, and it is a 2-Commutative Linear Representation System morphism
T: $((\ll R(a) \gg, S_\alpha, S_\beta), a, (0,0)) \rightarrow ((K^n, F_{\alpha s}, F_{\beta s}), \mathbf{e}_1, h_s)$.

By Corollary (3.12), the behavior of $((K^n, F_{\alpha s}, F_{\beta s}), \mathbf{e}_1, h_s)$ is a. It follows from the choice of $\{S_\alpha^i S_\beta^j a;\ 0 \leq i \leq \nu_{j+1} - 1,\ 0 \leq j \leq k-1\}$ and the determination of map T that $((K^n, F_{\alpha s}, F_{\beta s}), \mathbf{e}_1, h_s)$ is the Quasi-reachable Standard System.

3-C Partial Realization Theory

This appendix appears proofs for theorems and propositions stated in Section 3.4. See Appendices 3-A and 3-B for details of notions and notations.

3-C.1 Pointed $\{\alpha, \beta\}$-Actions

Here, let $P(\leq l_1, \leq m_1)$ be $P(\leq l_1, \leq m_1) := \sum_{i=0}^{l_1} \sum_{j=0}^{m_1} \lambda(i,j) z_\alpha^i z_\beta^j$.

Moreover, let $G_{(l_1,m_1)} = G \cdot J_{(l_1,m_1)}$,
where $J_{(l_1,m_1)}$ is the canonical injection : $P(\leq l_1, \leq m_1) \rightarrow K[z_\alpha, z_\beta]$.

Definition 3-C.1 If a pointed $\{\alpha, \beta\}$-action $((X, F_\alpha, F_\beta), x^0)$ satisfies $X = \ll \{F_\alpha^i F_\beta^j x^0; i \leq l_1, j \leq m_1\} \gg$, then $((X, F_\alpha, F_\beta), x^0)$ is said to be (l_1, m_1)-quasi-reachable, where $\ll S \gg$ denotes the smallest linear space which contains a set S.

Remark: Note that $((X, F_\alpha, F_\beta), x^0)$ is (l_1, m_1)-quasi-reachable if and only if $G_{(l_1,m_1)} = G \cdot J_{(l_1,m_1)} : P(\leq l_1, \leq m_1) \rightarrow X$ is surjective, where G is the linear input map : $(K[z_\alpha, z_\beta], z_\alpha, z_\beta) \rightarrow (X, F_\alpha, F_\beta)$ corresponding to $((X, F_\alpha, F_\beta), x^0)$.

Proposition 3-C.2. If a linear subspace S of $P(\leq l_1+1, \leq m_1+1)$ satisfies the following two conditions, then there uniquely exists an ideal $\underline{S} \subseteq K[z_\alpha, z_\beta]$ such that $\underline{S} \cap P(\leq l_1+1, \leq m_1+1) = S$ and $P(\leq l_1+1, \leq m_1+1)/S$ is isomorphic to $K[z_\alpha, z_\beta]/\underline{S}$. Moreover, a pointed $\{\alpha, \beta\}$-action $((K[z_\alpha, z_\beta]/\underline{S}, \dot{z}_\alpha, \dot{z}_\beta), \mathbf{1}+\underline{S})$ is (l_1, m_1)-quasi-reachable, where $\dot{z}_\alpha$ is given by $\dot{z}_\alpha(\lambda+\underline{S}) = z_\alpha\lambda+\underline{S}$ for $\lambda \in K[z_\alpha, z_\beta]$, and $\dot{z}_\beta$ is given by $\dot{z}_\beta(\lambda+\underline{S}) = z_\beta\lambda+\underline{S}$ for $\lambda \in K[z_\alpha, z_\beta]$.

Condition 1: $\lambda \in P(\leq l_1+1, \leq m_1+1) \cap S$ implies $z_\alpha\lambda \in S$ and $z_\beta\lambda \in S$.

Condition 2: There exist coefficients $\lambda_{l_1}(l,m) \in K$ such that $z_\alpha^{l_1+1}z_\beta - \sum_{l=0}^{l_1}\sum_{m=0}^{m_1}\lambda_{l_1}(l,m)z_\alpha^l z_\beta^m \in S$ and $\lambda_{m_1}(l,m) \in K$ such that $z_\alpha z_\beta^{m_1+1} - \sum_{l=0}^{l_1}\sum_{m=0}^{m_1}\lambda_{m_1}(l,m)z_\alpha^l z_\beta^m \in S$.

Proof. Let $J_{(l_1,m_1,l_1+1,m_1+1)} : P(\leq l_1, \leq m_1) \to P(\leq l_1+1, \leq m_1+1)$ be the canonical injection and $\pi_S : P(\leq l_1+1, \leq m_1+1) \to P(\leq l_1+1, \leq m_1+1)/S$ be the canonical surjection. Then **condition 2** implies that a composition map $\pi_S \cdot J_{(l_1,m_1,l_1+1,m_1+1)}$ is surjective. And **condition 1** implies that $\dot{z}_\alpha, \dot{z}_\beta$ can be defined uniquely by setting $\dot{z}_\alpha(\lambda_1+S) = z_\alpha\lambda_1+S$, $\dot{z}_\beta(\lambda_2+S) = z_\beta\lambda_2+S$, where λ_1, $\lambda_2 \in K[z_\alpha, z_\beta]$.

Therefore, $\dot{z}_\alpha$ and $\dot{z}_\beta$ satisfy $\dot{z}_\alpha, \dot{z}_\beta \in P(\leq l_1+1, \leq m_1+1)/S$ and $\dot{z}_\alpha\dot{z}_\beta = \dot{z}_\beta\dot{z}_\alpha$. Hence, $((P(\leq l_1+1, \leq m_1+1)/S, \dot{z}_\alpha\dot{z}_\beta), \mathbf{1}+S)$ is a pointed $\{\alpha, \beta\}$-action and it is (l_1, m_1)-quasi-reachable.

Setting $G_{(l_1+1,m_1+1)} := G \cdot J_{(l_1+1,m_1+1)}$, ker $G_{(l_1+1,m_1+1)} = S$ holds and $\underline{S} :=$ ker G satisfies $\underline{S} \cap P(\leq l_1+1, \leq m_1+1) = S$. Since G is a linear input map, $\underline{S}$ is an invariant subspace under z_α and z_β. Moreover, the surjectivity of G implies that $((P(\leq l_1+1, \leq m_1+1)/S, \dot{z}_\alpha\dot{z}_\beta), \mathbf{1}+S)$ is isomorphic to $((K[z_\alpha, z_\beta]/\underline{S}, \dot{z}_\alpha, \dot{z}_\beta), \mathbf{1}+\underline{S})$. Therefore, $((K[z_\alpha, z_\beta]/\underline{S}, \dot{z}_\alpha, \dot{z}_\beta), \mathbf{1}+\underline{S})$ is (l_1, m_1)-quasi-reachable. The uniqueness of $\underline{S}$ is proved by means of the uniqueness of $\dot{z}_\alpha, \dot{z}_\beta$ and G.

3-C.2 $\{\alpha, \beta\}$-Actions with a Readout Map

Set $F(\mathbf{l}_2 \times \mathbf{m}_2, Y) := \{$a function : $\mathbf{l}_2 \times \mathbf{m}_2 \to Y\}$, let $P_{(l_2,m_2)}$ be the canonical surjection : $F(N \times N, Y) \to F(\mathbf{l}_2 \times \mathbf{m}_2, Y); a \mapsto [; (i,j) \mapsto a(i,j)]$, and define $\underline{S}_\alpha$ and $\underline{S}_\beta$ by setting

$$\underline{S}_\alpha : F(\mathbf{l}_2 \times \mathbf{m}_2, Y) \to F((\mathbf{l}_2 - \mathbf{1}) \times \mathbf{m}_2, Y); a \mapsto \underline{S}_\alpha a[; (i,j) \mapsto a(i+1,j)]$$

and

$$\underline{S}_\beta : F(\mathbf{l}_2 \times \mathbf{m}_2, Y \to F(\mathbf{l}_2 \times (\mathbf{m}_2 - \mathbf{1}), Y); a \mapsto \underline{S}_\beta a[; (i,j) \mapsto a(i,j+1)].$$

Definition 3-C.3. If a $\{\alpha, \beta\}$-action with a readout map $((X, F_\alpha, F_\beta), h)$ satisfies the condition that $hF_\alpha^i F_\beta^j x = 0$ implies $x = 0$ for any $i \leq l_2$, $j \leq m_2$, then $((X, F_\alpha, F_\beta), h)$ is called (l_2, m_2)-distinguishable for some integers l_2, m_2.

Remark: Note that $((X, F_\alpha, F_\beta), h)$ is (l_2, m_2)-distinguishable if and only if a linear map $H_{(l_2,m_2)} := P_{(l_2,m_2)} \cdot H$ is injective, where H is a linear observation map corresponding to $((X, F_\alpha, F_\beta), h)$ and $P_{(l_2,m_2)}$ is the canonical surjection : $F(N \times N, Y) \to F(\mathbf{l}_2 \times \mathbf{m}_2, Y)$.

Remark: Note that $((K^n, F_\alpha, F_\beta), h)$ is distinguishable if and only if it is $(n-1, n-1)$-distinguishable.

Proposition 3-C.4 If a subspace Z of $F((\mathbf{l}_2 + \mathbf{1}) \times (\mathbf{m}_2 + \mathbf{1}), Y)$ satisfies the following two conditions, then there uniquely exists a $\{\alpha, \beta\}$-action (X, S_α, S_β) such that the map $P_{(l_2,m_2)}|_X : X \to Z$ is isomorphism, where $P_{(l_2,m_2)}|_X$ is a restriction of the canonical surjection $P_{(l_2,m_2)}$: $F(N \times N, Y) \to F(\mathbf{l}_2 \times \mathbf{m}_2, Y)$ to X, and a $\{\alpha, \beta\}$-action with a readout map $((X, S_\alpha, S_\beta), (0,0))$ is (l_2, m_2)-distinguishable.

Condition 3: The composition map $\pi \cdot j : Z \xrightarrow{j} F((\mathbf{l}_2 + \mathbf{1}) \times (\mathbf{m}_2 + \mathbf{1}), Y) \xrightarrow{\pi} F(\mathbf{l}_2 \times \mathbf{m}_2, Y)$ is injective, where j is the canonical injection and π is the canonical surjection.

Condition 4: im $(\underline{S}_\alpha \cdot j) \subseteq$ im $(j \cdot \pi_\alpha)$ holds in the sense of $F(\mathbf{l}_2 \times (\mathbf{m}_2 + \mathbf{1}), Y)$ and im $(\underline{S}_\beta \cdot j) \subseteq$ im $(j \cdot \pi_\beta)$ holds in the sense of $F((\mathbf{l}_2 + \mathbf{1}) \times \mathbf{m}_2, Y)$, where π_α is the canonical surjection: $F((\mathbf{l}_2 + \mathbf{1}) \times (\mathbf{m}_2 + \mathbf{1}), Y) \to F(\mathbf{l}_2 \times (\mathbf{m}_2 + \mathbf{1}), Y)$ and π_β is the canonical surjection : $F((\mathbf{l}_2 + \mathbf{1}) \times (\mathbf{m}_2 + \mathbf{1}), Y) \to F((\mathbf{l}_2 + \mathbf{1}) \times \mathbf{m}_2, Y)$.

Proof. By **conditions 3** and **4**, we can define $F_\alpha(z) = (\pi_\alpha \cdot j)^{-1} \underline{S}_\alpha \cdot j(z)$ and $F_\beta(z) = (\pi_\beta \cdot j)^{-1} \underline{S}_\beta \cdot j(z)$ for any $z \in Z$.

Then $F_\alpha, F_\beta \in L(Z)$ and $F_\alpha F_\beta = F_\beta F_\alpha$ hold. Hence, $((Z, F_\alpha, F_\beta), \overline{(0,0)})$ is a distinguishable $\{\alpha, \beta\}$-action with a readout map, where $\overline{(0,0)}$ is the map : $Z \to Y; a \mapsto a(0,0)$. Injectivity of $\pi_\alpha \cdot j$ and $\pi_\beta \cdot j$ imply that $((Z, F_\alpha, F_\beta), \overline{(0,0)})$ is (l_2, m_2)-distinguishable. It follows that the linear observation map H corresponding to $((Z, F_\alpha, F_\beta), \overline{(0,0)})$ is injective. Set $X :=$ im H, a map $H^{-1} : X \to Z$ is clearly the restriction of the map $P_{(l_2,m_2)}$: $F(N \times N, Y) \to F(\mathbf{l}_2 \times \mathbf{m}_2, Y)$ to X.

An equation $\overline{(0,0)} = (0,0) \cdot H$ implies that $((X, S_\alpha, S_\beta), (0,0))$ is isomorphic to $((Z, F_\alpha, F_\beta), \overline{(0,0)})$ in the sense of $\{\alpha, \beta\}$-action with a readout map. Therefore, $((X, S_\alpha, S_\beta), (0,0))$ is (l_2, m_2)-distinguishable. The uniqueness of X is demonstrated by the uniqueness of F_α, F_β and H.

3-C.3 Partial Realization Problem

For a 2-Commutative Linear Representation System $\sigma = ((X, F_\alpha, F_\beta), x^0, h)$ and some $l_1, m_1, l_2, m_2 \in N$.

For a given $\underline{a} \in F(\mathbf{L} \times \mathbf{M}, Y)$, if there exist l_1, m_1 and $l_2, m_2 \in N$ such that $l_1 + l_2 < L$, $m_1 + m_2 < M$ and its partial realization σ is (l_1, m_1)-quasi-reachable and (l_2, m_2)-distinguishable, then σ is said to be a natural partial realization of $\underline{a}$. For a partial finite-sized image $\underline{a} \in F(\mathbf{L} \times \mathbf{M}, Y)$, the matrix $H_{\underline{a}\ (l_1,m_1,L-l_1,M-m_1)}$ is said to be a finite-sized Hankel matrix of $\underline{a}$.

For the definition of $H_{\underline{a}\ (l_1,m_1,L-l_1,M-m_1)}$, see comments between Definition (3.23) and Theorem (3.24).

We can consider a partial linear input/output map

$$A_{(l_1,m_1,L-l_1,M-m_1)} : P(\leq l_1, \leq m_1) \to F((\mathbf{L}-\mathbf{l_1}) \times (\mathbf{M}-\mathbf{m}_1), Y)$$

for $\underline{a} \in F(\mathbf{L} \times \mathbf{M}, Y)$. The linear input/output map $A : (K[z_\alpha, z_\beta], z_\alpha, z_\beta) \to (F(N \times N, Y), S_\alpha, S_\beta)$ can be regarded as $a \in F(N \times N, Y)$ in the same way as in Appendix 3-A.

Lemma 3-C.5. Let $A_{(l_1,m_1,L-l_1,M-m_1)}$ be the partial linear input/output map corresponding to $\underline{a} \in F(\mathbf{L} \times \mathbf{M}, Y)$. Then the following diagrams commute:

$$\begin{array}{ccc} P(\leq l_1, \leq m_1) & \xrightarrow{A_{(l_1,m_1,L-l_1,M-m_1)}} & F((\mathbf{L}-\mathbf{l}_1) \times (\mathbf{M}-\mathbf{m}_1), Y) \\ {\scriptstyle \underline{i}}\downarrow & & \downarrow{\scriptstyle \pi} \\ P(\leq l_1+i, \leq m_1+j) & \xrightarrow{A_{(l_1+i,m_1+j,L-l_1-i,M-m_1-j)}} & F((\mathbf{L}-\mathbf{l}_1-\mathbf{i}) \times (\mathbf{M}-\mathbf{m}_1-\mathbf{j}), Y), \end{array}$$

where $\underline{i}$ is canonical injection and π is canonical surjection.

$$\begin{array}{ccc} P(\leq l_1, \leq m_1) & \xrightarrow{A_{(l_1,m_1,L-l_1,M-m_1)}} & F((\mathbf{L}-\mathbf{l}_1) \times (\mathbf{M}-\mathbf{m}_1), Y) \\ {\scriptstyle z_\alpha^i z_\beta^j}\downarrow & & \downarrow{\scriptstyle \underline{S}_\alpha^i \underline{S}_\beta^j} \\ P(\leq l_1+i, \leq m_1+j) & \xrightarrow{A_{(l_1+i,m_1+j,L-l_1-i,M-m_1-j)}} & F((\mathbf{L}-\mathbf{l}_1-\mathbf{i}) \times (\mathbf{M}-\mathbf{m}_1-\mathbf{j}), Y) \end{array}$$

Proof. These diagrams can be found by direct calculation.

Proposition 3-C.6. Let $A_{(\ ,\ ,\ ,\ )}$ be the partial linear input/output map corresponding to $\underline{a} \in F(\mathbf{L} \times \mathbf{M}, Y)$ and l_1, l_2, m_1 and m_2 be any integers such that $0 \leq l_2 \leq l_1$ and $0 \leq m_2 \leq m_1$. If

Ⓐ: $\operatorname{im} A_{(l_2+1,m_2,L-l_2-1,M-m_2)} = \operatorname{im} A_{(l_2,m_2,L-l_2-1,M-m_2)}$ and
Ⓑ: $\operatorname{im} A_{(l_2,m_2+1,L-l_2,M-m_2-1)} = \operatorname{im} A_{(l_2,m_2,L-l_2,M-m_2-1)}$ hold, then
$\operatorname{im} A_{(l_2+1,m_2+1,L-l_2-1,M-m_2-1)} = \operatorname{im} A_{(l_2+1,m_2,L-l_2-1,M-m_2-1)} =$
$\operatorname{im} A_{(l_2,m_2+1,L-l_2-1,M-m_2-1)} = \operatorname{im} A_{(l_2,m_2,L-l_2-1,M-m_2-1)}$ holds.

Proof. Let l_1', l_2', m_1' and m_2' be any integers such that $0 \leq l_2' \leq l_1'$ and $0 \leq m_2' \leq m_1'$.

Then we note that the following fact *) is trivial:

*) : $\operatorname{im} A_{(l_2',m_2',L-l_1',M-m_1')} = \operatorname{im} A_{(l_1',m_1',L-l_1',M-m_1')}$ is equivalent to $\operatorname{im} A_{(l_2',m_2',l,m)} = \operatorname{im} A_{(l_1',m_1',l,m)}$ for any l, m such that $0 \leq l \leq L - l_1'$ and $0 \leq m \leq M - m_1'$.

Assume that condition Ⓐ holds. Then we can show that the following equation holds:
① : im $A_{(l_2+l,m_2,L-l_2-l,M-m_2)} = \text{im } A_{(l_2,m_2,L-l_2-l,M-m_2)}$
for any $l \in N$ $(1 \leq l \leq L-l_2)$.
We prove ① by induction.
By assumption, ① holds for $l=1$.
Let's assume it holds for $l=k$, i.e., assume that im $A_{(l_2+k,m_2,L-l_2-k,M-m_2)} =$ im $A_{(l_2,m_2,L-l_2-k,M-m_2)}$. Then for any $i,j \in N$ such that $0 \leq i \leq L-l_2-k$ and $0 \leq j \leq M-m_2$, there exist coefficients $\{\lambda_\alpha(p,q) \in K; p,q \in N\}$ such that $\underline{a}(l_2+k+i, m_2+j) = \sum_{q=0}^{m_2}\sum_{p=0}^{l_2}\lambda_\alpha(p,q)\underline{a}(p+i,q+j)$. Therefore, for any $i,j \in N$ such that $0 \leq i \leq L-l_2-k-1$ and $0 \leq j \leq M-m_2$, it follows that $\underline{a}(l_2+k+i+1, m_2+j) = \sum_{q=0}^{m_2}\sum_{p=1}^{l_2+1}\lambda_\alpha(p-1,q)\underline{a}(p+i,q+j)$.
Hence im $A_{(l_2+k+1,m_2,L-l_2-k-1,M-m_2)} = \text{im } A_{(l_2,m_2,L-l_2-k-1,M-m_2)}$.
Thus, ① is proved.
Next, assume that condition Ⓑ holds. Then we will prove that the following equation holds:
② : im $A_{(l_2,m_2+m,L-l_2,M-m_2-m)} = \text{im } A_{(l_2,m_2,L-l_2,M-m_2-m)}$
for any $m \in N$ $(1 \leq m \leq M-m_2)$.
We prove ② by induction.
By assumption, ② holds for $m=1$.
Let's assume it holds for $m=k$, i.e., assume that im $A_{(l_2,m_2+k,L-l_2,M-m_2-k)} =$ im $A_{(l_2,m_2,L-l_2,M-m_2-k)}$. Then for any $i,j \in N$ such that $0 \leq i \leq L-l_2$ and $0 \leq j \leq M-m_2-k$, there exist coefficients $\{\lambda_\beta(p,q) \in K; p,q \in N\}$ such that $\underline{a}(l_2+i, m_2+k+j) = \sum_{q=0}^{m_2}\sum_{p=0}^{l_2}\lambda_\beta(p,q)\underline{a}(p+i,q+j)$. Therefore, for any $i,j \in N$ such that $0 \leq i \leq L-l_2$ and $0 \leq j \leq M-m_2-k-1$, it follows that $\underline{a}(l_2+i, m_2+k+j+1) = \sum_{q=1}^{m_2+1}\sum_{p=0}^{l_2}\lambda_\beta(p,q-1)\underline{a}(p+i,q+j)$ holds.
Hence im $A_{(l_2,m_2+k+1,L-l_2,M-m_2-k-1)} = \text{im } A_{(l_2,m_2,L-l_2,M-m_2-k-1)}$ holds.
Thus, ② is proved.
Next, assuming that condition Ⓐ holds, we will prove that
③ : im $A_{(l_2+1,m_2+1,L-l_2-1,M-m_2-1)} = \text{im } A_{(l_2,m_2+1,L-l_2-1,M-m_2-1)}$ holds.
By Ⓐ, for any $i,j \in N$ such that
$0 \leq i \leq L-l_2-1, 0 \leq j \leq M-m_2$, there exist coefficients $\{\lambda'(p,q) \in K; p,q \in N\}$ such that $\underline{a}(l_2+1+i, m_2+j) = \sum_{q=0}^{m_2}\sum_{p=0}^{l_2}\lambda'(p,q)\underline{a}(p+i,q+j)$.
Therefore, for any $i,j \in N$ such that $0 \leq i \leq L-l_2-1, 0 \leq j \leq M-m_2-1$, $\underline{a}(l_2+1+i, m_2+1+j) = \sum_{q=m_1-m_2+1}^{m_1+1}\sum_{p=l_1-l_2}^{l_1}\lambda'(p+l_2-l_1, q+m_2-m_1-1)\underline{a}(p+i,q+j)$.
Hence, ③ : im $A_{(l_2+1,m_2+1,L-l_2-1,M-m_2-1)} = \text{im } A_{(l_2,m_2+1,L-l_2-1,M-m_2-1)}$ is proved.
By ①, ②, ③ and the fact *) described in the top of this proof, we can derive the following equations:
im $A_{(l_2+1,m_2+1,L-l_2-1,M-m_2-1)} = \text{im } A_{(l_2+1,m_2,L-l_2-1,M-m_2-1)}$
$= \text{im } A_{(l_2,m_2+1,L-l_2-1,M-m_2-1)} = \text{im } A_{(l_2,m_2,L-l_2-1,M-m_2-1)}$.
Hence, Proposition (3-C.6) has been proved.

Proposition 3-C.7. Let $A_{(\ ,\ ,\ ,\ )}$ be the partial linear input/output map corresponding to $\underline{a} \in F(\mathbf{L} \times \mathbf{M}, Y)$ and l_1, l_2, m_1 and m_2 be any integers such that $0 \leq l_2 \leq l_1$ and $0 \leq m_2 \leq m_1$. If

Ⓐ : ker $A_{(l_2, m_2+1, L-l_2, M-m_2-1)} =$ ker $A_{(l_2, m_2+1, L-l_2-1, M-m_2-1)}$ and
Ⓑ : ker $A_{(l_2+1, m_2, L-l_2-1, M-m_2)} =$ ker $A_{(l_2+1, m_2, L-l_2-1, M-m_2-1)}$ hold,

then ker $A_{(l_2, m_2, L-l_2, M-m_2)} =$ ker $A_{(l_2, m_2, L-l_2-1, M-m_2)} =$
ker $A_{(l_2, m_2, L-l_2, M-m_2-1)} =$ ker $A_{(l_2, m_2, L-l_2-1, M-m_2-1)}$ holds.

Proof. Let l_1', l_2', m_1' and m_2' be any integers such that $0 \leq l_2' \leq l_1'$ and $0 \leq m_2' \leq m_1'$.
Then the following fact **) is trivial :
**) : ker $A_{(l_2', m_2', L-l_2', M-m_2')} =$ ker $A_{(l_2', m_2', L-l_1', M-m_1')}$ is equivalent to ker $A_{(l, m, L-l_2', M-m_2')} =$ ker $A_{(l, m, L-l_1', M-m_1')}$ for any l, m such that $0 \leq l \leq l_2'$ and $0 \leq m \leq m_2'$.
Assume that condition Ⓐ holds. Then by induction, we can prove
① : ker $A_{(l, m_2+1, L-l, M-m_2-1)} =$ ker $A_{(l, m_2+1, L-l-1, M-m_2-1)}$
for any $l \in N$ $(0 \leq l \leq l_2)$.
By assumption, ① holds for $l = 0$. Let's assume that it holds for $l = k$, i.e., assume that ker $A_{(k, m_2+1, L-k, M-m_2-1)} =$ ker $A_{(k, m_2+1, L-k-1, M-m_2-1)}$.
Then for any $i, j \in N$ such that $0 \leq i \leq k$, $0 \leq j \leq m_2 + 1$, there exist coefficients $\{\lambda_\alpha(p, q) \in K; p, q \in N\}$ such that
$\underline{a}(L - k + i, M - m_2 - 1 + j) = \sum_{q=0}^{M-m_2-1} \sum_{p=0}^{L-k-1} \lambda_\alpha(p, q)\underline{a}(p + i, q + j)$.
Therefore, for any $\bar{i}, \bar{j} \in N$ such that $0 \leq \bar{i} \leq k-1$, $0 \leq \bar{j} \leq m_2+1$, it follows that $\underline{a}(L-k+\bar{i}+1, M-m_2-1+\bar{j}) = \sum_{q=0}^{M-m_2-1} \sum_{p=1}^{L-k-1} \lambda_\alpha(p-1, q)\underline{a}(p+ i, q + j)$ holds.
Hence ker $A_{(k+1, m_2+1, L-k-1, M-m_2-1)} =$ ker $A_{(k+1, m_2+1, L-k-1-1, M-m_2-1)}$ also holds.
Thus, ① has been proved.
Next, assume that condition Ⓑ holds, and we will show that the following equation holds:
② : ker $A_{(l_2+1, m, L-l_2-1, M-m)} =$ ker $A_{(l_2+1, m, L-l_2-1, M-m-1)}$
for any $m \in N$ $(0 \leq m \leq m_2)$.
Assume that condition Ⓑ holds. Then by induction, we will prove ②.
The proof of ② proceeds in the same way as ①.
Next, assuming that condition Ⓐ holds, we are able to prove that
③ : ker $A_{(l_2, m_2, L-l_2, M-m_2-1)} =$ ker $A_{(l_2, m_2, L-l_2-1, M-m_2-1)}$.
By Ⓐ, for any $i, j \in N$ which satisfy that
$0 \leq i \leq l_2, 0 \leq j \leq m_2 + 1$, there exist coefficients $\{\lambda'(p, q) \in K; p, q \in N\}$ such that
$\underline{a}(L - l_2 + i, M - m_2 - 1 + j) = \sum_{q=0}^{M-m_2-1} \sum_{p=0}^{L-l_2-1} \lambda'(p, q)\underline{a}(p + i, q + j)$.
Therefore,
$\underline{a}(L-l_2+i-1, M-m_2-1+j) = \sum_{q=0}^{M-m_2-1} \sum_{p=1}^{L-l_2} \lambda'(p-1, q)\underline{a}(p+i, q+j)$
holds for $i, j \in N$ $(0 \leq i \leq l_2,\ 0 \leq j \leq m_2)$.

Consequently, ③ : ker $A_{(l_2,m_2,L-l_2,M-m_2-1)}$ = ker $A_{(l_2,m_2,L-l_2-1,M-m_2-1)}$ holds.
By ①, ②, ③ and the fact **), we obtain the following equations:
ker $A_{(l_2,m_2,L-l_2,M-m_2)}$ = ker $A_{(l_2,m_2,L-l_2-1,M-m_2)}$
= ker $A_{(l_2,m_2,L-l_2,M-m_2-1)}$ = ker $A_{(l_2,m_2,L-l_2-1,M-m_2-1)}$.
Hence, this proposition has been proved.

Lemma 3-C.8. For a partial linear input/output map $A_{(\ ,\ ,\ ,\ )}$ corresponding to
$\underline{a} \in F(\mathbf{L} \times \mathbf{M}, Y)$ and a 2-Commutative Linear Representation System $\sigma = ((X, F_\alpha, F_\beta), x^0, h)$, the following 1) and 2) hold, where $G_{(l,m)} := G \cdot J_{(l,m)}$, $H_{(L-l,M-m)} := P_{(L-l,M-m)} \cdot H$ for the linear input map G corresponding to x^0 and the linear observation map H corresponding to h and $A_{(l,m,L-l,M-m)} := H_{(L-l,M-m)} \cdot G_{(l,m)}$.

1) σ is a partial realization of $\underline{a}$ if and only if the following diagram commutes for any l and m such that $0 \le l < L,\ 0 \le m < M$.
2) σ is a natural partial realization of $\underline{a}$ if and only if the following diagram commutes, where $G_{(l,m)}$ is surjective and $H_{(L-l,M-m)}$ is injective for some l and m such that $0 \le l < L,\ 0 \le m < M$.

Proof. These can be obtained by the definition of partial and natural partial realization.

3-C.9. Proof of Theorem (3.24)
We prove the theorem by rewriting the conditions of partial Hankel matrix in Theorem (3.24) to a partial linear input/output map $A_{(\ ,\ ,\ ,\ )}$ corresponding to $\underline{a} \in F(\mathbf{L} \times \mathbf{M}, Y)$. By using Proposition (3-C.6) and (3-C.7), the conditions of the Hankel matrix can be equivalently transformed into the following equations:

(1) im $A_{(l_1+1,m_1,L-l_1-1,M-m_1)}$ = im $A_{(l_1,m_1,L-l_1-1,M-m_1)}$
(2) im $A_{(l_1,m_1+1,L-l_1,M-m_1-1)}$ = im $A_{(l_1,m_1,L-l_1,M-m_1-1)}$
(3) ker $A_{(l_1+1,m_1,L-l_1-1,M-m_1)}$ = ker $A_{(l_1+1,m_1,L-l_1-1,M-m_1-1)}$
(4) ker $A_{(l_1,m_1+1,L-l_1,M-m_1-1)}$ = ker $A_{(l_1,m_1+1,L-l_1-1,M-m_1-1)}$

We can prove the theorem by using the above four equations. First, we will show the necessity of the equations.

Let $\sigma = ((X, F_\alpha, F_\beta), x^0, h)$ be a natural partial realization of $\underline{a} \in F(\mathbf{L} \times \mathbf{M}, Y)$. Then σ is (l_1, m_1)-quasi-reachable, and $(L - l_1 - 1, M - m_1 - 1)$-distinguishable.

Let G be the linear input map corresponding to x^0 and H be the linear output map corresponding to h, and let l', m', L', M' $(l_1 \le l', m_1 \le m', L' \le L - l_1 - 1, M' \le M - m_1 - 1)$ be integers. Then $G_{(l',m')} := G \cdot J_{(l',m')}$ is onto and $H_{(L',M')} := P_{(L',M')} \cdot H$ is one-to-one.

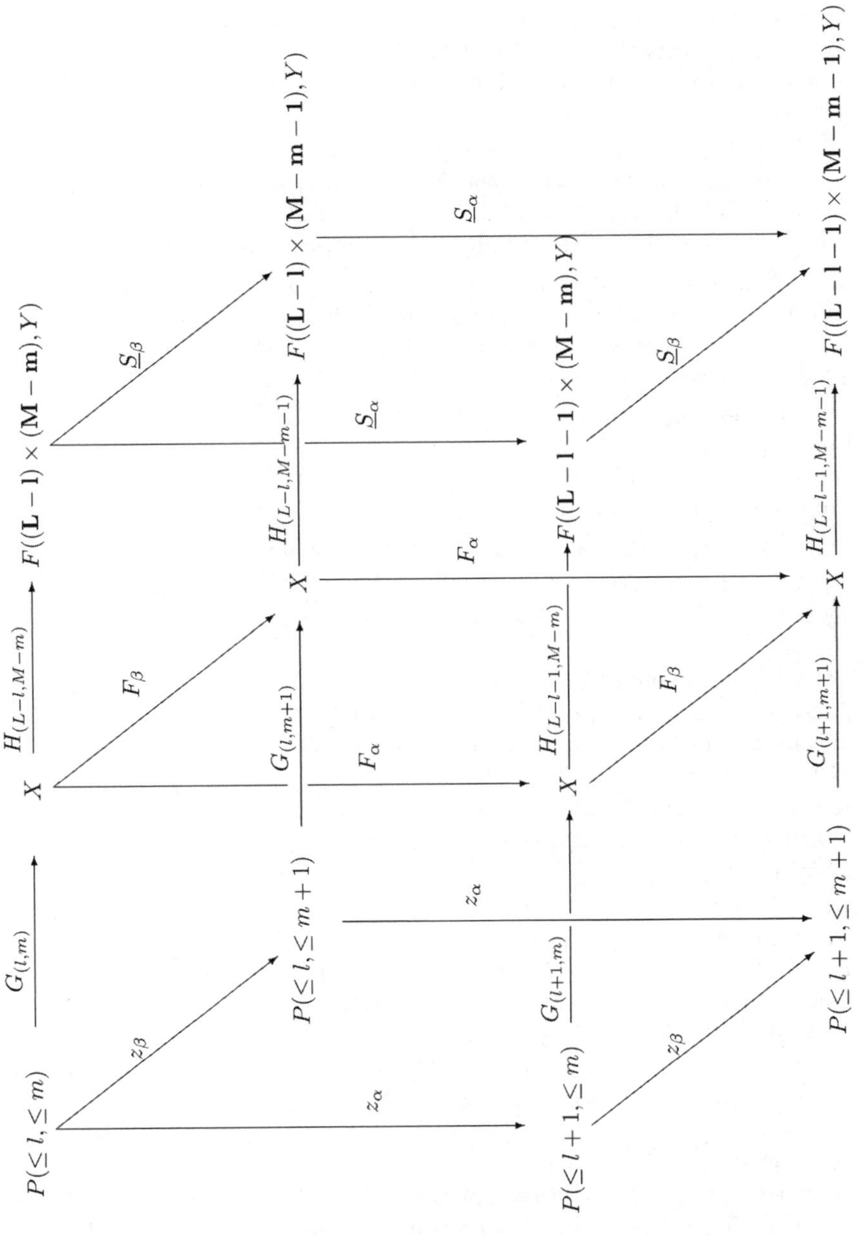

Therefore, $A_{(l',m',L',M')} := H_{(L',M')} \cdot G_{(l',m')}$ satisfies equations (1) through (4). Now, we will show sufficiency.

(1), (2) and Proposition (3-C.6) imply that the following equation (5) holds:

(5) : im $A_{(l_1+1,m_1+1,L-l_1-1,M-m_1-1)}$ = im $A_{(l_1+1,m_1,L-l_1-1,M-m_1-1)}$
= im $A_{(l_1,m_1+1,L-l_1-1,M-m_1-1)}$ = im $A_{(l_1,m_1,L-l_1-1,M-m_1-1)}$.
Equations (3), (4) and Proposition (3-C.7) imply that the following equation (6) holds.
(6) : ker $A_{(l_1,m_1,L-l_1,M-m_1)}$ = ker $A_{(l_1,m_1,L-l_1-1,M-m_1)}$
= ker $A_{(l_1,m_1,L-l_1,M-m_1-1)}$ = ker $A_{(l_1,m_1,L-l_1-1,M-m_1-1)}$.
Set $S :=$ ker $A_{(l_1+1,m_1+1,L-l_1-1,M-m_1-1)}$ and $Z :=$ im $A_{(l_1,m_1,L-l_1,M-m_1)}$.
Then equations (3) and (4) imply that a composition map $\pi \cdot j$:
$Z \xrightarrow{j} F((\mathbf{L}-\mathbf{l_1}) \times (\mathbf{M}-\mathbf{m_1}), Y) \xrightarrow{\pi} F((\mathbf{L}-\mathbf{l_1}-\mathbf{1}) \times (\mathbf{M}-\mathbf{m_1}-\mathbf{1}), Y)$ is injective, where π and j are the same as in Proposition (3-C.4).
Therefore, Z satisfies **condition 3** in Proposition (3-C.4). Equation (5) implies that there exist λ and $\lambda' \in P(\leq l_1, \leq m_1)$ such that
$A_{(l_1+1,m_1+1,L-l_1-1,M-m_1-1)}(z_\alpha^l z_\beta^{m_1+1}) = A_{(l_1,m_1,L-l_1-1,M-m_1)}(\lambda)$,
$A_{(l_1+1,m_1+1,L-l_1-1,M-m_1-1)}(z_\alpha^{l_1+1} z_\beta^m) = A_{(l_1,m_1,L-l_1-1,M-m_1)}(\lambda')$
for any $l, m \in N$ $(0 \leq l \leq l_1+1, 0 \leq m \leq m_1+1)$.
By Lemma (3-C.5), we find that
$A_{(l_1+1,m_1+1,L-l_1-1,M-m_1-1)}(z_\alpha^l z_\beta^{m_1+1} - \underline{i}\lambda) = 0$ and
$A_{(l_1+1,m_1+1,L-l_1-1,M-m_1-1)}(z_\alpha^{l_1+1} z_\beta^m - \underline{i}\lambda') = 0$. Thus, $z_\alpha^l z_\beta^{m_1+1} - \underline{i}\lambda$ and $z_\alpha^{l_1+1} z_\beta^m - \underline{i}\lambda' \in S$. Therefore, this S satisfies **condition 2** in Proposition (3-C.2).
Let $\underline{j}$ be the canonical injection
$\underline{j} : A_{(l_1+1,m_1+1,L-l_1-1,M-m_1-1)} \rightarrow F((\mathbf{L}-\mathbf{l_1}-\mathbf{1}) \times (\mathbf{M}-\mathbf{m_1}-\mathbf{1}), Y)$ and π_α and π_β be the same as in Proposition (3-C.4).
$B := (\underline{j})^{-1} \cdot \pi_\alpha \pi_\beta \cdot j : Z \rightarrow$ im $A_{(l_1+1,m_1+1,L-l_1-1,M-m_1-1)}$ is a bijective linear map by **condition 2** in Proposition (3-C.2).
When we consider the bijective linear map
$A^b :=$ im $A^b_{(l_1+1,m_1+1,L-l_1-1,M-m_1-1)}$:
$P(\leq l_1+1, \leq m_1+1)/S \rightarrow$ im $A_{(l_1+1,m_1+1,L-l_1-1,M-m_1-1)}$ associated with the linear map $A_{(l_1+1,m_1+1,L-l_1-1,M-m_1-1)}$:
$P(\leq l_1+1, \leq m_1+1) \rightarrow F((\mathbf{L}-\mathbf{l_1}-\mathbf{1}) \times (\mathbf{M}-\mathbf{m_1}-\mathbf{1}), Y)$, equation (2) implies that the linear map $B^{-1} \cdot A^b$ is a bijective linear map
$B^{-1} \cdot A^b : P(\leq l_1+1, \leq m_1+1)/S \rightarrow Z$.
For any $\lambda \in P(\leq l_1+1, \leq m_1) \cap S$, $A_{(l_1+1,m_1,L-l_1-1,M-m_1)}(\lambda) = 0$ holds by the injectivity of $B^{-1} \cdot A^b$. Therefore, im $A_{(l_1+1,m_1+1,L-l_1-1,M-m_1-1)}(z_\beta \lambda) = \underline{S_\beta}$ im $A_{(l_1+1,m_1,L-l_1-1,M-m_1)}(\lambda) = 0$ holds by Lemma (3-C.5). This implies that $z_\beta \lambda \in S$.
Similarly, we can assert that any $\lambda \in P(\leq l_1+1, \leq m_1) \cap S$ implies $z_\alpha \lambda \in S$. Therefore, S satisfies **condition 1** in Proposition (3-C.2). Then Proposition (3-C.2) implies that a pointed $\{\alpha, \beta\}$-action $((P(\leq l_1+1, \leq m_1+1)/S, \dot{z}_\alpha, \dot{z}_\beta), \mathbf{1}+S)$ is (l_1, m_1)-quasi-reachable.
Let $\underline{j_\alpha}$ be a canonical injection : im $A_{(l_1,m_1,L-l_1-1,M-m_1)} \rightarrow F((\mathbf{L}-\mathbf{l_1}-\mathbf{1}) \times (\mathbf{M}-\mathbf{m_1}), Y)$. Then (6) implies that $B_\alpha := (\underline{j_\alpha})^{-1} \pi_\alpha j$ is the bijective linear operator : $Z \rightarrow$ im $A_{(l_1,m_1,L-l_1-1,M-m_1)}$. Hence, equation (1) implies that

there exists $x \in \text{im } A_{(l_1,m_1,L-l_1-1,M-m_1)}$ such that $\underline{j}_{\alpha}(x) = \underline{S}_{\alpha} \cdot j(z)$ for any $z \in Z$. Moreover, by the surjectivity of B_α, there exists $z' \in Z$ such that $x = B_\alpha(z')$.
Therefore, $\underline{S}_{\alpha} \cdot j(z) = \underline{j_\alpha}(x) = \underline{j_\alpha}B_\alpha(z') = \pi_\alpha j(z')$, which implies that im $(\underline{S}_{\alpha} \cdot j) \subseteq$ im $(\pi_\alpha \cdot j)$.
Similarly, we can find that im $(\underline{S}_{\beta} \cdot j) \subseteq$ im $(\pi_\beta \cdot j)$.
It follows that Z satisfies **condition 4** in Proposition (3-C.4) and $((Z, F_\alpha, F_\beta), \underline{(0,0)})$ is $(L - l_1 - 1, M - m_1 - 1)$-distinguishable.
We can also show that $B^{-1} \cdot A^b : (P(\leq l_1 + 1, \leq m_1 + 1)/S, \dot{z}_\alpha, \dot{z}_\beta) \to (Z, F_\alpha, F_\beta)$ is a $\{\alpha, \beta\}$-morphism and that a 2-Commutative Linear Representation System
$\sigma_1 = ((P(\leq l_1 + 1, \leq m_1 + 1)/S, \dot{z}_\alpha, \dot{z}_\beta), \mathbf{1} + S, \underline{(0,0)} \cdot B^{-1} \cdot A^b)$ is isomorphic to a 2-Commutative Linear Representation System
$\sigma_2 = ((Z, F_\alpha, F_\beta), B^{-1} \cdot A^b(\mathbf{1} + S), \underline{(0,0)})$. It follows that σ_1 and σ_2 are the natural partial realizations of $\underline{a} \in F(\mathbf{L} \times \mathbf{M}, Y)$. Therefore, there exist the natural partial realizations of $\underline{a} \in F(\mathbf{L} \times \mathbf{M}, Y)$.

Lemma 3-C.10. Two canonical 2-Commutative Linear Representation Systems are isomorphic if and only if their behavior is the same.

Proof. This lemma can be derived from Theorem (3.5) and Corollary (3.12).

3-C.11. Proof of Theorem (3.25)
Let $A_{(\ ,\ ,\ ,\ )}$ be the partial linear input/output map corresponding to $\underline{a} \in F(\mathbf{L} \times \mathbf{M}, Y)$. In order to prove necessity, we assume the existence of the natural partial realization of $\underline{a}$. Suppose Theorem (3.25) holds for integers l_1, l_1', m_1 and m_1' that are mutually different. The following relations hold for l and m in place of l_1, l_1', m_1 and m_1':

(1) im $A_{(l+1,m,L-l-1,M-m)} =$ im $A_{(l,m,L-l-1,M-m)}$
(2) im $A_{(l,m+1,L-l,M-m-1)} =$ im $A_{(l,m,L-l,M-m-1)}$
(3) ker $A_{(l+1,m,L-l-1,M-m)} =$ ker $A_{(l+1,m,L-l-1,M-m-1)}$
(4) ker $A_{(l,m+1,L-l,M-m-1)} =$ ker $A_{(l,m+1,L-l-1,M-m-1)}$

First, we need to prove the necessity. We assume that the above four relations hold for $l = l_1$, $m = m_1$ and $l = l_1'$, $m = m_1'$. Then Propositions (3-C.6) and (3-C.7) imply that the dimension of $Z :=$ im $A_{(l_1,m_1,L-l_1,M-m_1)}$ is equivalent to the dimension of $Z' :=$ im $A_{(l_1',m_1',L-l_1',M-m_1')}$.

Let σ and σ' be the natural partial realizations of $\underline{a}$ whose state spaces are Z and Z' respectively and which can be obtained by the same procedure as in (3-C.9). Then σ is clearly isomorphic to σ' and the behavior of σ is equivalent to that of σ' by Lemma (3-C.10). This implies that the behavior of the natural partial realization is always the same regardless of different integers l_1 and l_1'. Therefore, the natural partial realization of $\underline{a}$ is unique modulo isomorphism by Lemma (3-C.10).

Next, we will demonstrate the sufficiency by contraposition. We assume that there does not exist a natural partial realization of $\underline{a} \in F(\mathbf{L} \times \mathbf{M}, Y)$.

Then minimum dimensional partial realization σ of $\underline{a}$ is (l_1, m_1)-quasi-reachable and (l_2, m_2)-distinguishable for $l_1 + l_2 \geq L$ and $m_1 + m_2 \geq M$. It cannot be quasi-reachable within $(l_1 - 1, m_1 - 1)$ and cannot be distinguishable within $(l_2 - 1, m_2 - 1)$. Next, there exists a state x in σ such that x can be reachable by a set $\{F_\alpha^i F_\beta^j x^0; i + j = l_1 + m_1\}$. The remaining data of $F((\mathbf{L} - \mathbf{i}) \times (\mathbf{M} - \mathbf{j}), Y)$ with $i + j = i_1 + m_1 - 1$ cannot determine a new state $F_\alpha x$ or $F_\beta x$ because of $L - l_1 - 1 < l_2$ or $M - m_1 - 1 < m_2$. Therefore, we cannot determine the transition matrix F_α or F_β uniquely by (l_2, m_2)-distinguishability. This implies that the minimum dimensional realization of $\underline{a}$ is not unique.

3-C.12. Proof of Theorem (3.26)
Let's take the natural partial realization $\sigma_2 = ((Z, F_\alpha, F_\beta), B^{-1} \cdot A^b(1 + S), \underline{(0,0)})$ of $\underline{a} \in F(\mathbf{L} \times \mathbf{M}, Y)$ given in (3-C.9). From σ_2, we can obtain the Quasi-reachable Standard System $\sigma_s = ((K^n, F_{\alpha s}, F_{\beta s}), \mathbf{e}_1, h_s)$ from in the same manner as Theorem (3.21) for a realization procedure.

4 Structures of 2-Commutative Linear Representation Systems

In this chapter, we will discuss the state structure of 2-Commutative Linear Representation Systems which is an extension of the linear system theory. Based on the results in Chapter 3, we will also discuss an effective encoding of two-dimensional images.

The results of Chapter 3 can be summarized as follows: First, we established a foundation for a new realization theory of two-dimensional images. It was shown that there exist canonical (quasi-reachable and distinguishable) 2-Commutative Linear Representation Systems which realize, that is, faithfully describe any two-dimensional image, and any canonical 2-Commutative Linear Representation Systems with the same behavior are isomorphic to each other. Finite-dimensional 2-Commutative Linear Representation Systems were investigated with the following results:

> Representation of the isomorphic class of finite-dimensional 2-Commutative Linear Representation Systems are characterized by the Quasi-reachable Standard System and the Distinguishable Standard System. Necessary and sufficient conditions for two-dimensional images to be behaviors of finite-dimensional 2-Commutative Linear Representation Systems are given by the rank condition of the Hankel matrix and the rationality of formal power series in two variables. A procedure to obtain the Quasi-reachable standard System from a given two-dimensional image was described.

Finally, partial realization problems were dealt with the following results:

> There always exists a minimal dimensional 2-Commutative Linear Representation System which realizes any given finite-sized two-dimensional image. The necessary and sufficient condition for the uniquely existence of a minimal dimensional 2-Commutative Linear Representation System which realizes a given two-dimensional image is given as a condition of finite-sized Hankel matrix. It is also shown that our algorithm provides partialy the same behavior for a given finite-sized two-dimensional image.

We now turn to clarifying the structure of finite-dimensional 2-Commutative Linear Representation Systems, something has not been done before. The structure problem can be stated as follows:

Find a 2-Commutative Linear Representation System which has a simpler state space and a simpler state transition than in the class of finite-dimensional 2-Commutative Linear Representation Systems which have quad the same behavior.

4.1 Structure Theory of 2-Commutative Linear Representation Systems

In this section, we propose a new structure theory which extends the structure theory of linear systems to 2-Commutative Linear Representation Systems. A new 2-Commutative Linear Representation System called the Invariant Standard System will be introduced. It is characterized by a simple state space and a simple transition matrix. By virtue of our proposed structure theory, the transition matrix is composed of fewer parameters than in the usual theory. We will also show that there uniquely exists an Invariant Standard System in the isomorphic class of canonical 2-Commutative Linear Representation Systems with the same behavior.

Recall the following fact: The structure theory of linear systems implies that the state space of the systems is divided into the direct sum of sub-state spaces and its transition matrix is block-diagonalized by the direct sum decomposition.

Definition 4.1. A canonical n-dimensional 2-Commutative Linear Representation System $\sigma_I = ((K^n, F_{\alpha I}, F_{\beta I}), \mathbf{e}_1, h_I)$ shown in Figure 4.1 is called an Invariant Standard System with an index $\nu = (\nu_1, \nu_2, \cdots, \nu_k)$, where $n = \sum_{i=1}^{k} \nu_i$. A $\{\alpha, \beta\}$-action $(K^n, F_{\alpha I}, F_{\beta I})$ in the system is given as follows: $\mathbf{B}_i \in K^n$ of $F_{\beta I}$ is given by $\mathbf{B}_i := [\mathbf{b}_i^1, \cdots, F_{\alpha I}^{\ell_i - 1}\mathbf{b}_i^1, \cdots, \mathbf{b}_i^k, \cdots, F_{\alpha I}^{\ell_k - 1}\mathbf{b}_i^k]$, where $\mathbf{b}_i^1 \in K^n$ is given by $\mathbf{b}_i^1 = [\mathbf{b}_i^{11}, \mathbf{b}_i^{12}, \cdots, \mathbf{b}_i^{1i}, \mathbf{e}_1^s, \mathbf{0}, \cdots, \mathbf{0}]^T$,
$\mathbf{b}_i^{11} := [\mathbf{0}, \widetilde{\mathbf{b}_i^{11}}, \mathbf{0}]^T \in K^{\nu_1}, \widetilde{\mathbf{b}_i^{11}} \in K^{\ell_i}$, $\mathbf{b}_i^{12} := [\mathbf{0}, \widetilde{\mathbf{b}_i^{12}}, \mathbf{0}]^T \in K^{\nu_2}, \widetilde{\mathbf{b}_i^{12}} \in K^{\ell_i}$, $\cdots$, $\mathbf{b}_i^{1i} := [\mathbf{0}, \widetilde{\mathbf{b}_i^{1i}}, \mathbf{0}]^T \in K^{\nu_i}, \widetilde{\mathbf{b}_i^{1i}} \in K^{\ell_i}, \mathbf{e}_1^s \in K^{\nu_{i+1}}$.

For the linear space $K^n$and its subspace K^ℓ, $\mathbf{e}_1 \in K^n$, $\mathbf{e}_i^s \in K^\ell$ and $\mathbf{e}_1^s \in K^\ell$ are defined thus:

$$\mathbf{e}_1 = [\overbrace{1, 0, \cdots, 0}^{n}]^T, \ \mathbf{e}_1^s = [\overbrace{1, 0, \cdots, 0}^{\ell}]^T, \ \mathbf{e}_i^s = [\overbrace{0, \cdots, 0, \underset{i}{1}, 0, \cdots, 0}^{\ell}]^T.$$

Theorem 4.2. For any n-dimensional canonical 2-Commutative Linear Representation System $((K^n, F_\alpha, F_\beta), x^0, h)$, there uniquely exists an Invariant Standard System $((K^n, F_{\alpha I}, F_{\beta I}), \mathbf{e}_1, h_I)$ which is isormorphic to it.

Proof. For this proof, see (4-A.12) in Appendix 4-A.

$$
F_i = \left[\begin{array}{c|c|c|c|c|ccc|c|c}
\leftarrow \ell_i & \rightarrow & \leftarrow \ell_{i+1} & & \rightarrow & & \cdot\ \cdot & & \leftarrow \ell_k & \rightarrow \\
0 \cdot 0\,0 & & & & 0 & & & & & 0 \\
1\ 0 \cdot 0 & & & & \vdots & & & & & \vdots \\
0\ \cdot & \alpha^{i} & & & & & & & & \\
\cdot\ 0\,1\,0 & & & & \vdots & & & & & \\
0\ \cdot\ 0\,1 & & & & 0 & & & & & \\
\hline
0\,0 & 1 & 0 \cdot 0\,0 & & & & & & & \\
0\,0 & 0 & 1\ 0\,0\,0 & \alpha^{i+1} & & & & & & \\
\cdot\ 0 & 0 & 0\ \cdot & & & & & & & \\
0\,0 & \cdot & \cdot\ 0\,1\,0 & & & & & & & \\
0\,0 & 0 & 0\ \cdot\ 0\,1 & & & & & & & \\
\hline
 & & & 1 & 0 \cdot \cdot\ 0\,0 & & & & & \\
 & & & 0 & 1 & & & & & \\
 & & & 0 & \cdot\ 1 & & & & & \\
 & & & \cdot & 0\ \cdot & & & & & \\
 & & & 0 & \cdot\ 0\,1\,0 & & & & & \vdots \\
 & & & \vdots & \cdot\ 0\,0\,1 & & & & & 0 \\
\hline
 & & & & 0\,1 & & & 0\ \ 0\,0 & & \\
 & & & & 0 & & & 1\ \ 0\,0 & \alpha^{k} & \\
 & & & & \cdot & & & 0\ \ 1\,0 & & \\
 & & & 0 & 0 & & & 0\ \ 0\,1 & &
\end{array}\right]
$$

$$
F_{\alpha I} = \begin{bmatrix}
F_1 & \mathbf{0} & \cdots & \cdots & \mathbf{0} \\
\mathbf{0} & F_2 & \mathbf{0} & \ddots & \vdots \\
\vdots & \mathbf{0} & \ddots & \ddots & \vdots \\
\vdots & \ddots & \ddots & \ddots & \mathbf{0} \\
\mathbf{0} & \cdots & \cdots & \mathbf{0} & F_k
\end{bmatrix}, \text{ where } F_i \in K^{\nu_i \times \nu_i}.
$$

$F_{\beta I} = [\mathbf{B}_1, \mathbf{B}_2, \mathbf{B}_3, \cdots, \mathbf{B}_k]$, where

$\mathbf{B}_i = [\mathbf{b}_i^1, F_{\alpha I}\mathbf{b}_i^1, \cdots F_{\alpha I}^{\ell_i - 1}\mathbf{b}_i^1, \mathbf{b}_i^2, \cdots, F_{\alpha I}^{\ell_{i+1}-1}\mathbf{b}_i^2, \cdots \mathbf{b}_i^k, \cdots F_{\alpha I}^{\ell_k - 1}\mathbf{b}_i^k]$
$\in K^{n \times \nu_i}$ for $1 \le i \le k$.
Moreover, $\phi_i(F_{\alpha I})\mathbf{b}_i^1 = \mathbf{0}$ $(1 \le i \le k)$ holds and $\mathbf{b}_i^j$ is given by
$\mathbf{b}_i^j = \chi_i(F_{\alpha I})\chi_{i+1}(F_{\alpha I}) \cdots \chi_{i+j-2}(F_{\alpha I})\mathbf{b}_i^1$ for j $(2 \le j \le k)$.

Fig. 4.1. The Invariant Standard System σ_I defined in Definition (4.1)

Remark 1: The Invariant Standard System $((K^n, F_{\alpha I}, F_{\beta I}), \mathbf{e}_1, h_I)$ with an index $\nu = (\nu_1, \nu_2, \cdots, \nu_k)$ has the following properties:

(1) $K^n = K^{\nu_1} \oplus K^{\nu_2} \oplus \cdots \oplus K^{\nu_k}$, $\bar{B_{ij}} \in L(K^{\nu_j}, K^{\nu_i})$, where $\bar{B_{ij}}$ is given by $\mathbf{B}_i = [\bar{B_{i1}}, \bar{B_{i2}}, \cdots, \bar{B_{ik}}]^T \in K^{n \times \nu_i}$.
(2) For i $(1 \le i \le k)$, the minimal polynomial $\phi_i(\lambda)$ of F_i is represented as $\phi_i(\lambda) = \chi_i(\lambda)\chi_{i+1}(\lambda) \cdots \chi_k(\lambda)$. For every $\phi_i(\lambda)$ $(1 \le i \le k-1)$, $\phi_i(\lambda)$ can be divided by the minimal polynomial $\phi_{i+1}(\lambda)$ of F_{i+1} (submatrix of $F_{\alpha I}$) and $\phi_i(\lambda)/\phi_{i+1}(\lambda) = \chi_i(\lambda)$ holds. Especially, by Lemma (4-A.2), the

minimal polynomial of $F_{\alpha I}$ is $\phi_1(\lambda)$. $\phi_i(\lambda)$ can be expressed by $\phi_i(\lambda) = \lambda^{\nu_i} - c^i_{i\nu_i}\lambda^{\nu_i-1} - c^i_{i\nu_i-1}\lambda^{\nu_i-2} - \cdots - c^i_{i2}\lambda - c^i_{i1}$ for i $(1 \leq i \leq k)$. For c^i_{ij} $(1 \leq j \leq \nu_i)$, see Definition (3.14). α^i in Figure 4.1 is represented as $\alpha^i = [\alpha^i_1, \alpha^i_2, \cdots, \alpha^i_{\ell_i-1}, \alpha^i_{\ell_i}]^T$ for $\chi_i(\lambda) = \lambda^{\ell_i} - \alpha^i_{\ell_i}\lambda^{\ell_i-1} - \alpha^i_{\ell_i-1}\lambda^{\ell_i-2} - \cdots - \alpha^i_2\lambda - \alpha^i_1$.

(3) A $\{\alpha, \beta\}$-action $(K^n, F_{\alpha I}, F_{\beta I})$ in the Invariant Standard System σ_I can be determined by $(n + \nu_1)$ parameters.

Remark 2: Structure Theorem (4.2) is a direct extension of the structure theory for linear systems [(8.1) in Kalman, et al., 1969] to 2-Commutative Linear Representation systems.

Remark 3: Gantmacher [1959] discussed the commutativity of two matrices, the number of parameters in the two matrices was given by $n + \nu_1 + 3\nu_2 + \cdots + (2k-1)\nu_k$. In our theory, the number of parameters, that is, $n + \nu_1$ is decreased by introducing quasi-reachabiliy into the commutativity of two matrices.

4.2 Structure Theory and a Coding Theory of Two-Dimensional Images

In the preceding section, we obtained a structure theory of the 2-Commutative Linear Representation Systems which contains a state structure having fewer parameters than in the existing theory. Therefore, we can now discuss an effective encoding of two-dimensional images.
This can be considered as encoding of channel. Note that an encoding of information source corresponds to the partial realization algorithm discussed in Chapter 3, where for a given finite-sized two-dimensional image, we derived the algorithm to obtain the Quasi-reachable Standard System $\sigma_s = ((K^n, F_{\alpha s}, F_{\beta s}), \mathbf{e}_1, h_s)$, which has the same partial behavior as the given image.

Let $\phi_i(\lambda)$ be the minimal polynomial of the matrix F_i in Figure 4.1. Then it can be expressed by $\phi_i(\lambda) = \lambda^{\nu_i} - c^i_{i\nu_i}\lambda^{\nu_i-1} - c^i_{i\nu_i-1}\lambda^{\nu_i-2} - \cdots - c^i_{i2}\lambda - c^i_{i1}$ given in Remark 1 of Theorem (4.2).

For c^j_{ml} in Definition (3.14), we use the following notations for the subsequent discussions:
$\mathbf{C}^i := [C^i_1, C^i_2, \cdots, C^i_i, \mathbf{0}]^T \in K^n$, $C^i_j := [c^i_{j1}, c^i_{j2}, \cdots, c^i_{j\nu_j}]^T \in K^{\nu_i}$ for i, j $(1 \leq i \leq k,\ 1 \leq j \leq i)$.

4.3. Procedure for effective encoding.
Let $\sigma_s = ((K^n, F_{\alpha s}, F_{\beta s}), \mathbf{e}_1, h_s)$ be the Quasi-reachable Standard System. Then the Invariant Standard System $\sigma_I = ((K^n, F_{\alpha I}, F_{\beta I}), \mathbf{e}_1, h_I)$ which is isomorphic to σ_s is derived by the following procedure:

① $F_{\alpha I}$, the transition matrix, can be derived as follows:
Since $\phi_i(\lambda) = \lambda^{\nu_i} - c^i_{i\nu_i}\lambda^{\nu_i-1} - c^i_{i\nu_i-1}\lambda^{\nu_i-2} - \cdots - c^i_{i2}\lambda - c^i_{i1}$,
$\phi_k(\lambda) = \chi_k(\lambda)$, $\phi_{k-1}(\lambda) = \chi_{k-1}(\lambda)\phi_k(\lambda), \cdots$,
$\phi_i(\lambda) = \chi_i(\lambda)\chi_{i+1}(\lambda)\cdots\chi_k(\lambda), \cdots$,, $\phi_1(\lambda) = \chi_1(\lambda)\chi_2(\lambda)\cdots\chi_k(\lambda)$
hold, and $\chi_i(\lambda) = \phi_i(\lambda)/\phi_{i+1}(\lambda)$ $(k-1 \geq i \geq 1)$ can be found in turn.
Then a companion form F_i of the polynomial $\chi_i(\lambda)$ can be found, and leads finally to $F_{\alpha I}$.

② The isomorphism $T_s : \sigma_s \to \sigma_I$ expressed by T_s
$= [\mathbf{t}^1_1, F_{\alpha I}\mathbf{t}^1_1, \cdots, F_{\alpha I}^{\nu_1-1}\mathbf{t}^1_1, \mathbf{t}^1_2, \cdots, F_{\alpha I}^{\nu_2-1}\mathbf{t}^1_2, \cdots, \mathbf{t}^1_k \cdots, F_{\alpha I}^{\nu_k-1}\mathbf{t}^1_k] \in K^{n\times n}$
is derived in this way:
(1) Set $\mathbf{t}^1_1 = [\mathbf{t}^{11}_1, \mathbf{0}]^T \in K^n$, $\mathbf{t}^{11}_1 = \mathbf{e}^s_1 := [1, 0, \cdots, 0]^T \in K^{\nu_1}$.
(2) Let $\mathbf{t}^1_2 = [\mathbf{t}^{11}_2, \mathbf{t}^{12}_2, \mathbf{0}, \cdots, \mathbf{0}]^T \in K^n$ be $\mathbf{t}^{11}_2 = [\mathbf{b}^{11}_1, \mathbf{0}, \cdots, \mathbf{0}]^T \in K^{\nu_1}$,
$\widetilde{\mathbf{b}^{11}_1} \in K^{\ell_1}$, $\mathbf{t}^{12}_2 := \mathbf{e}^s_1 \in K^{\nu_2}$, where
$\widetilde{\mathbf{b}^{11}_1} = \{[\phi_2(F_1)\mathbf{e}^s_1, \cdots, \phi_2(F_1)\mathbf{e}^s_{\ell_1}]^T[\phi_2(F_1)\mathbf{e}^s_1, \cdots, \phi_2(F_1)\mathbf{e}^s_{\ell_1}]\}^{-1}$
$\times [\mathbf{e}^s_1, \cdots, \mathbf{e}^s_{\ell_1}]^T[\mathbf{t}^1_1, \cdots, F_1^{\nu_1-1}\mathbf{t}^1_1]C^2_1$.
(3) Set $\mathbf{t}^1_3 = [\mathbf{t}^{11}_3, \mathbf{t}^{12}_3, \mathbf{t}^{13}_3, \mathbf{0}]^T \in K^n$ be $\mathbf{t}^{11}_3 = (a^{11}_2 + \mathbf{b}^{11}_2)^T \in K^{\nu_1}$,
$\mathbf{t}^{12}_3 = (a^{12}_2 + \mathbf{b}^{12}_2)^T \in K^{\nu_2}$ and $\mathbf{t}^{13}_3 = \mathbf{e}^s_1 \in K^{\nu_3}$.
Then $a^1_2 := [\mathbf{t}^1_2, F_1\mathbf{t}^1_2, \cdots, F_1^{\ell_1-1}\mathbf{t}^1_2]\ \widetilde{\mathbf{b}^{11}_1}$ becomes
$a^1_2 = [a^{11}_2, a^{12}_2, \mathbf{0}, \mathbf{0}]^T \in K^n$. Moreover $\widetilde{\mathbf{b}^1_2} \in K^{\ell_2}$ of $\mathbf{b}^1_2 := [\mathbf{0}, \widetilde{\mathbf{b}^1_2}, \mathbf{0}]^T \in K^{\nu_1}$
and $\widetilde{\mathbf{b}^2_2} \in K^{\ell_2}$ of $\mathbf{b}^2_2 := [\mathbf{0}, \widetilde{\mathbf{b}^2_2}, \mathbf{0}]^T \in K^{\nu_2}$ are calculated in this way: $\widetilde{\mathbf{b}^1_2} =$
$\{[\phi_3(F_1)\mathbf{e}^s_{\ell_1+1}, \cdots, \phi_3(F_1)\mathbf{e}^s_{\ell_1+\ell_2}]^T[\phi_3(F_1)\mathbf{e}^s_{\ell_1+1}, \cdots, \phi_3(F_1)\mathbf{e}^s_{\ell_1+\ell_2}]\}^{-1}$
$\times [\mathbf{e}^s_{\ell_1+1}, \cdots, \mathbf{e}^s_{\ell_1+\ell_2}]^T\{[\mathbf{t}^1_1, \cdots, F_1^{\nu_1-1}\mathbf{t}^1_1]C^3_1$
$+ [\mathbf{t}^1_2, \cdots, F_1^{\nu_2-1}\mathbf{t}^1_2]C^3_2 - \phi_3(F_1)a^1_2\}$,
$\widetilde{\mathbf{b}^2_2} = \{[\phi_3(F_2)\mathbf{e}^s_1, \cdots, \phi_3(F_2)\mathbf{e}^s_{\ell_2}]^T[\phi_3(F_2)\mathbf{e}^s_1, \cdots, \phi_3(F_2)\mathbf{e}^s_{\ell_2}]\}^{-1}$
$\times [\mathbf{e}^s_1, \cdots, \mathbf{e}^s_{\ell_2}]^T\{[\mathbf{t}^1_2, \cdots, F_2^{\nu_2-1}\mathbf{t}^1_2]C^3_2 - \phi_3(F_2)a^{12}_2\}$.
(4) For i $(4 \leq i \leq k)$, let $\mathbf{t}^1_i = [\mathbf{t}^{11}_i, \mathbf{t}^{12}_i, \cdots, \mathbf{t}^{1i}_i, \mathbf{0}]^T \in K^n$ be
$\mathbf{t}^{11}_i = (a^{11}_{i-1} + \mathbf{b}^{11}_{i-1})^T \in K^{\nu_1}$, $\mathbf{t}^{12}_i = (a^{12}_{i-1} + \mathbf{b}^{12}_{i-1})^T \in K^{\nu_2}$, $\cdots$,
$\mathbf{t}^{1i}_i = \mathbf{e}^s_1 \in K^{\nu_i}$.
For $\mathbf{b}^j_{i-1} := [\mathbf{0}, \widetilde{\mathbf{b}^j_{i-1}}, \mathbf{0}]^T \in K^{\nu_j}, \widetilde{\mathbf{b}^j_{i-1}} \in K^{\ell_j}$ $(1 \leq j \leq i-1)$ are given as follows:
$\widetilde{\mathbf{b}^j_{i-1}}$
$= \{[\phi_i(F_j)\mathbf{e}^s_{\ell_j+\cdots+\ell_{i-2}+1}, \cdots, \phi_3(F_2)\mathbf{e}^s_{\ell_j+\cdots+\ell_{i-2}+\ell_{i-1}}]^T$
$\times[\phi_i(F_j)\mathbf{e}^s_{\ell_j+\cdots+\ell_{i-2}+1}, \cdots, \phi_3(F_2)\mathbf{e}^s_{\ell_j+\cdots+\ell_{i-2}+\ell_{i-1}}]\}^{-1}$
$\times[\mathbf{e}^s_{\ell_j+\cdots+\ell_{i-2}+1}, \cdots, \mathbf{e}^s_{\ell_j+\cdots+\ell_{i-2}+\ell_{i-1}}]^T \times \{[\mathbf{t}^{1j}_j, \cdots, F_j^{\nu_j-1}\mathbf{t}^{1j}_j]C^i_j$
$+ [\mathbf{t}^{1j}_{j+1}, \cdots, F_j^{\nu_{j+1}-1}\mathbf{t}^{1j}_{j+1}]C^i_{j+1} + \cdots + [\mathbf{t}^{1j}_{i-1}, \cdots, F_j^{\nu_{j+1}-1}\mathbf{t}^{1j}_{i-1}]C^i_{i-1}$
$- \phi_i(F_j)a^{1j}_{i-1}\}$.
Then $a^1_{i-1} := \mathbf{B}_1\mathbf{t}^{11}_{i-1} + \mathbf{B}_2\mathbf{t}^{12}_{i-1} + \cdots + \mathbf{B}_{i-2}\mathbf{t}^{1i-2}_{i-1}$ becomes
$a^1_{i-1} = [a^{11}_{i-1}, a^{12}_{i-1}, \cdots, a^{1i-1}_{i-1}, \mathbf{0}, \cdots, \mathbf{0}]^T \in K^n$, $a^{11}_{i-1} \in K^{\nu_1}$, $a^{12}_{i-1} \in K^{\nu_2}$,
$\cdots$, $a^{1i-1}_{i-1} \in K^{\nu_{i-1}}$.

③ Let $F_{\beta I}$ be $F_{\beta I} := T_s F_{\beta s} T_s^{-1}$.
④ Let h_I be $h_I := h_s T_s^{-1}$.

Proof. See (4-A.13) in Appendix 4-A.

Remark 1: It is possible to obtain the Quasi-reachable Standard System from a given actual finite-sized two-dimensional image using the partial realization algorithm discussed in Theorem (3.26). The procedure in (4.3) is executed to obtain the Invariant Standard System from the Quasi-reachable Standard System. Therefore, the Invariant Standard System can be found starting from the finite-sized two-dimensional image.

The partial realization algorithm provides a method of encoding an information source in image processing. On the other hand, the procedure discussed in this chapter for obtaining the Invariant Standard System can be interpreted as an effective encoding of channel.

Example 4.4. Consider the geometric pattern depicted in Figure 4.2 as an example of encoding a two-dimensional image.

This image can be encoded as in Figure 7.5.

Let K be $N/3N$ which is the quotient field modulo the prime number 3, and let the set Y of output values be K. Then the image is transformed into the following two-dimensional array:

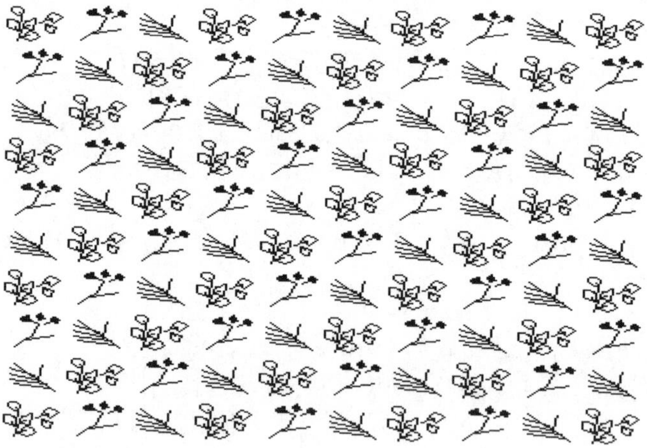

Fig. 4.2. The geometrical pattern in Example (4.4)

Fig. 4.3. The coding list for Example (4.4)

$$\begin{array}{cccccccccc}
1&0&2&1&0&2&1&0&2&1\\
0&2&1&0&2&1&0&2&1&0\\
2&1&0&2&1&0&2&1&0&2\\
1&0&2&1&0&2&1&0&2&1\\
0&2&1&0&2&1&0&2&1&0\\
2&1&0&2&1&0&2&1&0&2\\
1&0&2&1&0&2&1&0&2&1\\
0&2&1&0&2&1&0&2&1&0\\
2&1&0&2&1&0&2&1&0&2\\
1&0&2&1&0&2&1&0&2&1
\end{array}$$

Applying the partial realization algorithm noted in Theorem (3.26) yields the following Quasi-reachable Standard System $\sigma_s = ((K^3, F_{\alpha s}, F_{\beta s}), \mathbf{e}_1, h_s)$:

$$F_{\alpha s} = \begin{bmatrix} 0\,2\,2 \\ 1\,2\,1 \\ 0\,0\,1 \end{bmatrix},\; F_{\beta s} = \begin{bmatrix} 0\,2\,2 \\ 0\,1\,1 \\ 1\,1\,1 \end{bmatrix},\; \mathbf{e}_1 = \begin{bmatrix} 1 \\ 0 \\ 0 \end{bmatrix},\; h_s = \begin{bmatrix} 1\,0\,0 \end{bmatrix}.$$

Based on the algorithm for an effective encoding algorithm in (4.3), the Invariant Standard System $\sigma_I = ((K^3, F_{\alpha I}, F_{\beta I}), \mathbf{e}_1, h_I)$ is derived in this way.

① From $F_{\alpha s}$, $\phi_1(\lambda) = \lambda^2 + \lambda + 1$, $\phi_2(\lambda) = \lambda + 2$ can be obtained, which leads to $\chi_1(\lambda) = \phi_1(\lambda)/\phi_2(\lambda) = \lambda + 2$ and $\chi_2(\lambda) = \phi_2(\lambda) = \lambda + 2$. Thus we arrive at the following $F_{\alpha I}$:

$$F_{\alpha I} = \begin{bmatrix} 1\,0\,0 \\ 1\,1\,0 \\ 0\,0\,1 \end{bmatrix}$$

② For the 2-Commutative Linear Representation System morphism $T_s : \sigma_s \to \sigma_I$ given by $T_s = [\mathbf{t}_1^1, F_{\alpha I}\mathbf{t}_1^1, \mathbf{t}_2^1] \in K^{3\times 3}$, where $\mathbf{t}_1^1 = \mathbf{e}_1^s \in K^2$,$\mathbf{t}_2^1 = [\mathbf{t}_2^{11}, \mathbf{t}_2^{12}, \mathbf{0}]^T$, $\mathbf{t}_2^{11} = [\mathbf{b}_1^{11}, \mathbf{0}]^T \in K^2$,$\mathbf{b}_1^{11} \in K$, $\mathbf{t}_2^{12} = \mathbf{e}_1^s \in K$. Then $\mathbf{b}_1^{11} = 1$. Therefore, T_s and T_s^{-1} are represented in this way:

$$T_s = \begin{bmatrix} 1\,1\,1 \\ 0\,1\,0 \\ 0\,0\,1 \end{bmatrix}, \qquad T_s^{-1} = \begin{bmatrix} 1\,2\,2 \\ 0\,1\,0 \\ 0\,0\,1 \end{bmatrix}$$

③ Set $F_{\beta I} = T_s F_{\beta s} T_s^{-1}$.

④ Set $h_I = h_s T_s^{-1}$.

Finally, we have $\sigma_I = ((K^3, F_{\alpha I}, F_{\beta I}), \mathbf{e}_1, h_I)$, where

$$F_{\alpha I} = \begin{bmatrix} 1\,0\,0 \\ 1\,1\,0 \\ 0\,0\,1 \end{bmatrix},\; F_{\beta I} = \begin{bmatrix} 1\,0\,0 \\ 0\,1\,1 \\ 1\,0\,0 \end{bmatrix},\; h_I = \begin{bmatrix} 1\,2\,2 \end{bmatrix}.$$

Example 4.5. Consider the geometric pattern depicted in Figure 4.4 as an example of encoding a two-dimensional image.

This image can be encoded as in Figure 4.5.

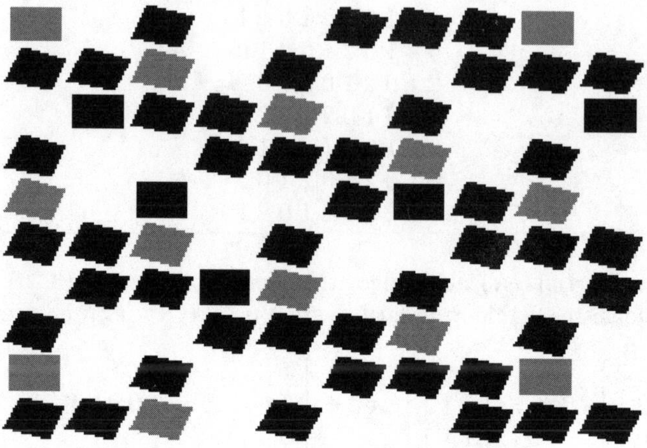

Fig. 4.4. The geometrical pattern in Example (4.5)

Fig. 4.5. The coding list for Example (4.5)

Let K be $N/3N$ which is the quotient field modulo the prime number 3, and let the set Y of output values be K. Then the image is transformed into the following two-dimensional array.

1	2	0	2	2	0	0	0	1	2
0	0	1	2	0	2	2	0	0	0
2	0	0	0	1	2	0	2	2	0
0	2	2	0	0	0	1	2	0	2
1	2	0	2	2	0	0	0	1	2
0	0	1	2	0	2	2	0	0	0
2	0	0	0	1	2	0	2	2	0
0	2	2	0	0	0	1	2	0	2
1	2	0	2	2	0	0	0	1	2
0	0	1	2	0	2	2	0	0	0

Applying the partial realization algorithm described in Theorem (3.26), we find the Quasi-reachable Standard System $\sigma_s = ((K^4, F_{\alpha s}, F_{\beta s}), \mathbf{e}_1, h_s)$, where

$$F_{\alpha s} = \begin{bmatrix} 0\,0\,1\,2 \\ 1\,0\,2\,0 \\ 0\,1\,1\,1 \\ 0\,0\,0\,1 \end{bmatrix}, \; F_{\beta s} = \begin{bmatrix} 0\,2\,0\,1 \\ 0\,0\,1\,2 \\ 0\,1\,2\,1 \\ 1\,1\,1\,0 \end{bmatrix}, \; \mathbf{e}_1 = \begin{bmatrix} 1 \\ 0 \\ 0 \\ 0 \end{bmatrix}, \; h_s = \begin{bmatrix} 1\,0\,2\,2 \end{bmatrix}.$$

Based on the algorithm for an effective encoding in (4.3), the Invariant Standard System $\sigma_I = ((K^4, F_{\alpha I}, F_{\beta I}), \mathbf{e}_1, h_I)$ is derived in this way:

① From $F_{\alpha s}$, $\phi_1(\lambda) = \lambda^3 - 2\lambda^2 - 2\lambda - 1$, $\phi_2(\lambda) = \lambda - 1$ is obtained, and then $\chi_1(\lambda) = \phi_1(\lambda)/\phi_2(\lambda) = \lambda^2 - 2$ and $\chi_2(\lambda) = \phi_2(\lambda) = \lambda - 1$. Finally, we get the following $F_{\alpha I}$:

$$F_{\alpha I} = \begin{bmatrix} 0\,2\,0\,0 \\ 1\,0\,0\,0 \\ 0\,1\,1\,0 \\ 0\,0\,0\,1 \end{bmatrix}$$

② The 2-Commutative Linear Representation System morphism $T_s : \sigma_s \to \sigma_I$ is represented as $T_s = [\mathbf{t}_1^1, F_{\alpha I}\mathbf{t}_1^1, F_{\alpha I}^2\mathbf{t}_1^1, \mathbf{t}_2^1] \in K^{4\times 4}$, where $\widetilde{\mathbf{t}_1^1} = \mathbf{e}_1^s \in K^3$, $\widetilde{\mathbf{t}_2^1} = [\mathbf{t}_2^{11}, \mathbf{t}_2^{12}, \mathbf{0}]^T$, $\mathbf{t}_2^{11} = [\mathbf{b}_1^{11}, \mathbf{0}]^T \in K^3$, $\widetilde{\mathbf{b}_1^{11}} \in K^2$, $\mathbf{t}_2^{12} = \mathbf{e}_1^s \in K$. Then $\widetilde{\mathbf{b}_1^{11}} = [1, 1]^T$. Therefore, T_s and T_s^{-1} is expressed in this way:

$$T_s = \begin{bmatrix} 1\,0\,2\,1 \\ 0\,1\,0\,1 \\ 0\,0\,1\,0 \\ 0\,0\,0\,1 \end{bmatrix}, \qquad T_s^{-1} = \begin{bmatrix} 1\,0\,1\,2 \\ 0\,1\,0\,2 \\ 0\,0\,1\,0 \\ 0\,0\,0\,1 \end{bmatrix}$$

③ Set $F_{\beta I} = T_s F_{\beta s} T_s^{-1}$.

④ Set $h_I = h_s T_s^{-1}$.

Consequently, we have $\sigma_I = ((K^4, F_{\alpha I}, F_{\beta I}), \mathbf{e}_1, h_I)$, where

$$F_{\alpha I} = \begin{bmatrix} 0\,2\,0\,0 \\ 1\,0\,0\,0 \\ 0\,1\,1\,0 \\ 0\,0\,0\,1 \end{bmatrix}, \; F_{\beta I} = \begin{bmatrix} 1\,2\,0\,0 \\ 1\,1\,0\,0 \\ 0\,1\,2\,0 \\ 1\,1\,2\,1 \end{bmatrix}, \; h_I = \begin{bmatrix} 1\,0\,0\,1 \end{bmatrix}.$$

4.3 Historical Notes and Concluding Remarks

We have proposed a structure problem for 2-Commutative Linear Representation Systems which is an extension of linear systems proposed in [Kalman, et al., 1969]. Making use of the Invariant Standard System which has the simplest structure in the class of 2-Commutative Linear Representation Systems, we investigated the structure problem and obtained new results. The Invariant Standard System has a companion form and it is a representative of the isomorphic class. In addition, the system has the following properties:

① The state space of the Invariant Standard System $\sigma_I = ((K^n, F_{\alpha I}, F_{\beta I}), \mathbf{e}_1, h_I)$, is represented as the direct sum of the invariant subspaces under the transition matrix $F_{\alpha I}$.
② The invariant subspaces can be characterized by their minimal polynomials.
③ The two transition matrices $F_{\alpha I}$ and $F_{\beta I}$ are characterized by a minimum number of parameters, and the system makes the position of these parameters in $F_{\alpha I}$ and $F_{\beta I}$ clear.

Properties ① and ② are the same as those in the structure theorem for discrete-time linear systems. However, in regard to property ③, the number of parameters which determine commutative matrices is fewer than the number found in previous analysis [Gantmacher, 1959]. This reduction in number was achieved by introducing quasi-reachability into our systems. Moreover, the number of parameters is not only less; it is minimal. This property makes the coding of channel for two-dimensional images clear. Our results provide a new algebraic encoding method. Note that the algorithm for obtaining the system from a given two-dimensional image corresponds to an encoding of information source.

Appendix to Chapter 4

In this appendix, we provide proofs for the developments in the preceding sections. First, we present some facts which will be needed for the proofs.

Consider a pointed $\{\alpha, \beta\}$-action $((X, F_\alpha, F_\beta), x^0)$ which can be expressed by the following equations:

$$\begin{cases} x(i+1, j) = F_\alpha x(i, j) \\ x(i, j+1) = F_\beta x(i, j) \\ x(0, 0) \quad\; = x^0, \end{cases}$$

for any $i, j \in N$, where $x(i, j) \in X$.

Now consider the quasi-reachable standard form $((K^n, F_{\alpha s}, F_{\beta s}), \mathbf{e}_1)$ which is the pointed $\{\alpha, \beta\}$-action of the Quasi-reachable Standard System $((K^n, F_{\alpha s}, F_{\beta s}), \mathbf{e}_1, h_s)$ with a vector index $\nu = (\nu_1, \nu_2, \cdots, \nu_k)$ discussed in Definition (3.14) and Definition (3-B.9).

Lemma 4-A.1. Let $((K^n, F_{\alpha s}, F_{\beta s}), \mathbf{e}_1)$ be the quasi-reachable standard form with a vector index $\nu = (\nu_1, \nu_2, \cdots, \nu_k)$. Then the polynomial $\phi_1(\lambda)$ is a minimal polynomial of the matrix $F_{\alpha s}$.

Proof. It is apparent that $\phi_1(F_{\alpha s})\mathbf{e}_1 = \mathbf{0}$ because of the selection of the independent vectors in the Quasi-reachable Standard System, where $\mathbf{e}_1 := [10\cdots0]^T \in K^n$.
$\phi_1(F_{\alpha s})\mathbf{e}_2 = \phi_1(F_{\alpha s})F_{\alpha s}\mathbf{e}_1 = F_{\alpha s}\phi_1(F_{\alpha s})\mathbf{e}_1 = \mathbf{0}$,

$$\phi_1(F_{\alpha s})\mathbf{e}_3 = \phi_1(F_{\alpha s})F_{\alpha s}^2\mathbf{e}_1 = F_{\alpha s}^2\phi_1(F_{\alpha s})\mathbf{e}_1 = \mathbf{0},$$

$$\vdots$$

$$\phi_1(F_{\alpha s})\mathbf{e}_{\nu_1+1} = \phi_1(F_{\alpha s})F_{\beta s}\mathbf{e}_1 = F_{\beta s}\phi_1(F_{\alpha s})\mathbf{e}_1 = \mathbf{0},$$

$$\vdots$$

$$\phi_1(F_{\alpha s})\mathbf{e}_{\nu_1+\nu_2} = \phi_1(F_{\alpha s})F_{\beta s}F_{\alpha s}^{\nu_2-1}\mathbf{e}_1 = F_{\beta s}F_{\alpha s}^{\nu_2-1}\phi_1(F_{\alpha s})\mathbf{e}_1 = \mathbf{0},$$

We obtain that $\phi_1(F_{\alpha s})\mathbf{e}_{\nu_1+\nu_2+\cdots+\nu_k} = \mathbf{0}$ in a similar manner.

Therefore, we can show that $\phi_1(F_{\alpha s})\mathbf{e}_i = \mathbf{0}$ for any i $(1 \le i \le n)$. We insist that $\phi_1(\lambda)$ is an annihilating polynomial of $F_{\alpha s}$.

Next, we will show that $\phi_1(\lambda)$ is a minimal polynomial of $F_{\alpha s}$.

Let $\phi(\lambda)$ be a minimal polynomial of $F_{\alpha s}$ such that the degree of $\phi(\lambda)$ is less than the degree of $\phi_1(\lambda)$. Since $\phi(\lambda)$ of $F_{\alpha s}$ is a minimal polynomial, $\phi(F_{\alpha s})\mathbf{e}_1 = \mathbf{0}$ holds. But $\phi_1(\lambda)$ is a minimal polynomial such that $\phi_1(F_{\alpha s})\mathbf{e}_1 = \mathbf{0}$ by definition of the quasi-reachable standard form. This means that the assumption of $\phi(\lambda)$ is inconsistent with $\phi_1(\lambda)$.

Lemma 4-A.2. Let $((K^n, F_{\alpha s}, F_{\beta s}), \mathbf{e}_1)$ be the quasi-reachable standard form with a vector index $\nu = (\nu_1, \nu_2, \cdots, \nu_k)$. Let i be an integer such that $1 \le i \le k$ holds and let $\phi_i(\lambda)$ be the minimal polynomial of F_i in $F_{\alpha I}$. Then $\phi_i(\lambda)$ is divided by $\phi_{i+1}(\lambda)$ for i $(1 \le i \le k-1)$. Therefore, there exist polynomials $\{\chi_i(\lambda) : 1 \le i \le k\}$ such that $\phi_i(\lambda) = \chi_i(\lambda)\chi_{i+1}(\lambda)\cdots\chi_k(\lambda)$. Furthermore, $\phi_i(\lambda)$ can be expressed by $\phi_i(\lambda) = \lambda^{\nu_i} - c^i_{i\nu_i}\lambda^{\nu_i-1} - c^i_{i\nu_i-1}\lambda^{\nu_i-2} - \cdots - c^i_{i2}\lambda - c^i_{i1}$ by using $\{c^i_{ij};\ 1 \le j \le i\}$ in Definition (3.14).

Proof. Each $\nu_i \times \nu_i$ $(1 \le i \le k)$ submatrix of block-diagonal part in transition matrix $F_{\alpha s}$ has a companion form of a polynomial $\phi_i(\lambda)$ $(1 \le i \le k)$. Since $F_{\alpha s}$ is a block upper triangular matrix, the characteristic polynomial of $F_{\alpha s}$ can be expressed as a product of the characteristic polynomial of each block-diagonal matrix $\phi_1(\lambda)\phi_2(\lambda)\cdots\phi_k(\lambda)$.

We assume that $\phi_i(\lambda)$ is not divided by $\phi_{i+1}(\lambda)$. This contradicts that $\phi_1(\lambda)$ is the minimal polynomial of $F_{\alpha s}$. Therefore, $\phi_i(\lambda)$ can be divided by $\phi_{i+1}(\lambda)$, and $\phi_i(\lambda)$ can be expressed as $\phi_i(\lambda) = \chi_i(\lambda)\chi_{i+1}(\lambda)\cdots\chi_k(\lambda)$. Since each $\phi_i(\lambda)$ is the characteristic polynomial of each block-diagonal matrix of $F_{\alpha s}$, $\phi_i(\lambda)$ can be expressed by using the coefficients $\{c^i_{ij};\ 1 \le j \le i\}$.

Remark: The characteristic polynomial of the transition matrix $F_{\alpha I}$ in the Invariant Standard System can also be expressed as the product of the characteristic polynomials of the block-diagonal matrices. Lemma (4-A.1) and Lemma (4-A.2) imply that the transition matrix $F_{\alpha s}$ in the Quasi-reachable Standard Sytem is isomorphic to the transition matrix $F_{\alpha I}$ in the Invariant Standard System.

Lemma 4-A.3. Let $\phi_i(\lambda)$ be a characteristic polynomial of F_i (submatrix of $F_{\alpha I}$) for any i $(1 \le i \le k)$. Then rank $\phi_j(F_i) = \nu_i - \nu_j$ holds for $j \ge i$.

Proof. A direct calculation leads to this lemma.

Next we consider the other transition matrix $F_{\beta I} \in K^{n \times n}$ in the Invariant Standard System such that $F_{\alpha I} F_{\beta I} = F_{\beta I} F_{\alpha I}$.

Lemma 4-A.4. Two matrices $F_{\alpha I}$ and $F_{\beta I}$ satisfy the commutativity $F_{\alpha I} F_{\beta I} = F_{\beta I} F_{\alpha I}$ if and only if $F_{\beta I}$ can be expressed as $F_{\beta I} = [\mathbf{B}_1, \mathbf{B}_2, \cdots, \mathbf{B}_k]$ and $\chi_i(F_{\alpha I})\chi_{i+1}(F_{\alpha I}) \cdots \chi_k(F_{\alpha I})\mathbf{b}_i^1 = \mathbf{0}$ $(1 \le i \le k)$ holds for $\mathbf{B}_i = [\mathbf{b}_i^1, F_{\alpha I}\mathbf{b}_i^1, \cdots, F_{\alpha I}^{\ell_i - 1}\mathbf{b}_i^1, \mathbf{b}_i^2, \cdots, F_{\alpha I}^{\ell_{i+1}-1}\mathbf{b}_i^2, \cdots, \mathbf{b}_i^k, F_{\alpha I}\mathbf{b}_i^k, \cdots, F_{\alpha I}^{\ell_k - 1}\mathbf{b}_i^k]$, where $\mathbf{b}_i^j$ is given by $\mathbf{b}_i^j = \chi_i(F_{\alpha I})\chi_{i+1}(F_{\alpha I}) \cdots \chi_{i+j-2}(F_{\alpha I})\mathbf{b}_i^1$ for j $(2 \le j \le k)$.

Proof. This lemma can be obtained by direct calculations.

Here we set the following notations:

Let $\phi_i(\lambda)$ be the minimal polynomial of the matrix F_i in Figure 4.1. Then it can be expressed by $\phi_i(\lambda) = \lambda^{\nu_i} - c^i_{i\nu_i}\lambda^{\nu_i - 1} - c^i_{i\nu_i - 1}\lambda^{\nu_i - 2} - \cdots - c^i_{i2}\lambda - c^i_{i1}$. For c^i_{jl} in Definition (3.14), we use the following notations in the subsequent discussions: $\mathbf{C}^i := [C^i_1, C^i_2, \cdots, C^i_i, \mathbf{0}]^T \in K^n$, $C^i_j := [c^i_{j1}, c^i_{j2}, \cdots, c^i_{j\nu_j}]^T \in K^{\nu_i}$ for i, j $(1 \le i \le k,\ 1 \le j \le i)$.

Lemma 4-A.5. Let $(K^n, F_{\alpha s}, F_{\beta s})$ be the $\{\alpha, \beta\}$-action in the quasi-reachable standard form $((K^n, F_{\alpha s}, F_{\beta s}), \mathbf{e}_1)$ with a vector index $\nu = (\nu_1, \nu_2, \cdots, \nu_k)$ and let $(K^n, F_{\alpha I}, F_{\beta I})$ be the $\{\alpha, \beta\}$-action in Lemma (4-A.4).

Then T_s is the linear $\{\alpha, \beta\}$-morphism $T_s : (K^n, F_{\alpha s}, F_{\beta s}) \to (K^n, F_{\alpha I}, F_{\beta I})$, i.e., $F_{\alpha I} T_s = T_s F_{\alpha s}$ and $F_{\beta I} T_s = T_s F_{\beta s}$ if and only if the following three conditions hold:

① T_s is given by $T_s = [\mathbf{e}_1, \cdots, F_{\alpha I}^{\nu_1 - 1}\mathbf{e}_1, \mathbf{b}_1^1, \cdots, F_{\alpha I}^{\nu_2 - 1}\mathbf{b}_1^1, F_{\beta I}\mathbf{b}_1^1, \cdots, F_{\alpha I}^{\nu_3 - 1} F_{\beta I}\mathbf{b}_1^1, \cdots, F_{\beta I}^{k-2}\mathbf{b}_1^1, \cdots, F_{\alpha I}^{\nu_k - 1} F_{\beta I}^{k-2}\mathbf{b}_1^1]$, where $\mathbf{b}_1^1$ is given by $\mathbf{b}_1^1 = \mathbf{B}_1\mathbf{e}_1$.

② For any j $(1 \le j \le k)$, the following equation holds: $[\mathbf{e}_1, \cdots, F_{\alpha I}^{\nu_1 - 1}\mathbf{e}_1, \mathbf{b}_1^1, \cdots, F_{\alpha I}^{\nu_2 - 1}\mathbf{b}_1^1, F_{\beta I}\mathbf{b}_1^1, \cdots, F_{\alpha I}^{\nu_3 - 1}F_{\beta I}\mathbf{b}_1^1, \cdots, F_{\beta I}^{j-2}\mathbf{b}_1^1, \cdots, F_{\alpha I}^{\nu_j - 1}F_{\beta I}^{j-2}\mathbf{b}_1^1]$ $[C_1^{j+1}, C_2^{j+1}, \cdots, C_j^{j+1}]^T = \phi_{j+1}(F_{\alpha I})F_{\beta I}^{j-1}\mathbf{b}_1^1$.

③ $T_s\mathbf{c}^{k+1} = F_{\beta I}^{k-1}\mathbf{b}_1^1$.

Proof. Direct calculations provide this lemma.

Remark : T_s is a regular matrix and a linear $\{\alpha, \beta\}$-morphism $T_s : (K^n, F_{\alpha s}, F_{\beta s}) \to (K^n, F_{\alpha I}, F_{\beta I})$, i.e., $F_{\alpha I} T_s = T_s F_{\alpha s}$ and $F_{\beta I} T_s = T_s F_{\beta s}$ if and only if T_s preserves the quasi-reachability of $((K^n, F_{\alpha s}, F_{\beta s}), \mathbf{e}_1)$.

Henceforth, we will only seek out T_s which is a regular matrix. In fact, it will be evident that T_s is a right upper triangular matrix from the Remarks of Lemmas (4-A.7) and (4-A.8) and Proposition (4-A.10).

Here, let T_s which satisfies Lemma 4-A.5 be $T_s = [\mathbf{t}_1^1, \cdots, F_{\alpha I}^{\nu_1 - 1}\mathbf{t}_1^1, \mathbf{t}_2^1, \cdots, F_{\alpha I}^{\nu_2 - 1}\mathbf{t}_2^1, \mathbf{t}_3^1, \cdots, F_{\alpha I}^{\nu_3 - 1}\mathbf{t}_3^1, \cdots, \mathbf{t}_k^1, \cdots, F_{\alpha I}^{\nu_k - 1}\mathbf{t}_k^1]$, where $\mathbf{t}_1^1 = \mathbf{e}_1$, $\mathbf{t}_j^1 = F_{\beta I}^{j-2}\mathbf{b}_1^1$ for j $(2 \le j \le k)$.

Lemma 4-A.6. Let T_s be a regular matrix and the $\{\alpha,\beta\}$-morphism $T_s : (K^n, F_{\alpha s}, F_{\beta s}) \to (K^n, F_{\alpha I}, F_{\beta I})$. Then $F_{\alpha I}$ can be uniquely determined from $F_{\alpha s}$, and $F_{\alpha I}$ has an independent ν_1 parameters.

Proof. Note that the characteristic polynomial of $F_{\alpha s}$ is $\phi_1(\lambda)\phi_2(\lambda)\cdots\phi_k(\lambda)$, and each $\phi_i(\lambda)$ can be expressed as $\phi_i(\lambda) = \chi_i(\lambda)\chi_{i+1}(\lambda)\cdots\chi_k(\lambda)$ for i $(1 \le i \le k)$. If we select the polynomials $\phi_k(\lambda)$, $\phi_{k-1}(\lambda)$, $\cdots$, $\phi_2(\lambda)$ and $\phi_1(\lambda)$, we can calculate $\chi_k(\lambda)$, $\chi_{k-1}(\lambda)$, $\cdots$, $\chi_2(\lambda)$ and $\chi_1(\lambda)$ in turn. This means that $F_{\alpha I}$ can be uniquely determined. The polynomial $\chi_i(\lambda)$ can be determined by ℓ_i parameters for i $(1 \le i \le k)$. The polynomials $\chi_k(\lambda)$, $\chi_{k-1}(\lambda)$, $\cdots$, $\chi_2(\lambda)$ and $\chi_1(\lambda)$ can be determined by $\ell_k + \ell_{k-1} + \cdots + \ell_2 + \ell_1 = \nu_1$ parameters. Therefore, the matrix $F_{\alpha I}$ is determined by ν_1 parameters.

Remark: Let Lemma 4-A.6 hold. Since $\chi_i(\lambda)$ is a characteristic polynomial of $F_i \in K^{\nu_i \times \nu_i}$, $K^{\nu_i} = K^{\ell_i} \oplus K^{\ell_{i+1}} \oplus + \cdots \oplus K^{\ell_k}$ and $K^n = K^{\nu_1} \oplus K^{\nu_2} \oplus + \cdots \oplus K^{\nu_k}$ hold, $\mathbf{b}_i^1$ $(1 \le i \le k)$ in Lemma 4-A.4 has non-zero values of $K^{\nu_i} = K^{\ell_i} \oplus K^{\ell_{i+1}} \oplus + \cdots \oplus K^{\ell_k}$ in every subspace K^{ν_j} $(1 \le j \le k)$, while the others become zero vectors.

Lemma 4-A.7. Let T_s be one to one, onto and the $\{\alpha,\beta\}$-morphism $T_s : (K^n, F_{\alpha s}, F_{\beta s}) \to (K^n, F_{\alpha I}, F_{\beta I})$. Then $\mathbf{b}_1^1 \in K^n$ of $F_{\beta I}$ is determined by ℓ_1 minimal parameters.

Proof. Let the given conditions hold. According to Lemma 4-A.2 and condition ② in Lemma 4-A.5, $[\mathbf{e}_1, \cdots, F_{\alpha I}^{\nu_1 - 1}\mathbf{e}_1]C_1^2 = \phi_2(F_{\alpha I})\mathbf{b}_1^1 = \chi_2(F_{\alpha I})$ $\chi_3(F_{\alpha I})\cdots\chi_k(F_{\alpha I})\mathbf{b}_1^1$ holds. By $\chi_i(\lambda)$ $(2 \le i \le k)$ being a characteristic polynomial of F_i (submatrix of $F_{\alpha I}$) and the Remark in Lemma (4-A.6), the vector C_1^2 depends only on elements of $\mathbf{b}_1^1$ which belongs to the subspace K^{ℓ_1} in K^{ν_1} and does not depend on the other subspace of $\mathbf{b}_1^1$. Hence, the vector $\mathbf{b}_1^1$ can be determined by ℓ_1 parameters.

Remark: In the constrained conditions given in Lemma (4-A.7), we will seek out a $\{\alpha,\beta\}$-morphism $T_s : (K^n, F_{\alpha s}, F_{\beta s}) \to (K^n, F_{\alpha I}, F_{\beta I})$. Then we can set $\mathbf{b}_1^1 \in K^n$ as $\mathbf{b}_1^1 = [\mathbf{b}_1^{11}, \mathbf{e}_1^s, \mathbf{0}, \cdots, \mathbf{0}]^T$, where $\mathbf{b}_1^{11} = [\widetilde{\mathbf{b}_1^{11}}, \mathbf{0}]^T \in K^{\nu_1}$, $\widetilde{\mathbf{b}_1^{11}} \in K^{\ell_1}$ and $\mathbf{e}_1^s \in K^{\nu_2}$. Using $[\mathbf{e}_1, \cdots, F_{\alpha I}^{\nu_1 - 1}\mathbf{e}_1]C_1^2 = \phi_2(F_{\alpha I})\mathbf{b}_1^1$, we obtain $[\mathbf{e}_1^s, \cdots, F_1^{\nu_1 - 1}\mathbf{e}_1^s]C_1^2 = \phi_2(F_1)[\widetilde{\mathbf{b}_1^{11}}, \mathbf{0}]^T$.

Hence, $\widetilde{\mathbf{b}_1^{11}}$ is given by $\widetilde{\mathbf{b}_1^{11}} = \{[\phi_2(F_1)\mathbf{e}_1^s, \cdots, \phi_2(F_1)\mathbf{e}_{\ell_1}^s]^T[\phi_2(F_1)\mathbf{e}_1^s, \cdots, \phi_2(F_1)\mathbf{e}_{\ell_1}^s]\}^{-1} \times [\mathbf{e}_1^s, \cdots, \mathbf{e}_{\ell_1}^s]^T[\mathbf{e}_1^s, \cdots, F_1^{\nu_1 - 1}\mathbf{e}_1^s]C_1^2$, where $\mathbf{b}_1^{11} = [\widetilde{\mathbf{b}_1^{11}}, \mathbf{0}]^T \in K^{\nu_1}$, $\widetilde{\mathbf{b}_1^{11}} \in K^{\ell_1}$ and $\mathbf{e}_1^s \in K^{\nu_2}$. If we refer to the Remark on the form of T_s in Lemma (4-A.5), then we can say that all the column vectors from the first to the $(\nu_1 + \nu_2)$-th column of T_s are mutually independent.

Lemma 4-A.8 Let T_s be a regular matrix and the $\{\alpha,\beta\}$-morphism $T_s : (K^n, F_{\alpha s}, F_{\beta s}) \to (K^n, F_{\alpha I}, F_{\beta I})$.

And let $\mathbf{b}_1^1 \in K^n$ be $\mathbf{b}_1^1 = [\mathbf{b}_1^{11}, \mathbf{e}_1^s, \mathbf{0}, \cdots, \mathbf{0}]^T$, where $\mathbf{b}_1^{11} = [\widetilde{\mathbf{b}_1^{11}}, \mathbf{0}]^T \in K^{\nu_1}$, $\widetilde{\mathbf{b}_1^{11}} \in K^{\ell_1}$ and $\mathbf{e}_1^s \in K^{\nu_2}$. Then $\mathbf{b}_2^1 \in K^n$ of $F_{\beta I}$ is determined by $2 \times \ell_2$ minimal parameters.

Proof. Lemma (4-A.5) implies that
$[\mathbf{t}_1^1, \cdots, F_{\alpha I}^{\nu_1 - 1}\mathbf{t}_1^1, \mathbf{t}_2^1, \cdots, F_{\alpha I}^{\nu_2 - 1}\mathbf{t}_2^1][C_1^3, C_2^3]^T = \phi_3(F_{\alpha I})F_{\beta I}\mathbf{b}_1^1$.

On the other hand, $\mathbf{b}_1^1 = [\mathbf{b}_1^{11}, \mathbf{e}_1^s, \mathbf{0}, \cdots, \mathbf{0}]^T$ satisfies $F_{\beta I}\mathbf{b}_1^1 = [\mathbf{B}_1, \mathbf{B}_2]\mathbf{t}_2^1 = [\mathbf{b}_1^1, \cdots, F_1^{\nu_1 - 1}\mathbf{b}_1^1]\mathbf{b}_1^{11} + \mathbf{b}_2^1$.

Here, set $a_2^1 := [\mathbf{b}_1^1, \cdots, F_1^{\nu_1 - 1}\mathbf{b}_1^1]\ \mathbf{b}_1^{11}$. Then $a_2^1 = [a_2^{11}, a_2^{12}, \mathbf{0}, \cdots, \mathbf{0}]^T \in K^n$ holds, where $a_2^{11} \in K^{\nu_1}$ and $a_2^{12} \in K^{\nu_2}$.
Using Lemma (4-A.2), the above equation
$[\mathbf{t}_1^1, \cdots, F_{\alpha I}^{\nu_1 - 1}\mathbf{t}_1^1, \mathbf{t}_2^1, \cdots, F_{\alpha I}^{\nu_2 - 1}\mathbf{t}_2^1][C_1^3, C_2^3]^T = \phi_3(F_{\alpha I})F_{\beta I}\mathbf{b}_1^1$ becomes
$[\mathbf{t}_1^1, \cdots, F_{\alpha I}^{\nu_1 - 1}\mathbf{t}_1^1, \mathbf{t}_2^1, \cdots, F_{\alpha I}^{\nu_2 - 1}\mathbf{t}_2^1][C_1^3, C_2^3]^T - \phi_3(F_{\alpha I})a_2^1 = \phi_3(F_{\alpha I})\mathbf{b}_2^1 = \chi_3(F_{\alpha I})\chi_4(F_{\alpha I}) \cdots \chi_k(F_{\alpha I})\mathbf{b}_2^1$.

Since $\chi_i(\lambda)$ $(3 \leq i \leq k)$ is a characteristic polynomial of F_i (submatrix of $F_{\alpha I}$) from the above equation and the Remark in Lemma (4-A.5), the vector $[C_1^2, C_2^3]^T \in K^{\nu_1 + \nu_2}$ depends only on the elements of $\mathbf{b}_2^1 \in K^n$ which belong to the subspace K^{ℓ_2} in K^{ν_1} and K^{ν_2}. It does not depend on the other subspaces of $\mathbf{b}_2^1 \in K^n$. Hence the vector $\mathbf{b}_2^1$ can be determined by $2 \times \ell_2$ parameters.
Remark: Let $\mathbf{b}_1^1$ satisfy Lemma 4-A.8, then
$T_s\mathbf{e}_{\nu_1 + \nu_2 + 1} = \mathbf{t}_3^1 = F_{\beta I}\mathbf{b}_1^1 = a_2^1 + \mathbf{b}_2^1$ holds.
Under the constraints given in Lemma (4-A.8), we look for a $\{\alpha, \beta\}$-morphism $T_s : (K^n, F_{\alpha s}, F_{\beta s}) \to (K^n, F_{\alpha I}, F_{\beta I})$ which is a regular matrix. Then we can set $\mathbf{b}_2^1 \in K^n$ as $\mathbf{b}_2^1 = [\mathbf{b}_2^{11}, \mathbf{b}_2^{12}, \mathbf{e}_1^s, \mathbf{0}]^T$,
where $\mathbf{b}_2^{11} := [\mathbf{0}, \widetilde{\mathbf{b}_2^{11}}, \mathbf{0}]^T \in K^{\nu_1}, \widetilde{\mathbf{b}_2^{11}} \in K^{\ell_2}$, $\mathbf{b}_2^{12} := [\mathbf{0}, \widetilde{\mathbf{b}_2^{12}}, \mathbf{0}]^T \in K^{\nu_2}$, $\widetilde{\mathbf{b}_2^{12}} \in K^{\ell_2}$ and $\mathbf{e}_1^s \in K^{\nu_3}$ hold.
Referring to the Remark on the form of T_s in Lemma (4-A.5), we will say that all the column vectors from the first column to the $(\nu_1 + \nu_2 + \nu_3)$-th column of T_s are mutually independent.
Using
$[\mathbf{t}_1^1, \cdots, F_{\alpha I}^{\nu_1 - 1}\mathbf{t}_1^1, \mathbf{t}_2^1, \cdots, F_{\alpha I}^{\nu_2 - 1}\mathbf{t}_2^1][C_1^3, C_2^3]^T = \phi_3(F_{\alpha I})F_{\beta I}\mathbf{b}_1^1$
$= \phi_3(F_{\alpha I})(a_2^1 + \mathbf{b}_2^1)$, $\mathbf{b}_2^1 \in K^n$ as $\mathbf{b}_2^1 = [\mathbf{b}_2^{11}\mathbf{b}_2^{12}\mathbf{e}_1^s, 0]^T$ is given as follows:
$\widetilde{\mathbf{b}_2^{11}}$ is given by
$\widetilde{\mathbf{b}_2^{11}} = \{[\phi_3(F_1)\mathbf{e}_{\ell_1+1}^s, \cdots, \phi_3(F_1)\mathbf{e}_{\ell_1+\ell_2}^s]^T[\phi_3(F_1)\mathbf{e}_{\ell_1+1}^s, \cdots, \phi_3(F_1)\mathbf{e}_{\ell_1+\ell_2}^s]\}^{-1}$
$\times [\mathbf{e}_{\ell_1+1}^s, \cdots, \mathbf{e}_{\ell_1+\ell_2}^s]^T\{[\mathbf{e}_1^s, \cdots, F_1^{\nu_1 - 1}\mathbf{e}_1^s]C_1^3 + [\mathbf{t}_2^{11}, \cdots, F_1^{\nu_1 - 1}\mathbf{t}_2^{11}]$
$C_2^3 - \phi_3(F_1)a_2^{11}\}$
$\widetilde{\mathbf{b}_2^{12}}$ is given by
$\widetilde{\mathbf{b}_2^{12}} = \{[\phi_3(F_2)\mathbf{e}_1^s, \cdots, \phi_3(F_2)\mathbf{e}_{\ell_2}^s]^T[\phi_3(F_2)\mathbf{e}_1^s, \cdots, \phi_3(F_2)\mathbf{e}_{\ell_2}^s]\}^{-1}$
$\times [\mathbf{e}_1^s, \cdots, \mathbf{e}_{\ell_2}^s]^T\{[\mathbf{t}_2^{12}, \cdots, F_2^{\nu_2 - 1}\mathbf{t}_2^{12}]C_2^3 - \phi_3(F_2)a_2^{11}\}$

Next, we will find $\mathbf{b}_3^1 \in K^n$.
By the relations of $T_s\mathbf{e}_{\nu_1+\nu_2+\nu_3+1} = \mathbf{t}_4^1 = F_{\beta I}^2\mathbf{b}_1^1 = F_{\beta I}(a_2^1 + \mathbf{b}_2^1) = F_{\beta I}\mathbf{t}_3^1$ and by the Remark in Lemma (4-A.8), $F_{\beta I}^2\mathbf{b}_1^1 = a_3^1 + \mathbf{b}_3^1$ holds, where $a_3^1 :=$

$\mathbf{B}_1\mathbf{t}_3^{11} + \mathbf{B}_2\mathbf{t}_3^{12}$. Then we can obtain $a_3^1 = [a_3^{11}, a_3^{12}, a_3^{13}, \mathbf{0}, \cdots, \mathbf{0}]^T \in K^n$, where $a_3^{11} \in K^{\nu_1}, a_3^{12} \in K^{\nu_2}$ and $a_3^{13} \in K^{\nu_3}$.

According to ② in Lemma (4-A.5), $\mathbf{b}_3^1$ is given by the following relation:
$[\mathbf{t}_1^1, \cdots, F_{\alpha I}^{\nu_1-1}\mathbf{t}_1^1, \mathbf{t}_2^1, \cdots, F_{\alpha I}^{\nu_2-1}\mathbf{t}_2^1, \mathbf{t}_3^1, \cdots, F_{\alpha I}^{\nu_3-1}\mathbf{t}_3^1][C_1^4, C_2^4, C_3^4]^T$
$= \phi_4(F_{\alpha I})F_{\beta I}^2\mathbf{b}_1^1 = \phi_4(F_{\alpha I})(a_3^1 + \mathbf{b}_3^1)$.

Then the vector $\mathbf{b}_3^1 \in K^n$ is determined by $3 \times \ell_3$ parameters in the same way as in Lemma 4-A.8.

Since $a_3^1 + \mathbf{b}_3^1 = F_{\beta I}^2\mathbf{b}_1^1 = T_s\mathbf{e}_{\nu_1+\nu_2+\nu_3+1}$ holds, set $\mathbf{b}_3^1$ as
$\mathbf{b}_3^1 = [\mathbf{b}_3^{11}, \mathbf{b}_3^{12}, \mathbf{b}_3^{13}, \mathbf{e}_1^s, \mathbf{0}, \cdots, \mathbf{0}]^T$ for all the column vectors from the first to the $(\nu_1+\nu_2+\nu_3)$-th column of T_s are mutually independent, where
$\mathbf{b}_3^{11} := [\mathbf{0}, \widetilde{\mathbf{b}_3^{11}}, \mathbf{0}] \in K^{\nu_1}, \widetilde{\mathbf{b}_3^{11}} \in K^{\ell_3}$, $\mathbf{b}_3^{12} := [\mathbf{0}, \widetilde{\mathbf{b}_3^{12}}, \mathbf{0}] \in K^{\nu_2}, \widetilde{\mathbf{b}_3^{12}} \in K^{\ell_3}$ and $\mathbf{b}_3^{13} := [\mathbf{0}, \widetilde{\mathbf{b}_3^{13}}, \mathbf{0}] \in K^{\nu_3}, \widetilde{\mathbf{b}_3^{13}} \in K^{\ell_3}$. $\widetilde{\mathbf{b}_3^{11}}$, $\widetilde{\mathbf{b}_3^{12}}$ and $\widetilde{\mathbf{b}_3^{13}} \in K^{\ell_3}$ are given as:

$$\widetilde{\mathbf{b}_3^{11}} = \{[\phi_4(F_1)\mathbf{e}_{\ell_1+\ell_2+1}^s, \cdots, \phi_4(F_1)\mathbf{e}_{\ell_1+\ell_2+\ell_3}^s]^T[\phi_4(F_1)\mathbf{e}_{\ell_1+\ell_2+1}^s, \cdots, \phi_4(F_1)\mathbf{e}_{\ell_1+\ell_2+\ell_3}^s]\}^{-1} \times [\mathbf{e}_{\ell_1+\ell_2+1}^s, \cdots, \mathbf{e}_{\ell_1+\ell_2+\ell_3}^s]^T\{[\mathbf{t}_1^{11}, \cdots, F_1^{\nu_1-1}\mathbf{t}_1^{11}]C_1^4 + [\mathbf{t}_2^{11}, \cdots, F_1^{\nu_1-1}\mathbf{t}_2^{11}]C_2^4 + [\mathbf{t}_3^{11}, \cdots, F_1^{\nu_1-1}\mathbf{t}_3^{11}]C_3^4 - \phi_4(F_1)a_3^{11}\}$$

$$\widetilde{\mathbf{b}_3^{12}} = \{[\phi_4(F_2)\mathbf{e}_{\ell_2+1}^s, \cdots, \phi_4(F_2)\mathbf{e}_{\ell_2+\ell_3}^s]^T[\phi_4(F_2)\mathbf{e}_{\ell_2+1}^s, \cdots, \phi_4(F_2)\mathbf{e}_{\ell_2+\ell_3}^s]\}^{-1} \times [\mathbf{e}_{\ell_2+1}^s, \cdots, \mathbf{e}_{\ell_2+\ell_3}^s]^T\{[\mathbf{t}_2^{12}, \cdots, F_2^{\nu_2-1}\mathbf{t}_2^{12}]C_2^4 + [\mathbf{t}_3^{12}, \cdots, F_2^{\nu_3-1}\mathbf{t}_3^{12}]C_3^4 - \phi_4(F_2)a_3^{12}\}$$

$$\widetilde{\mathbf{b}_3^{13}} = \{[\phi_4(F_3)\mathbf{e}_1^s, \cdots, \phi_4(F_3)\mathbf{e}_{\ell_3}^s]^T[\phi_4(F_3)\mathbf{e}_1^s, \cdots, \phi_4(F_3)\mathbf{e}_{\ell_3}^s]\}^{-1} \times [\mathbf{e}_1^s, \cdots, \mathbf{e}_{\ell_3}^s]^T\{[\mathbf{t}_3^{13}, \cdots, F_2^{\nu_3-1}\mathbf{t}_3^{13}]C_3^4 - \phi_4(F_3)a_3^{13}\}$$

Thus we have determined $\mathbf{b}_3^1$ as $\mathbf{b}_3^1 = [\mathbf{b}_3^{11}, \mathbf{b}_3^{12}, \mathbf{b}_3^{13}, \mathbf{e}_1^s, \mathbf{0}, \cdots, \mathbf{0}]^T \in K^n$.
We have also obtained a_3^1 as $a_3^1 = [a_3^{11}, a_3^{12}, a_3^{13}, \mathbf{0}, \cdots, \mathbf{0}]^T \in K^n$.

In the same way, we can determine $\mathbf{b}_i^1$ and $F_{\beta I}^{i-1}\mathbf{b}_1^1 = \mathbf{t}_{i+1}^1 = a_i^1 + \mathbf{b}_i^1$ for i $(4 \le i \le k-1)$ such that all column vectors from the first column to the $(\nu_1+\nu_2+\nu_3+\cdots+\nu_k)$-th column of T_s are mutually independent, where a_i^1 is given by the equation $a_i^1 := B_1\mathbf{t}_i^{11} + B_2\mathbf{t}_i^{12} + \cdots + B_{i-1}\mathbf{t}_i^{1i-1}$, and a_i^1 becomes $a^i = [a^{i1}a^{i2}, \cdots, a^{ii}, \mathbf{0}, \cdots, \mathbf{0}]^T \in K^n, a^{i1} \in K^{\nu_1}, a^{i2} \in K^{\nu_2}, \cdots, a^{ii} \in K^{\nu_i}$.

Consider $\widetilde{\mathbf{b}_i^{1j}}$ $(1 \le j \le i-1)$. They belong to
$\mathbf{b}_i^1 = [\mathbf{b}_i^{11}, \mathbf{b}_i^{12}, \cdots, \mathbf{b}_i^{1i}, \mathbf{e}_1^s, \mathbf{0}, \cdots, \mathbf{0}]^T \in K^n$ as the vector, where $\mathbf{b}_i^{11}, \mathbf{b}_i^{12}, \cdots$, and $\mathbf{b}_i^{1i}$ are represented as follows:
$\mathbf{b}_i^{11} := [\mathbf{0}, \widetilde{\mathbf{b}_i^{11}}, \mathbf{0}] \in K^{\nu_1}, \widetilde{\mathbf{b}_i^{11}} \in K^{\ell_i}$, $\mathbf{b}_i^{12} := [\mathbf{0}, \widetilde{\mathbf{b}_i^{12}}, \mathbf{0}] \in K^{\nu_2}, \widetilde{\mathbf{b}_i^{12}} \in K^{\ell_i}$, $\cdots$,
$\mathbf{b}_i^{1j} := [\mathbf{0}, \widetilde{\mathbf{b}_i^{1j}}, \mathbf{0}] \in K^{\nu_j}, \widetilde{\mathbf{b}_i^{1j}} \in K^{\ell_i}$ and $\mathbf{e}_1^s \in K^{\nu_{i+1}}$.

Therefore, the number of parameters in $\mathbf{b}_i^1$ becomes $i \times \ell_i$, which is apparently minimal.

Next, we determine $\mathbf{b}_k^1$, which appears in $F_{\beta I} = [\mathbf{B}_1, \mathbf{B}_2, \cdots, \mathbf{B}_k]$ and $\mathbf{B}_i = [\mathbf{b}_i^1 F_{\alpha I}\mathbf{b}_i^1, \cdots, F_{\alpha I}^{\ell_i - 1}\mathbf{b}_i^1, \mathbf{b}_i^2, \cdots, F_{\alpha I}^{\ell_{i+1}-1}\mathbf{b}_i^2, \cdots, \mathbf{b}_i^k F_{\alpha I}\mathbf{b}_i^k, \cdots, F_{\alpha I}^{\ell_k - 1}$ $\mathbf{b}_i^k]$ for i $(1 \leq i \leq k)$. $\mathbf{b}_i^j$ is given by
$\mathbf{b}_i^j = \chi_i(F_{\alpha I})\chi_{i+1}(F_{\alpha I}) \cdots \chi_{i+j-2}(F_{\alpha I})\mathbf{b}_i^1$ for i, j $(1 \leq i \leq k,\ 2 \leq j \leq \ell_i)$.
In order to determine $\mathbf{b}_k^1$, we introduce Lemma 4-A.9.

Lemma 4-A.9. Let T_s be a regular matrix and a $\{\alpha, \beta\}$-morphism $T_s : (K^n, F_{\alpha s}, F_{\beta s}) \to (K^n, F_{\alpha I}, F_{\beta I})$. Also let $\mathbf{b}_i^1 \in K^n$ be
$\mathbf{b}_i^1 = [\mathbf{b}_i^{11}, \mathbf{b}_i^{12}, \cdots, \mathbf{b}_i^{1i}, \mathbf{e}_1^s, \mathbf{0}, \cdots, \mathbf{0}]^T \in K^n$ for i $(1 \leq i \leq k-1)$, where $\mathbf{b}_i^{11} := [\mathbf{0}, \widetilde{\mathbf{b}_i^{11}}, \mathbf{0}]^T \in K^{\nu_1}, \widetilde{\mathbf{b}_i^{11}} \in K^{\ell_i}$, $\mathbf{b}_i^{12} := [\mathbf{0}, \widetilde{\mathbf{b}_i^{12}}, \mathbf{0}]^T \in K^{\nu_2}$, $\widetilde{\mathbf{b}_i^{12}} \in K^{\ell_i}$, $\cdots$, $\mathbf{b}_i^{1i} := [\mathbf{0}, \widetilde{\mathbf{b}_i^{1i}}, \mathbf{0}]^T \in K^{\nu_i}, \widetilde{\mathbf{b}_i^{1i}} \in K^{\ell_i}, \mathbf{e}_1^s \in K^{\nu_{i+1}}$. Then $\mathbf{b}_k^1 \in K^n$ of $F_{\beta I}$ is determined by $k \times \ell_k$ minimal parameters.

Proof. According to ③ in Lemma (4-A.5) and Lemma (4-A.2), we obtain $T_s\mathbf{c}^{k+1} = F_{\beta I}^{k-1}\mathbf{b}_1^1 = F_{\beta I}\mathbf{t}_k^1 = \mathbf{B}_1\mathbf{t}_k^{11} + \mathbf{B}_2\mathbf{t}_k^{12} + \cdots + \mathbf{B}_{k-1}\mathbf{t}_k^{1k-1} + \mathbf{b}_k^1$.

Hence $\mathbf{b}_k^1 = T_s\mathbf{c}^{k+1} - a_k^1$ holds, where a_k^1 is given by $a_k^1 = \mathbf{B}_1\mathbf{t}_k^{11} + \mathbf{B}_2\mathbf{t}_k^{12} + \cdots + \mathbf{B}_{k-1}\mathbf{t}_k^{1k-1}$. If we consider Lemma (4-A.4) and the Remark in Lemma (4-A.5), we can conclude that the vector $\mathbf{b}_k^1$ can be determined by $k \times \ell_k$ parameters.

By summarizing from Lemma (4-A.4) through Lemma (4-A.9) and their Remarks, we arrive at the following proposition.

Proposition 4-A.10. If $\mathbf{b}_i^1$ $(1 \leq i \leq k-1)$ and $\mathbf{b}_k^1$ of $F_{\beta I}$ in $\{\alpha, \beta\}$-action $(K^n, F_{\alpha I}, F_{\beta, I})$ are determined by the following conditions (1) and (2), then $F_{\beta I}$ can be determined by the minimal parameters and the linear operator T_s is $\{\alpha, \beta\}$-morphism $T_s : (K^n, F_{\alpha s}, F_{\beta s}) \to (K^n, F_{\alpha I}, F_{\beta I})$; moreover, T_s is bijective.

(1) $\mathbf{b}_i^1 \in K^n$ is given by $\mathbf{b}_i^1 = [\mathbf{b}_i^{11}, \mathbf{b}_i^{12}, \cdots, \mathbf{b}_i^{1i}, \mathbf{e}_1^s, \mathbf{0}, \cdots, \mathbf{0}]^T \in K^n$ for i $(1 \leq i \leq k-1)$, where $\mathbf{b}_i^{11} := [\mathbf{0}, \widetilde{\mathbf{b}_i^{11}}, \mathbf{0}]^T \in K^{\nu_1}, \widetilde{\mathbf{b}_i^{11}} \in K^{\ell_i}$, $\mathbf{b}_i^{12} := [\mathbf{0}, \widetilde{\mathbf{b}_i^{12}}, \mathbf{0}]^T \in K^{\nu_2}$, $\widetilde{\mathbf{b}_i^{12}} \in K^{\ell_i}$, $\cdots$, $\mathbf{b}_i^{1i} := [\mathbf{0}, \widetilde{\mathbf{b}_i^{1i}}, \mathbf{0}]^T \in K^{\nu_i}, \widetilde{\mathbf{b}_i^{1i}} \in K^{\ell_i}, \mathbf{e}_1^s \in K^{\nu_{i+1}}$.
(2) $\mathbf{b}_k^1 \in K^n$ is given by $\mathbf{b}_k^1 = [\mathbf{b}_k^{11}, \mathbf{b}_k^{12}, \cdots, \mathbf{b}_k^{1k}]^T \in K^n$, where $\mathbf{b}_k^{1j} := [\mathbf{0}, \widetilde{\mathbf{b}_k^{1j}}, \mathbf{0}]^T \in K^{\nu_j}, \widetilde{\mathbf{b}_k^{1j}} \in K^{\ell_j}$ and $\mathbf{b}_k^{1k} := \widetilde{\mathbf{b}_k^{1k}} \in K^{\ell_k}$ for j $(1 \leq j \leq k-1)$.

Remark: $F_{\beta I}$ of $\{\alpha, \beta\}$-action $(K^n, F_{\alpha I}, F_{\beta I})$ obtained in Proposition (4-A.10) can be determined by n $(\ell_1 + 2 \times \ell_2 + 3 \times \ell_3 + \cdots + k \times \ell_k = n)$ parameters. Moreover, note that $(K^n, F_{\alpha I}, F_{\beta I})$ is the same as $\{\alpha, \beta\}$-action of the invariant standard form $((K^n, F_{\alpha I}, F_{\beta I}), \mathbf{e}_1)$.

At this point, we will make the relation between the Quasi-reachable Standard Systems and the Invariant Standard Systems clear.

Theorem 4-A.11. For any Quasi-reachable Standard System $((K^n, F_{\alpha s}, F_{\beta s}), \mathbf{e}_1, h_s)$, there exists a unique Invariant Standard System $((K^n, F_{\alpha I}, F_{\beta I}), \mathbf{e}_1, h_I)$ which is isomorphic to it.

Proof. Let $\sigma_s = ((K^n, F_{\alpha s}, F_{\beta s}), \mathbf{e}_1, h_s)$ be the Quasi-reachable Standard System with the index $\nu = (\nu_1, \nu_2, \cdots, \nu_k)$. Let $(K^n, F_{\alpha I}, F_{\beta I})$ be the $\{\alpha, \beta\}$-action which satisfies Proposition (4-A.10). For the regular matrix T_s which is the $\{\alpha, \beta\}$-morphism $T_s : (K^n, F_{\alpha s}, F_{\beta s}) \to (K^n, F_{\alpha I}, F_{\beta I})$ obtained in Proposition (4-A.10), set $h_I := h_s T_s^{-1}$. Since $T_s \mathbf{e}_1 = \mathbf{e}_1$ holds, T_s is the 2-Commutative Linear Representation System morphism $T_s : ((K^n, F_{\alpha s}, F_{\beta s}), \mathbf{e}_1, h_s) \to ((K^n, F_{\alpha I}, F_{\beta I}), \mathbf{e}_1, h_I)$. Since T_s is bijective, $((K^n, F_{\alpha I}, F_{\beta I}), \mathbf{e}_1, h_I)$ is canonical, and is also the Invariant Standard System with the index $\nu = (\nu_1, \nu_2, \cdots, \nu_k)$.

Since we take out the minimal parts which depends on $((K^n, F_{\alpha s}, F_{\beta s}), \mathbf{e}_1, h_s)$ in the process of obtaining $F_{\beta I}$ of Lemma (4-A.9), the Invariant Standard System $((K^n, F_{\alpha I}, F_{\beta I}), \mathbf{e}_1, h_I)$ is clearly unique.

4-A.12. Proof of Theorem (4.2)
Let $((K^n, F_\alpha, F_\beta), \mathbf{e}_1, h)$ be any n-dimensional canonical 2-Commutative Linear Representation System. Then Theorem (3.15) implies that there exists a unique Quasi-reachable Standard System $((K^n, F_{\alpha s}, F_{\beta s}), \mathbf{e}_1, h_s)$ and a unique 2-Commutative Linear Representation System morphism $T : ((K^n, F_\alpha, F_\beta), \mathbf{e}_1, h) \to ((K^n, F_{\alpha s}, F_{\beta s}), \mathbf{e}_1, h_s)$ which is isomorphic to it. Let T_s be the 2-Commutative Linear Representation System morphism $T_s : ((K^n, F_{\alpha s}, F_{\beta s}), \mathbf{e}_1, h_s) \to ((K^n, F_{\alpha I}, F_{\beta I}), \mathbf{e}_1, h_I)$ which was obtained in the proof of Theorem (4-A.11). Then $T_s{\cdot}T$ is a 2-Commutative Linear Representation System morphism $T_s{\cdot}T : ((K^n, F_\alpha, F_\beta), \mathbf{e}_1, h) \to ((K^n, F_{\alpha I}, F_{\beta I}), \mathbf{e}_1, h_I)$. Clearly, $((K^n, F_{\alpha I}, F_{\beta I}), \mathbf{e}_1, h_I)$ is unique for $((K^n, F_\alpha, F_\beta), \mathbf{e}_1, h)$.

Moreover, the behavior of $((K^n, F_{\alpha I}, F_{\beta I}), \mathbf{e}_1, h_I)$ is the same as that of $((K^n, F_\alpha, F_\beta), \mathbf{e}_1, h)$.

4-A.13. Proof of Procedure for effective encoding (4.3)
On the basis of the facts in Lemma (4-A.4) to Lemma (4-A.9) and the remarks in those lemmas, we can obtain the Invariant Standard System $((K^n, F_{\alpha I}, F_{\beta I}), \mathbf{e}_1, h_I)$ which has minimal parameters.

5 Design for Two-Dimensional Images

In Chapter 3, it was newly proposed that any two-dimensional geometrical pattern can be realized by a mathematical model called the 2-Commutative Linear Representation Systems. Many methods of image generation are intended primarily for the reality of images. Our proposed method is fundamentally different; it is a graphic generation method. This geometrical pattern generation is a new attempt at design patterns or artifacts which come only in our imaginations. All the calculations consisting of linear operations, this generation method is very suitable for computer algorithms. It is also of major importance that our model can describe any two-dimensional geometrical pattern and reproduce it exactly. The method can be applied to patterns for tableware or fabrics. A number of examples of applications will be shown to illustrate the effectiveness of this method.

Computer graphics have been used to present exact images of natural objects and phenomena. See [Mcormick and Jayaramamurthy, 1974]. Usually, generation algorithms for graphic arts are irregular and complex. On the contrary, for fancy articles, regular and simple patterns may be preferable. For example, Culik and Valenta, [1997] propose a designing method for simple colored images by means of an automaton. Meiszner, et al. [1998] treat an art of knitted fabrics in three-dimensional images. However, the methods presented in these papers are executed by means of non-linear operations.

There are generation methods [Heeger and Bergen, 1995], [Zhu, Wu and Mumford, 1998], [Bonet and Viola, 1997] which produce new or similar images from original image data. However, these methods rely on certain filter, which require actual visual data to produce images; they can not reproduce original images. On the contrary, our method is a 'true' image generator and does not need any actual image when it originates a pattern.

We know that any two-dimensional geometrical pattern can be modeled by the 2-Commutative Linear Representation system. For the simple design for two-dimensional images, a direct idea for two-dimensional image (geometrical pattern generation) is presented by using 2-Commutative Linear Representation Systems. Several examples of geometrical pattern generation are also presented.

5.1 2-Commutative Linear Representation Systems for Design

Any finite-sized two-dimensional pattern can usually be decomposed into two categories:

(1) Patterns with no periodicity;
(2) Patterns with periodicity. We will discuss first 2-Commutative Linear Representation Systems σ which realize patterns with no periodicity and then take up patterns with periodicity. The patterns treated will be $(L+1) \times (M+1)$-sized two-dimensional images for positive integers L, $M \in N$.

Lemma 5.1. For any $(L+1)\times(M+1)$-sized pattern $a \in F(\mathbf{L}\times\mathbf{M}, Y)$ with no periodicity, there exists a $(L+1)\times(M+1)$-dimensional 2-Commutative Linear Representation System $\sigma = ((K^{(L+1)\times(M+1)}, F_\alpha, F_\beta), x^0, h)$ which realizes it. See Figure 5.1 for the σ.

$$F_\alpha = \underbrace{\begin{bmatrix} F_{\alpha 0} & 0 & \cdots & 0 \\ 0 & F_{\alpha 0} & \ddots & \vdots \\ \vdots & \ddots & \ddots & 0 \\ 0 & \cdots & 0 & F_{\alpha 0} \end{bmatrix}}_{(L+1)\times(M+1)}, \quad F_\beta = \underbrace{\begin{bmatrix} 0 & \cdots & 0 & 0 \\ I_{M+1} & \ddots & \vdots & \vdots \\ 0 & \ddots & 0 & \vdots \\ 0 & 0 & I_{M+1} & 0 \end{bmatrix}}_{(L+1)\times(M+1)}$$

$$x^0 = \left.\begin{bmatrix} 1 \\ 0 \\ \vdots \\ 0 \end{bmatrix}\right\} (L+1)\times(M+1), \quad F_{\alpha 0} = \underbrace{\begin{bmatrix} 0 & \cdots & 0 & 0 \\ 1 & \ddots & \vdots & \vdots \\ 0 & \ddots & 0 & \vdots \\ 0 & 0 & 1 & 0 \end{bmatrix}}_{M+1},$$

I_{M+1} is a $(M+1) \times (M+1)$ identity matrix,

$$h = \underbrace{\begin{bmatrix} a_{0,0} \cdots a_{M,0} \, a_{0,1} \cdots a_{M,1} \cdots a_{0,L} \cdots a_{M,L} \end{bmatrix}}_{(L+1)\times(M+1)}$$

Fig. 5.1. The 2-Commutative Linear Representation System $\sigma = ((K^{(L+1)\times(M+1)}, F_\alpha, F_\beta), x^0, h)$ in Lemma (5.1)

Proof. This lemma can be obtained easily by direct calculations.

Remark: According to this lemma, we can easily treat a pattern with no periodicity.

5.2. Patterns with periodicity.
A pattern $a \in F(N^2, Y)$ with a period of l length in the vertical and m length in the horizontal directions may be expressed as the two-dimensional image given in Figure 5.1. Such pattern $a \in F(N^2, Y)$ is written by the two-dimensional image with a $l \times m$-period. For simplicity, in the figure, we write $a(i, j)$ as $a_{i,j}$ for $i, j \in N$.

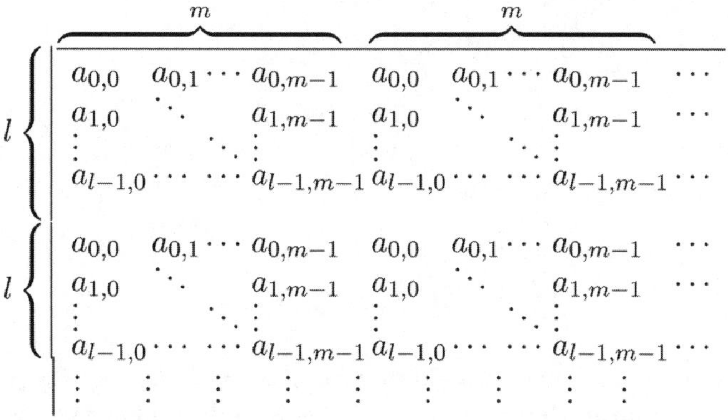

Fig. 5.2. The two-dimensional image with a $\ell \times m$-period

Proposition 5.3. *Any two-dimensional image with a $\ell \times m$-period can be realized by a 2-Commutative Linear Representation System $\sigma_p = ((K^{l \times m}, F_{\alpha p}, F_{\beta p}), x_p^0, h_p)$. See Figure 5.3 for the system σ_p.*

Proof. This can be obtained easily by direct calculations.

5.4. A direct sum of 2-Commutative Linear Representation System
We now introduce a direct sum $\sigma_1 \oplus \sigma_2$ for the 2-Commutative Linear Representation Systems σ_1 and σ_2, where $\sigma_1 = ((K^{n_1}, F_{\alpha 1}, F_{\beta 1}), x_1^0, h_1)$ and $\sigma_2 = ((K^{n_2}, F_{\alpha 2}, F_{\beta 2}), x_2^0, h_2)$.

The behavior of $\sigma_1 \oplus \sigma_2$ is given by $a_{\sigma_1} + a_{\sigma_2}$, namely, $a_{\sigma_1 \oplus \sigma_2} = a_{\sigma_1} + a_{\sigma_2}$ holds, where $\sigma_1 \oplus \sigma_2$ is expressed as follows:

$$\sigma_1 \oplus \sigma_2 = ((K^{n_1+n_2}, F_\alpha, F_\beta), x^0, h),$$

where

$$F_\alpha = \begin{bmatrix} F_{\alpha 1} & 0 \\ 0 & F_{\alpha 2} \end{bmatrix}, F_\beta = \begin{bmatrix} F_{\beta 1} & 0 \\ 0 & F_{\beta 2} \end{bmatrix}, x^0 = \begin{bmatrix} x_1^0 \\ x_2^0 \end{bmatrix}$$

and

$$h = \begin{bmatrix} h_1 \; h_2 \end{bmatrix}.$$

$$F_{\alpha p} = I_m \otimes F_\ell = \underbrace{\begin{bmatrix} F_\ell & 0 & \cdots & 0 \\ 0 & F_\ell & \ddots & \vdots \\ \vdots & \ddots & \ddots & 0 \\ 0 & \cdots & 0 & F_\ell \end{bmatrix}}_{\ell \times m}, \; F_{\beta p} = F_m \otimes I_\ell = \underbrace{\begin{bmatrix} 0 & \cdots & 0 & I_\ell \\ I_\ell & \ddots & \vdots & 0 \\ 0 & \ddots & 0 & \vdots \\ 0 & 0 & I_\ell & 0 \end{bmatrix}}_{\ell \times m}$$

$$x^0 = \left.\begin{bmatrix} 1 \\ 0 \\ \vdots \\ 0 \end{bmatrix}\right\} \ell \times m \;, \; F_q = \underbrace{\begin{bmatrix} 0 & \cdots & 0 & 1 \\ 1 & \ddots & \vdots & 0 \\ 0 & \ddots & 0 & \vdots \\ 0 & 0 & 1 & 0 \end{bmatrix}}_{q}, \text{ where } q = \ell \text{ or } m.$$

Fig. 5.3. The 2-Commutative Linear Representation System $\sigma_p = ((K^{\ell \times m}, F_{\alpha p}, F_{\beta p}), x_p^0, h_p)$ for patterns with a $\ell \times m$-period

To design two-dimensional images at high speed, we will use the direct sum of 2-Commutative Linear Representation Systems and number theory. First, we will state the following lemma, which is a special case of Dirichlet's theorem [Hardy and Wright, 1979]:

Lemma 5.5. Let p be a prime number such that $p = L \times M + 1$ for a positive integer L and for a fixed integer M. Then there are infinite primes p of the form.

Proof. This is Dirichlet's theorem itself.

5.6. Fermat's Lemma [Hardy and Wright, 1979] If p is prime and x is not divisible by p, then $x^{p-1} \equiv 1 \ (mod\ p)$ holds.

By virtue of Fermat's lemma (5.6), if an integer L satisfies $L < p$, then $x^L - 1 \equiv 0 \ (mod\ p)$ has L different solutions. Hence we can obtain the following lemma.

Lemma 5.7. Let p be a prime number such that $p = L \times M + 1$ for a positive integer L and a fixed integer M. Then $x^L - 1 \equiv (x - x_1)(x - x_2)(x - x_3) \cdots (x - x_L) \ (mod\ p)$ holds.

Proof. By the condition for the selection of the prime number p, we obtain $x^L - 1 = x^{\frac{p-1}{M}} - 1 = (x^{\frac{1}{M}})^{p-1} - 1 \equiv 0 \ (mod\ p)$.

Definition 5.8. The 2-Commutative Linear Representation System $\sigma_e = ((K^{\ell \times m}, F_{\alpha e}, F_{\beta e}), x_e^0, h_e)$ given in Figure 5.4 is called an Eigen Standard System.

$$F_{\alpha e}=\underbrace{\begin{bmatrix} F_{\ell e} & 0 & \cdots & 0 \\ 0 & F_{\ell e} & \ddots & \vdots \\ \vdots & \ddots & \ddots & 0 \\ 0 & \cdots & 0 & F_{\ell e} \end{bmatrix}}_{\ell \times m}, \; F_{\ell e} \in K^{\ell \times \ell},$$

$$F_{\beta e}=\underbrace{\begin{bmatrix} F_1 & 0 & \cdots & 0 \\ 0 & F_2 & \ddots & \vdots \\ \vdots & \ddots & \ddots & 0 \\ 0 & \cdots & 0 & F_m \end{bmatrix}}_{\ell \times m}, \; F_i \in K^{\ell \times \ell} \text{ for } i \; (1 \leq i \leq m),$$

$$F_{\ell e} = \underbrace{\begin{bmatrix} \alpha_1 & 0 & \ldots & 0 \\ 0 & \alpha_2 & 0 & \vdots \\ \vdots & \ddots & \ddots & 0 \\ 0 & \ldots & 0 & \alpha_\ell \end{bmatrix}}_{\ell}, \; x^\ell - 1 \equiv (x-\alpha_1)(x-\alpha_2)(x-\alpha_3)\cdots(x-\alpha_\ell) \; (mod \, p)$$

$$F_i = \underbrace{\begin{bmatrix} \beta_i & 0 & \ldots & 0 \\ 0 & \beta_i & 0 & \vdots \\ \vdots & \ddots & \ddots & 0 \\ 0 & \ldots & 0 & \beta_i \end{bmatrix}}_{\ell}, \; y^m - 1 \equiv (y-\beta_1)(y-\beta_2)(y-\beta_3)\cdots(y-\beta_m) \; (mod \, p)$$

$$x_e^0 = \left.\begin{bmatrix} 1 \\ 1 \\ \vdots \\ 1 \end{bmatrix}\right\} l \times m.$$

Fig. 5.4. The Eigen Standard System $\sigma_e = ((K^{\ell \times m}, F_{\alpha e}, F_{\beta e}), x_e^0, h_e)$, as given in Definition (5.8)

Theorem 5.9. Let us consider the 2-Commutative Linear Representation System σ_p which realizes any two-dimensional image with a $\ell \times m$ period, where σ_p is given by $\sigma_p = ((K^{l \times m}, F_{\alpha p}, F_{\beta p}), x_p^0, h_p)$ as we discussed in Proposition (5.3).

Then σ_p is isomorphic to the Eigen Standard System σ_e.

Proof. Take the following $(\ell \times m) \times (\ell \times m)$ matrix T_e:

$$T_e = [\, x_e^0 \; F_{\alpha e} x_e^0 \ldots \; F_{\alpha e}^{\ell-1} x_e^0 \; F_{\beta e} x_e^0 \; F_{\alpha e} F_{\beta e} x_e^0$$
$$\ldots \; F_{\alpha e}^{\ell-1} F_{\beta e} x_e^0 \ldots \; F_{\beta e}^{m-1} x_e^0 \ldots \; F_{\alpha e}^{\ell-1} F_{\beta e}^{m-1} x_e^0 \,].$$

Let h_e be $h_p := h_e T_e$. Then T_e is a 2-Commutative Linear Representation System morphism:

$$T_e : \sigma_p = ((K^{l\times m}, F_{\alpha p}, F_{\beta p}), x_p^0, h_p) \to \sigma_e = ((K^{\ell\times m}, F_{\alpha e}, F_{\beta e}), x_e^0, h_e).$$

Hence the behavior of σ_p is the same as the behavior of σ_e because of Corollary (3.12).

Remark: By Theorem (5.9), when we design any two-dimensional image, the Eigen Standard System σ_e can design that image more rapidly than it can be generated using σ_p.

5.10. Problem statement for finding a prime number
Consider the following problem in view of the need for rapid calculation in designing two-dimensional periodic geometrical patterns.

"Find a prime number p for given integers l_1 and l_2 such that $p - 1 = m_1 \times l_1 = m_2 \times l_2$ for some integers m_1 and m_2."

Remark: By virtue of Lemma (5.7), if such a prime number p is found for given integers l_1 and l_2, the polynomials $x^{l_1} - 1$ and $x^{l_2} - 1$ can be factorized simultaneously via modulo p.

5.11. Algorithm for finding a prime number p

1. Calculate the greatest common divisor g of l_1 and l_2.
2. Calculate the least common multiple l of l_1 and l_2.
3. Find the minimum prime number which satisfies $p = m_1 \times l_1 + 1 = m_2 \times l_2 + 1$ and $p \geq l_1 \times l_2$.

Remark: In this algorithm, the greatest common divisor can be obtained by using the well-known Euclidean algorithm, while the least common multiple is calculated by using the well-known relation $l \times g = l_1 \times l_2$. Upon finding the apparent minimum prime number, one only has to judge whether the given number is truly a prime or not.

5.12. Table of prime numbers
A prime number table which has been obtained by using the algorithm (5.11) is shown in Table 5.1. In this table, the positive integers l_1 and l_2 which are the period of two-dimensional images are restricted from 1 through 20.
For a given period of a two-dimensional image l_1 and l_2, the required minimum prime number is found at the cross point of the l_1 column and the l_2 row.

Example 5.13. Consider the following 2×2-periodic image.

Let K be $N/5N$, which is the quotient field modulo the prime number 5, and the set Y of output values be K^3, that is, the set of pixels, hues and rotations. The details of these sets are given in Figure 5.6. Given Figure 5.6, the readout map h_p is represents Figure 5.5 as follows:

Table 5.1. Prime number table for two-dimensional periodic images with period l_1 and l_2

-	20	19	18	17	16	15	14	13	12	11	l_1 / l_2
-	401	761	541	1021	401	421	281	521	241	661	20
-	-	419	2053	647	1217	571	1597	1483	229	419	19
9	109	-	379	307	433	271	379	937	397	199	18
8	73	73	-	307	1361	1021	239	443	409	1123	17
7	127	113	71	-	257	241	337	1249	193	353	16
6	73	73	43	37	-	241	211	1171	181	331	15
5	181	41	71	31	31	-	197	547	337	463	14
4	37	41	29	37	41	17	-	313	157	859	13
3	37	73	43	19	31	13	13	-	157	397	12
2	19	17	29	13	11	13	7	5	-	199	11
1	19	17	29	7	11	5	7	3	2	-	-
l_1 / l_2	9	8	7	6	5	4	3	2	1	-	-

l_1 \ l_2	10	9	8	7	6	5	4	3	2	1
20	201	181	241	281	181	101	101	61	41	41
19	191	2053	457	1597	229	191	229	229	191	191
18	181	163	433	127	109	181	73	73	37	19
17	1021	307	137	239	103	1021	137	103	103	103
16	241	433	193	113	97	241	97	97	97	17
15	151	181	241	211	151	151	61	61	31	31
14	211	127	113	113	127	71	113	43	29	29
13	131	937	313	547	79	131	53	79	53	53
12	181	109	97	337	73	61	61	37	37	13
11	331	199	89	463	67	331	89	67	23	23
10	101	181	241	71	61	61	41	31	31	11

$$h_p = \begin{bmatrix} 0\,1\,0\,1 \\ 0\,2\,0\,1 \\ 0\,0\,2\,1 \end{bmatrix}.$$

For this image, since $l = m = 2$, $F_2, F_{\alpha p}, F_{\beta p}$ and x^0 in Figure 5.3 are written as follows:

$$F_2 = \begin{bmatrix} 0\,1 \\ 1\,0 \end{bmatrix}, \qquad x^0 = \begin{bmatrix} 1 \\ 0 \\ 0 \\ 0 \end{bmatrix},$$

Fig. 5.5. The 2 × 2-periodic image for Example (5.13)

pixel	: 0:		1:			
hue	: 0:	normal	1:	dark	2:	light
rotation	: 0:	0°	1:	90°	2:	180°

Fig. 5.6. The coding list for Example (5.13)

$$F_{\alpha p} = I_2 \otimes F_2 = \begin{bmatrix} 0\,1\,0\,0 \\ 1\,0\,0\,0 \\ 0\,0\,0\,1 \\ 0\,0\,1\,0 \end{bmatrix}, \qquad F_{\beta p} = F_2 \otimes I_2 = \begin{bmatrix} 0\,0\,1\,0 \\ 0\,0\,0\,1 \\ 1\,0\,0\,0 \\ 0\,1\,0\,0 \end{bmatrix}.$$

For rapid design, we must express our image in the Eigen Standard System depicted in Figure 5.4. First, find the prime number p which can be factorized $x^l - 1$ and $x^m - 1$ simultaneously. Since $l = m = 2$, we find that $p = 5$ from Table 5.1. Since $x^2 - 1 \equiv (x-1)(x-4) \pmod 5$, we can set $\alpha_1 = 1, \alpha_2 = 4, \beta_1 = 1, \beta_2 = 4$. Then, from Figure 5.4, $F_{le}, F_1, F_2, x_e^0, F_{\alpha e}$ and F_{β_e} will be obtained as follows:

$$F_{le} = \begin{bmatrix} 1\,0 \\ 0\,4 \end{bmatrix}, \qquad F_1 = \begin{bmatrix} 1\,0 \\ 0\,1 \end{bmatrix}, \qquad F_2 = \begin{bmatrix} 4\,0 \\ 0\,4 \end{bmatrix}, \qquad x_e^0 = \begin{bmatrix} 1 \\ 1 \\ 1 \\ 1 \end{bmatrix},$$

$$F_{\alpha e} = \begin{bmatrix} 1\,0\,0\,0 \\ 0\,4\,0\,0 \\ 0\,0\,1\,0 \\ 0\,0\,0\,4 \end{bmatrix}, \qquad F_{\beta e} = \begin{bmatrix} 1\,0\,0\,0 \\ 0\,1\,0\,0 \\ 0\,0\,4\,0 \\ 0\,0\,0\,4 \end{bmatrix}.$$

From the proof of Theorem (5.9), the 2-Commutative Linear Representation System morphism T_e will be constructed as

$$T_e = \left[x_e^0, F_{\alpha e}x_e^0, F_{\beta e}x_e^0, F_{\alpha e}F_{\beta e}x_e^0\right] = \begin{bmatrix} 1\,1\,1\,1 \\ 1\,4\,1\,4 \\ 1\,1\,4\,4 \\ 1\,4\,4\,1 \end{bmatrix}.$$

Then the readout map of the Eigen Standard System h_e will be as follows:

$$h_e = \begin{bmatrix} 3\,2\,0\,0 \\ 2\,3\,4\,1 \\ 2\,4\,3\,1 \end{bmatrix}.$$

Example 5.14. Consider the following 3×2-periodic image.

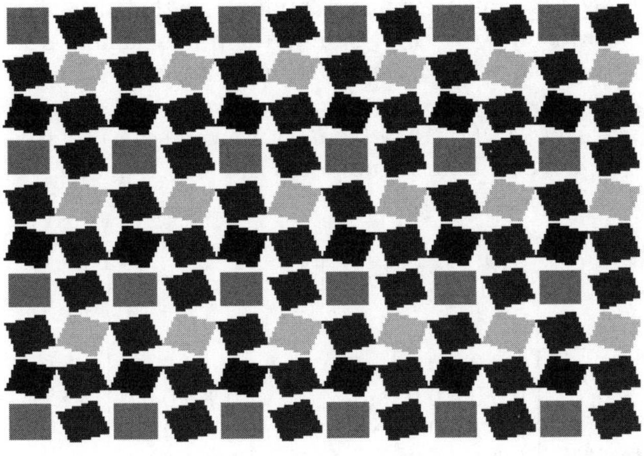

Fig. 5.7. The 2×2 periodic image for Example (5.14).

Fig. 5.8. The coding list for Example (5.14).

Let K be $N/7N$, which is the quotient field modulo the prime number 7, and let the set Y of output values be K. Given Figure 5.8, the readout map h_p

is represents Figure 5.7 as follows:

$$h_p = \begin{bmatrix} 1\,3\,5\,0\,2\,4 \end{bmatrix}.$$

For this image, since $l = 3, m = 2$, $F_2, F_3, F_{\alpha p}, F_{\beta p}$ and x^0 in Figure 5.3 are written as follows:

$$F_2 = \begin{bmatrix} 0\,1 \\ 1\,0 \end{bmatrix}, \qquad F_3 = \begin{bmatrix} 0\,0\,1 \\ 1\,0\,0 \\ 0\,1\,0 \end{bmatrix}, \qquad x^0 = \begin{bmatrix} 1 \\ 0 \\ 0 \\ 0 \\ 0 \\ 0 \end{bmatrix},$$

$$F_{\alpha p} = I_3 \otimes F_2 = \begin{bmatrix} 0\,1\,0\,0\,0\,0 \\ 1\,0\,0\,0\,0\,0 \\ 0\,0\,0\,1\,0\,0 \\ 0\,0\,1\,0\,0\,0 \\ 0\,0\,0\,0\,0\,1 \\ 0\,0\,0\,0\,1\,0 \end{bmatrix}, F_{\beta p} = F_3 \otimes I_2 = \begin{bmatrix} 0\,0\,0\,0\,1\,0 \\ 0\,0\,0\,0\,0\,1 \\ 1\,0\,0\,0\,0\,0 \\ 0\,1\,0\,0\,0\,0 \\ 0\,0\,1\,0\,0\,0 \\ 0\,0\,0\,1\,0\,0 \end{bmatrix}.$$

For rapid design, we must express our image in the Eigen Standard System depicted in Figure 5.4. First, we find the prime number p which can be factorized $x^l - 1$ and $x^m - 1$ simultaneously. Since $l = 3, m = 2$, we find that $p = 7$ from Table 5.1. Since $x^2 - 1 \equiv (x-1)(x-6) \pmod 7$ and $x^3 - 1 \equiv (x-1)(x-2)(x-4) \pmod 7$, we can set $\alpha_1 = 1, \alpha_2 = 6, \beta_1 = 1, \beta_2 = 2, \beta_3 = 4$. Then, from Figure 5.4, $F_{le}, F_1, F_2, F_3, x_e^0, F_{\alpha e}$ and F_{β_e} will be obtained as follows:

$$F_{le} = \begin{bmatrix} 1\,0 \\ 0\,6 \end{bmatrix}, \quad F_1 = \begin{bmatrix} 1\,0 \\ 0\,1 \end{bmatrix}, \quad F_2 = \begin{bmatrix} 2\,0 \\ 0\,2 \end{bmatrix}, \quad F_3 = \begin{bmatrix} 4\,0 \\ 0\,4 \end{bmatrix},$$

$$x_e^0 = \begin{bmatrix} 1 \\ 1 \\ 1 \\ 1 \\ 1 \\ 1 \end{bmatrix}, F_{\alpha e} = \begin{bmatrix} 1\,0\,0\,0\,0\,0 \\ 0\,6\,0\,0\,0\,0 \\ 0\,0\,1\,0\,0\,0 \\ 0\,0\,0\,6\,0\,0 \\ 0\,0\,0\,0\,1\,0 \\ 0\,0\,0\,0\,0\,6 \end{bmatrix}, \quad F_{\beta e} = \begin{bmatrix} 1\,0\,0\,0\,0\,0 \\ 0\,1\,0\,0\,0\,0 \\ 0\,0\,2\,0\,0\,0 \\ 0\,0\,0\,2\,0\,0 \\ 0\,0\,0\,0\,4\,0 \\ 0\,0\,0\,0\,0\,4 \end{bmatrix}.$$

From the proof of Theorem (5.9), the 2-Commutative Linear Representation System morphism T_e will be constructed as follows:

$$T_e = \begin{bmatrix} x_e^0, F_{\alpha e} x_e^0, F_{\beta e} x_e^0, F_{\alpha e} F_{\beta e} x_e^0, F_{\beta e}^2 x_e^0, F_{\alpha e} F_{\beta e}^2 x_e^0 \end{bmatrix}$$

$$= \begin{bmatrix} 1 & 1 & 1 & 1 & 1 & 1 \\ 1 & 6 & 1 & 6 & 1 & 6 \\ 1 & 1 & 2 & 2 & 4 & 4 \\ 1 & 6 & 2 & 5 & 4 & 3 \\ 1 & 1 & 4 & 4 & 2 & 2 \\ 1 & 6 & 4 & 3 & 2 & 5 \end{bmatrix}.$$

Then the readout map of the Eigen Standard System h_e will be obtained as follows:

$$h_e = \begin{bmatrix} 4 & 4 & 4 & 0 & 5 & 0 \end{bmatrix}.$$

5.2 Design Methods for Geometrical Patterns

To make a geometrical pattern like Figure 5.5 or 5.7 on a computer screen, the usual procedure or design technique is as follows:

1) A pattern for the screen is imaginated and decided on.
2) It is determined whether the desired pattern is periodic or non-periodic.
3) The design elements are produced manually at the place when they are to appear in the finished pattern. For a periodic pattern, one whole period is made. In the case of a non-periodic design, whole design elements are set manually at each place, and the geometrical pattern is now complete.
4) In the case of a periodic pattern, the period made in Step 3) is copied manually into every desired direction as need arises.

These steps are very laborious, tedious and monotonous; moreover, the error rate is high. If a given geometrical pattern is complex, the work of generating the pattern becomes even harder. Thus the usual method to a computer screen is not very efficient for getting a desired geometrical pattern.

Examples (5.13) and (5.14) show that our method is conducted mainly by computation using a computer program. Once the program based on the σ_e discussed in Theorem (5.9) is correctly made, any complex geometrical pattern can be generated on the screen automatically and with no error which is caused by executing the design process. The design procedure is as follows:

First, we make a corresponding table among picture's elements, thickness and numerals and make a two-dimensional image of numerals in the place of a designed pattern which comes to our minds. Then, this new design method may be executed by a program composed of the following procedures.

1) Input data for a pattern to be displayed on a screen.
2) Determine whether the desired pattern is non-periodic or periodic.
   a) For a non-periodic pattern, determine the 2-Commutative Linear Representation System σ (Lemma (5.1)).

b) For a periodic pattern, determine the 2-Commutative Linear Representation System σ_e (Theorem (5.9)).

3) According to a calculation of the behavior of the determined 2-Commutative Linear Representation System, show the design pattern on the screen by using the corresponding table of colors and numerals.

This program will make any designer see the designed pattern on the screen directly and confirm easily whether the pattern is good or not.

5.3 Historical Notes and Concluding Remarks

In Chapter 5, we have proposed a new method of designing geometrical patterns such as fancyworks without actual (visual) patterns. We showed that geometrical patterns can be concisely generated by a mathematical model that is called 2-Commutative Linear Representation System. Several examples demonstrated that this new design method is suitable for geometrical pattern generation. Because all the calculations are composed of linear operations, this method is applied by using computers, and designs are produced much more easily and efficiently than by manual methods.

The usual design of geometrical patterns may be done laboriously and monotonously. But, our method assists that computer-controlled machines can be practically devised by his final judging whether the generated geometrical pattern is good or not. Since any pattern in examples of Section 5.1 can be easily changed to any element, the computer-controlled machines will be able to make anyone a good designer of geometrical patterns easily.

6 Realization Theory of Three-Dimensional Images

Because of recent developments in computers, treating three-dimensional images has become easier, useful and important. The purpose of this chapter is to establish the foundations of a realization theory for three-dimensional images. To this end, we will replace the problem of describing three-dimensional images by the realization problem. A new mathematical model for three-dimensional images will be presented to accomplish this task.

Roughly speaking, the realization problem for three-dimensional images is stated as follows: <Find at least one mathematical model which realizes, that is, faithfully describes any given three-dimensional image.> The realization problem for two-dimensional images have already been established in principle in Chapter 3. It is stated as: <There exist more than two mathematical models which realize, that is, faithfully describe any given two-dimensional image, and those models are unique modulo isomorphism.>

In this chapter, the category of 3-Commutative Linear Representation Systems of the model for the three-dimensional image $a \in F(N^3, Y)$ is considered. We will establish the realization theory for three-dimensional image a in the same way as in the two-dimensional image case. We will find that the realization problem for three-dimensional images is an extension of that for two-dimensional ones. Finite-dimensional 3-Commutative Linear Representation Systems are important models from the point of view of actual implementation as computer programs or non-linear circuits. This chapter will also clarify how finite-dimensional 3-Commutative Linear Representation Systems are characterized.

In Section 6.1, the realization theory of three-dimensional images is stated. The proof of the theorem is provided in Appendix 6-A.

In Section 6.2, some concrete examples of 3-Commutative Linear Representation Systems will be presented.

In Section 6.3, we will derive the conditions when the three-dimensional infinite image admits to the finite-dimensional 3-Commutative Linear Representation System.

In Section 6.4, we will derive the condition when finite-sized three-dimensional image can be partially realized uniquely. The algorithm to get the 3-Commutative Linear Representation System from the given finite-sized image is also described.

There are many methods for treating three-dimensional images such as shape analysis or various techniques of computer graphics to indicate something on a screen [Nevatia, 1982], [Serra, 1982]. There is also a method of using oct-trees to encode 3-D objects [Jacking and Tanimoto, 1980].

The method presented in this chapter, however, is quite different from these. Our method depends on how characteristic rules are extracted preserving the connections of of each pixel in all three directions for three-dimensional image. Our realization theory can be regarded as a fundamental coding theory for three-dimensional images.

6.1 3-Commutative Linear Representation Systems

We define the model which realizes, that is, faithfully describes any three-dimensional image and state the realization theory.

Definition 6.1.

(1) A system given by the following equations is written as a collection $\sigma = ((X, F_\alpha, F_\beta, F_\gamma), x^0, h)$, and it is called a 3-Commutative Linear Representation System.

$$\begin{cases} x(i+1,j,k) = F_\alpha x(i,j,k) \\ x(i,j+1,k) = F_\beta x(i,j,k) \\ x(i,j,k+1) = F_\gamma x(i,j,k) \\ x(0,0,0) \quad = x^0 \\ \gamma(i,j,k) \quad = hx(i,j,k) \ , \end{cases}$$

for any $i, j, k \in N$, $x(i,j,k) \in X$ and $\gamma(i,j,k) \in Y$, where X is a linear space over the field K. F_α, F_β and F_γ are linear operators on X which satisfy $F_\alpha F_\beta = F_\beta F_\alpha, F_\beta F_\gamma = F_\gamma F_\beta$ and $F_\gamma F_\alpha = F_\alpha F_\gamma$. $x^0 \in X$ is the initial state and $h : X \to Y$ is the linear operator.

(2) Three-dimensional image $a_\sigma : N^3 \to Y; (i,j,k) \mapsto hF_\alpha^i F_\beta^j F_\gamma^k x^0$ is called the behavior of σ.

(3) For any $a \in F(N^3, Y)$, the 3-Commutative Linear Representation System σ which satisfies $a_\sigma = a$ is called a realization of a.

(4) A 3-Commutative Linear Representation System σ is called quasi-reachable if the linear hull of the reachable set $\{F_\alpha^i F_\beta^j F_\gamma^k x^0; i,j,k \in N\}$ equals X.

(5) A 3-Commutative Linear Representation System σ is called distinguishable if $hF_\alpha^i F_\beta^j F_\gamma^k x_1 = hF_\alpha^i F_\beta^j F_\gamma^k x_2$ for any $i,j,k \in N$ implies $x_1 = x_2$.

(6) A 3-Commutative Linear Representation System σ is called canonical if σ is quasi-reachable and distinguishable.

Remark 1: The $x(i,j,k)$ in the system equation of 3-Commutative Linear Representation System σ is the state which produces the image value at the position (i,j,k), while linear operator $h : X \to Y$ generates the image value $a_\sigma(i,j,k) = hx(i,j,k)$ at the position (i,j,k).

Remark 2: That σ is a faithful model for the image a implies that the 3-Commuta-tive Linear Representation System σ realizes a. If we find a 3-Commutative Linear Representation System σ which realizes the image a, we will be able to encode the image. Conversely, if we can calculate the behavior a_σ of the 3-Commutative Linear Representation System σ, decoding a_σ will be accomplished.

Remark 3: Notice that a canonical 3-Commutative Linear Representation System $\sigma = ((X, F_\alpha, F_\beta, F_\gamma), x^0, h)$ is a system with the most reduced state space X among systems that have the behavior a_σ. See Proposition (6-A.9), (6-A.10), (6-A.15), (6-A.18) and Definition (6-A.16) in Appendix 6-A .

Remark 4: The realization theorem for two-dimensional images has been already established by 2-Commutative Linear Representaion System $\sigma = ((X, F_\alpha, F_\beta), x^0, h)$ as the model which realizes a two-dimensional image, where X is a linear space over a field K. F_α and F_β are linear operators on X which satisfy $F_\alpha F_\beta = F_\beta F_\alpha$, $x^0 \in X$. h is a linear operator: $X \to Y$. The system equation of $\sigma = ((X, F_\alpha, F_\beta), x^0, h)$ is represented as:

$$\begin{cases} x(i+1,j) = F_\alpha x(i,j) \\ x(i,j+1) = F_\beta x(i,j) \\ x(0,0) \quad\; = x^0 \\ a_\sigma(i,j) \quad = hx(i,j) \;, \end{cases}$$

for any $i,j \in N$, where $x(i,j) \in X$ and $a_\sigma(i,j) \in Y$. In this way, we can observe that the model for three-dimensional images, the 3-Commutative Linear Representation System, is a direct extension of the model for two-dimensional images, the 2-Commutative Linear Representation System.

Definition 6.2. Let $\sigma_1 = ((X_1, F_{\alpha_1}, F_{\beta_1}, F_{\gamma_1}), x_1^0, h_1)$ and $\sigma_2 = ((X_2, F_{\alpha_2}, F_{\beta_2}, F_{\gamma_2}), x_2^0, h_2)$ be 3-Commutative Linear Representation Systems. Then the linear operator $T : X_1 \to X_2$ is called the linear representation system morphism $T : \sigma_1 \to \sigma_2$ if T satisfies $TF_{\alpha_1} = F_{\alpha_2}T$, $TF_{\beta_1} = F_{\beta_2}T$, $TF_{\gamma_1} = F_{\gamma_2}T$, $Tx_1^0 = x_2^0$ and $h_1 = h_2 T$. If $T : X_1 \to X_2$ is bijective, then $T : \sigma_1 \to \sigma_2$ is called an isomorphism.

Theorem 6.3. Realization Theorem of three-dimensional images

(1) Existence: For any image $a \in F(N^3, Y)$, there exist at least two canonical 3-Commutative Linear Representation Systems which realize a.
(2) Uniqueness: Let σ_1 and σ_2 be any two canonical 3-Commutative Linear Representation Systems which realize $a \in F(N^3, Y)$. Then there exists an isomorphism $T : \sigma_1 \to \sigma_2$.

Proof. See Propositions (6-A.10), (6-A.15) (6-A.18), Theorem (6-A.19), Corollary (6-A.20) and Remark of Lemma(6-A.24) in Appendix 6-A.

6.2 Definite Examples of Images Generated by Finite-Dimensional 3-Commutative Linear Representation Systems

In order to show some examples of three-dimensional images, we briefly introduce the finite-dimensional 3-Commutative Linear Representation System which can be dealt with by computer or non-linear circuits. The 3-Commutative Linear Representation System $\sigma = ((X, F_\alpha, F_\beta, F_\gamma), x^0, h)$ is called a finite (or n) -dimensional 3-Commutative Linear Representation System if the state space X is a finite (or n)-dimensional linear space.

Lemma 6.4. For any image $a \in F(N^3, Y)$, the following three conditions are equivalent to each other.

(1) a is the behavior of a finite-dimensional canonical 3-Commutative Linear Representation System.
(2) The quotient space $K[z_\alpha, z_\beta, z_\gamma]/_{=a}$ is finite-dimensional.
(3) The linear space generated by $\{S_\alpha^i S_\beta^j S_\gamma^k a \; : \; i, j, k \in N\}$ is finite-dimensional.

Note that $K[z_\alpha, z_\beta, z_\gamma]/_{=a}$ is a quotient space given by the following equivalence relation $a_1 = a_2 \Longleftrightarrow a_1(i,j,k) = a_2(i,j,k)$ for any $i, j, k \in N$.

Moreover, $S_\alpha, S_\beta, S_\gamma \in L(F(N^3, Y))$ are given by $S_\alpha a : N^3 \to Y; (i,j,k) \mapsto a(i+1,j,k)$, $S_\beta a : N^3 \to Y; (i,j,k) \mapsto a(i,j+1,k)$ and $S_\gamma a : N^3 \to Y); (i,j,k) \mapsto a(i,j,k+1)$.

Proof. This lemma is the direct consequence of RealizationTheorem (6.3).

Example 6.5. A three-dimensional canonical 3-Commutative Linear Representation System.
Let the state space be K^3 and let $F_\alpha, F_\beta, F_\gamma \in K^{3\times 3}, \mathbf{e}_1 \in K^3$ and $h \in K^{1\times 3}$ be:

$$F_\alpha = \begin{bmatrix} 0 & \alpha_1 & 0 \\ 1 & \alpha_2 & 0 \\ 0 & 1 & \alpha_3 \end{bmatrix}, \; F_\beta = \begin{bmatrix} \beta_1 & \beta_3 & 0 \\ \beta_2 & \beta_4 & 0 \\ 0 & \beta_2 & \beta_5 \end{bmatrix}, \; F_\gamma = \begin{bmatrix} \gamma_1 & \gamma_4 & 0 \\ \gamma_2 & \gamma_5 & 0 \\ \gamma_3 & \gamma_6 & \gamma_7 \end{bmatrix},$$

$$\mathbf{e}_1 = \begin{bmatrix} 1 \\ 0 \\ 0 \end{bmatrix}, \; h = \begin{bmatrix} h_1 & h_2 & h_3 \end{bmatrix},$$

where $\beta_4 = \alpha_1\beta_2$, $\beta_5 = \beta_1 + \alpha_2\beta_2$, $\beta_6 = \beta_2 + \alpha_3\beta_3$, $\beta_7 = \beta_1 + \alpha_3\beta_2 - \alpha_1\beta_3 - \alpha_2\alpha_3\beta_3 + \alpha_3^2\beta_3$, $\gamma_4 = \alpha_1\gamma_2$, $\gamma_5 = \gamma_1 + \alpha_2\gamma_2$, $\gamma_6 = \gamma_2 + \alpha_3\gamma_3$ and $\gamma_7 = \gamma_1 + \alpha_3\gamma_2 - \alpha_1\gamma_3 - \alpha_2\alpha_3\gamma_3 + \alpha_3^2\gamma_3$.

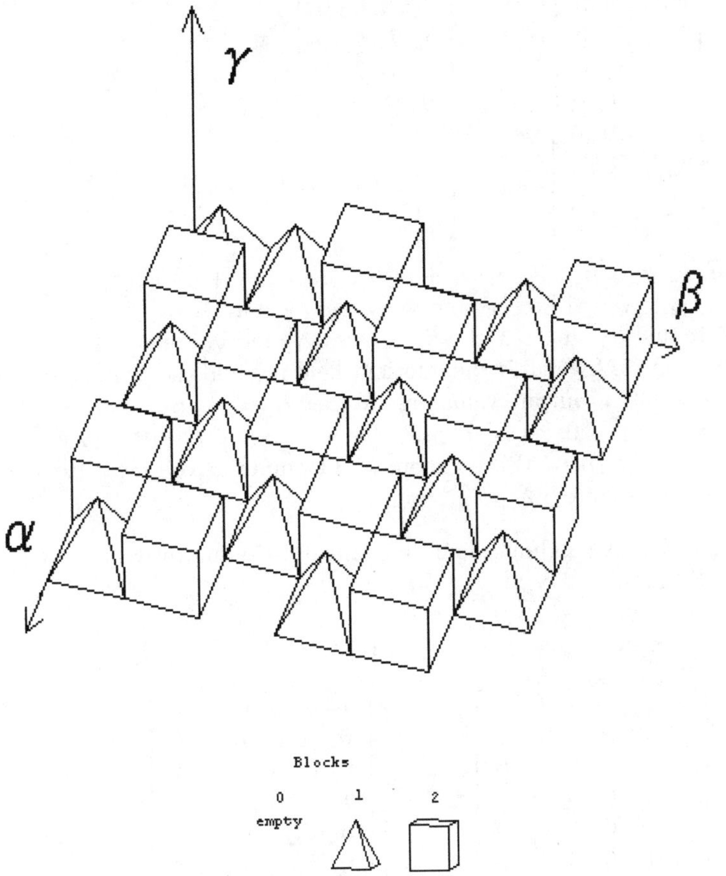

Fig. 6.1. A three-dimensional image generated by a three-dimensional canonical 3-Commutative Representation System σ

Note that $\alpha_1, \alpha_2, \alpha_3, \beta_1, \beta_2, \beta_3, \gamma_1, \gamma_2$ and γ_3 are arbitrary parameters. Let K be $N/3N$ which is the quotient field modulo the prime number 3, and let the set Y of output values be K. Set $\alpha_1 = 0, \alpha_2 = 1, \alpha_3 = 1, \beta_1 = 0, \beta_2 = 1, \beta_3 = 1, \gamma_1 = 0, \gamma_2 = 0, \gamma_3 = 0$ and $h = [1, 2, 2]$. Then $\sigma = ((K^3, F_\alpha, F_\beta, F_\gamma), \mathbf{e}_1, h)$ is a canonical 3-Commutative Linear Representation System. See Figure 6.1.

Example 6.6. A four-dimensional canonical 3-Commutative Linear Representation System.

Let the state space be K^4 and let $F_\alpha, F_\beta, F_\gamma \in K^{4\times4}, \mathbf{e}_1 \in K^4$ and $h \in K^{1\times4}$ be:

$$F_\alpha = \begin{bmatrix} 0 & \alpha_1 & 0 & 0 \\ 1 & \alpha_2 & 0 & 0 \\ 0 & 1 & \alpha_3 & 0 \\ 0 & 0 & 0 & \alpha_3 \end{bmatrix}, \; F_\beta = \begin{bmatrix} \beta_1 & \beta_3 & 0 & 0 \\ \beta_2 & \beta_4 & 0 & 0 \\ 0 & \beta_2 & \beta_5 & 0 \\ 0 & 0 & 0 & \beta_5 \end{bmatrix},$$

$$F_\gamma = \begin{bmatrix} \gamma_1 & \gamma_3 & 0 & 0 \\ \gamma_2 & \gamma_4 & 0 & 0 \\ 0 & \gamma_2 & \gamma_5 & \gamma_7 \\ 1 & \alpha_3 & \gamma_6 & \gamma_8 \end{bmatrix}, \; \mathbf{e}_1 = \begin{bmatrix} 1 \\ 0 \\ 0 \\ 0 \end{bmatrix}, \; h = \begin{bmatrix} h_1 & h_2 & h_3 & h_4 \end{bmatrix},$$

where $\beta_3 = \alpha_1\beta_2$, $\beta_4 = \beta_1 + \alpha_2\beta_2$, $\beta_5 = \beta_1 + \alpha_3\beta_2$, $\gamma_3 = \alpha_1\gamma_2$, $\gamma_4 = \gamma_1 + \alpha_2\gamma_2$, $\gamma_5 = \gamma_1 + \alpha_3\gamma_2$ and $\gamma_6 = \alpha_3\alpha_3 - \alpha_2\alpha_3 - \alpha_1$.

Note that α_1, α_2, α_3, β_1, β_2, γ_1, γ_2, γ_7 and γ_8 are arbitrary parameters. Let K be $N/5N$ which is the quotient field modulo the prime number 5, and let the set Y of output values be K. Set $\alpha_1 = 4$, $\alpha_2 = 3$, $\alpha_3 = 2$, $\beta_1 = 1$, $\beta_2 = 0$, $\gamma_1 = 0$, $\gamma_2 = 0$, $\gamma_7 = 0$, $\gamma_8 = 0$ and $h = [1, 2, 3, 4]$. Then $\sigma = ((K^4, F_\alpha, F_\beta, F_\gamma), \mathbf{e}_1, h)$ is a canonical 3-Commutative Linear Representation System. See Figure 6.2.

Example 6.7. A six-dimensional canonical 3-Commutative Linear Representation System.

Let the state space be K^6 and let $F_\alpha, F_\beta, F_\gamma \in K^{6\times 6}, \mathbf{e}_1 \in K^6$ and $h \in K^{1\times 6}$ be:

$$F_\alpha = \begin{bmatrix} \alpha_1 & 0 & 0 & 0 & 0 & 0 \\ 1 & \alpha_2 & 0 & 0 & 0 & 0 \\ 0 & 1 & \alpha_3 & 0 & 0 & 0 \\ 0 & 0 & 0 & \alpha_2 & 0 & 0 \\ 0 & 0 & 0 & 1 & \alpha_3 & 0 \\ 0 & 0 & 0 & 0 & 0 & \alpha_3 \end{bmatrix}, \; F_\beta = \begin{bmatrix} \beta_1 & 0 & 0 & 0 & 0 & 0 \\ 0 & \beta_1 & 0 & \beta_2 & 0 & 0 \\ 0 & 0 & \beta_1 & 0 & \beta_2 & 0 \\ 1 & \beta_4 & 0 & \beta_3 & 0 & 0 \\ 0 & 1 & \beta_5 & 0 & \beta_3 & 0 \\ 0 & 0 & 0 & 1 & \beta_6 & \beta_7 \end{bmatrix},$$

$$F_\gamma = \begin{bmatrix} 0 & 0 & 0 & 0 & 0 & 0 \\ 0 & 0 & 0 & 0 & 0 & 0 \\ 0 & 0 & 0 & 0 & 0 & 0 \\ 0 & 0 & 0 & 0 & 0 & 0 \\ 0 & 0 & 0 & 0 & 0 & 0 \\ 2 & \gamma_2 & \gamma_3 & 1 & \beta_6 & \gamma_1 \end{bmatrix}, \; \mathbf{e}_1 = \begin{bmatrix} 1 \\ 0 \\ 0 \\ 0 \\ 0 \\ 0 \end{bmatrix},$$

$$h = \begin{bmatrix} h_1 & h_2 & h_3 & h_4 & h_5 & h_6 \end{bmatrix},$$

where $\beta_4 = \alpha_2 - \alpha_1$, $\beta_5 = \alpha_3 - \alpha_1$, $\beta_6 = \alpha_3 - \alpha_2$, $\gamma_2 = 2(\alpha_3 - \alpha_1)$, $\gamma_3 = 2(\alpha_3 - \alpha_1)(\alpha_3 - \alpha_2)$, $2\beta_7 = 2\beta_1 + 1$ and $\gamma_1 = -2\beta_2(\alpha_3 - \alpha_1) - \beta_3 + \beta_7$. α_1, α_2, α_3, β_1, β_2 and β_3 are arbitrary parameters.

Let K be $N/3N$ which is the quotient field modulo the prime number 3, and let the set Y of output values be K. Set $\alpha_1 = 1$, $\alpha_2 = 1$, $\alpha_3 = 0$, $\beta_1 = 2$, $\beta_2 = 0$, $\beta_3 = 0$ and $h = [2, 0, 0, 1, 0, 1]$.

Then $\sigma = ((K^6, F_\alpha, F_\beta, F_\gamma), \mathbf{e}_1, h)$ is a canonical 3-Commutative Linear Representation System. See Figure 6.3.

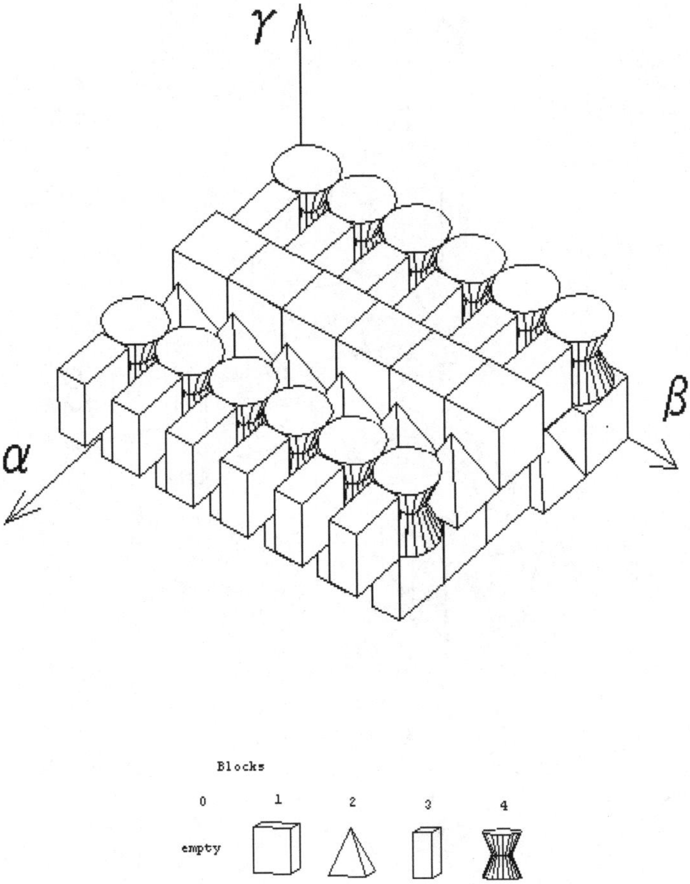

Fig. 6.2. A three-dimensional image generated by a four-dimensional canonical 3-Commutative Linear Representation System σ

6.3 Finite-Dimensional 3-Commutative Linear Representation Systems

This section investigates the structures of finite-dimensional 3-Commutative Linear Representation System based on the Realization Theorem (6.3). The matters discussed this section are significant for embodying finite-dimensional 3-Commutative Linear Representation Systems as computer programs or as non-linear circuits. The main results of this section are demonstrated by the following four steps. First, we present the conditions when a finite-dimensional 3-Commutative Linear Representation System is canonical. Sec-

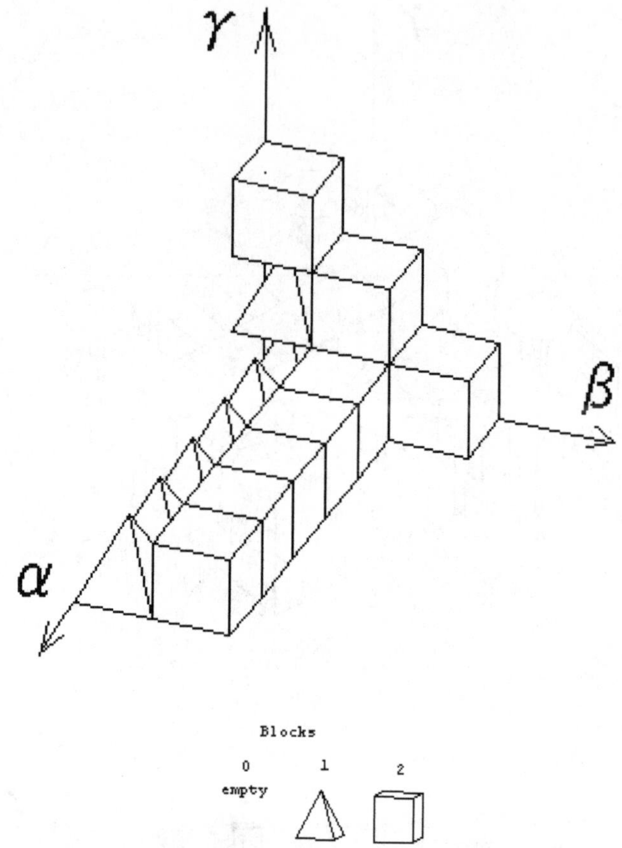

Fig. 6.3. A three-dimensional image generated by a six-dimensional canonical 3-Commutative Linear Representation System σ

ond, we derive the representation theorem for the finite-dimensional canonical 3-Commutative Linear Representation System, that is, there exists a unique Quasi-reachable Standard System as a representative in equivalence classes. Third, two necessary and sufficient conditions for an image to become the behavior of the finite-dimensional 3-Commutative Linear Representation System are provided. One is a rank condition for the Hankel matrix and the other is a condition for the rational function. Finally, the realization procedure is presented for obtaining the finite-dimensional canonical 3-Commutative Linear Representation System, in particular, the Quasi-reachable Standard System which realizes the given image. Proofs of these matters are provided in Appendix 6-B.

Corollary 6.7. *Let T be a 3-Commutative Linear Representation System morphism $T : \sigma_1 \to \sigma_2$. Then $a_{\sigma_1} = a_{\sigma_2}$ holds.*

Proof. A direct calculation based on the definition of the behavior and Linear Representation System morphism provides this corollary.

Definition 6.8. The linear input/output map $A : (K[z_\alpha, z_\beta, z_\gamma], z_\alpha, z_\beta, z_\gamma) \to (F(N^3, Y), S_\alpha, S_\beta, S_\gamma)$ that corresponds to any image $a \in F(N^3, Y)$ satisfies $A(z_\alpha^i z_\beta^j z_\gamma^k)(\tilde{i}, \tilde{j}, \tilde{k}) = (S_\alpha^i S_\beta^j S_\gamma^k a)(\tilde{i}, \tilde{j}, \tilde{k}) = a(i + \tilde{i}, j + \tilde{j}, k + \tilde{k})$ for $(i, j, k), (\tilde{i}, \tilde{j}, \tilde{k}) \in N^3$. This A is represented by the following infinite matrix H_a and it is called a Hankel matrix of a:

$$H_a = \begin{array}{c} \\ \\ \\ (\tilde{i}, \tilde{j}, \tilde{k}) \end{array} \begin{pmatrix} & & (i, j, k) & \\ & & \vdots & \\ & & \vdots & \\ \cdots & \cdots & a(\tilde{i} + i, \tilde{j} + j, \tilde{k} + k) & \end{pmatrix}$$

See Remark of Definition (6-A.4) on the linear input/output map A and Remark 2 of Proposition (6-A.9).

It is a fact on finite-dimensional linear spaces that n-dimensional linear spaces over the field K are isomorphic to K^n, and $L(K^n)$ is isomorphic to $K^{n \times n}$ [Halmos, 1958]. Therefore, without loss of generality, an n-dimensional 3-Commutative Linear Representation System is regarded as $\sigma = ((K^n, F_\alpha, F_\beta, F_\gamma), x^0, h)$, where $F_\alpha, F_\beta, F_\gamma \in K^{n \times n}, x^0 \in K^n$ and $h \in K^{t \times n}$.

Theorem 6.9. A 3-Commutative Linear Representation System such as $\sigma = ((K^n, F_\alpha, F_\beta, F_\gamma), x^0, h)$ is canonical if and only if the following conditions hold:

1) rank $[x^0, F_\alpha x^0, \cdots, F_\alpha^{n-1} x^0, F_\beta x^0, F_\beta F_\alpha x^0, \cdots, F_\beta F_\alpha^{n-2} x^0, \cdots,$ $F_\beta^{n-1} x^0, F_\gamma x^0, F_\gamma F_\alpha x^0, \cdots, F_\gamma F_\alpha^{n-1} x^0, \cdots, F_\gamma^{n-1} x^0] = n$
2) rank $[h^T, (hF_\alpha)^T, \cdots, (hF_\alpha^{n-1})^T, (hF_\beta)^T, \cdots, (hF_\beta^{n-1})^T,$ $(hF_\gamma)^T, \cdots, (hF_\gamma^{n-1})^T] = n$.

Proof. See Propositions (6-B.6) and (6-B.12) in Appendix 6-B.

Definition 6.10. A canonical 3-Commutative Linear Representation System canonical 3-Commutative Linear Representation System.
$\sigma_s = ((K^n, F_{\alpha s}, F_{\beta s}, F_{\gamma s}), \mathbf{e}_1, h_s)$ is called a Quasi-reachable Standard System with the vector index $\nu = (\nu_1, \nu_2, \nu_3, \cdots, \nu_p)$, where $\nu_i = (\nu_{i1}, \nu_{i2}, \cdots, \nu_{iq_i})$, if σ_s satisfies the following conditions:

1) $0 \leq \nu_{iq_i} \leq \cdots \leq \nu_{i2} \leq \nu_{i1}$ holds for any $i (1 \leq i \leq p)$.
   $n = \sum_{i=1}^{p} \nu_i$ and $\nu_i = \sum_{j=1}^{q_i} \nu_{ij}$ hold.

2) For any k $(1 \leq k \leq p)$, j $(1 \leq j \leq q_k)$ and i $(1 \leq i \leq \nu_{kj})$,
   $F_{\gamma s}^{k-1} F_{\beta s}^{j-1} F_{\alpha s}^{i-1} \mathbf{e}_1 = \mathbf{e}_{\nu_1+\nu_2+\cdots+\nu_{k-1}+\nu_{k1}+\nu_{k2}+\cdots+\nu_{kj-1}+i}$,
   $F_{\gamma s}^{k-1} F_{\beta s}^{j-1} F_{\alpha s}^{\nu_{kj}} \mathbf{e}_1 = \sum_{m=1}^{\nu_1+\nu_2+\cdots+\nu_{k-1}+\nu_{k1}+\nu_{k2}+\cdots+\nu_{kj}} c_m^{kj} \mathbf{e}_m$,
   $F_{\gamma s}^{k-1} F_{\beta s}^{q_k} \mathbf{e}_1 = \sum_{m=1}^{\nu_1+\nu_2+\cdots+\nu_k} c_m^{kq_k+1} \mathbf{e}_m$,
   $F_{\gamma s}^{p} \mathbf{e}_1 = \sum_{m=1}^{n} c_m^{p+1} \mathbf{e}_m$, where ν_0=0 and $\nu_{i0} = 0$ for any i.

3) $F_{\beta s} \mathbf{c}^{ij} = F_{\alpha s}^{\nu_{ij}-\nu_{ij+1}} \mathbf{c}^{ij+1}$ for i, j $(1 \leq i \leq p, 1 \leq j \leq q_i.)$.
   $F_{\gamma s} \mathbf{c}^{ij} = F_{\alpha s}^{\nu_{ij}-\nu_{ij+1}} \mathbf{c}^{ij+1}$ for i, j $(1 \leq i \leq p, 1 \leq j \leq q_i)$.
   $F_{\gamma s} \mathbf{c}^{ij} = F_{\alpha s}^{\nu_{ij}} F_{\beta s}^{j-q_{i+1}-1} \mathbf{c}^{i+1 q_{i+1}+1}$ for i, j $(1 \leq i \leq p-1,\ q_{i+1}+1 \leq j \leq q_i + 1)$.
   $F_{\gamma s} \mathbf{c}^{pj} = F_{\alpha s}^{\nu_{pj}} F_{\beta s}^{j-1} \mathbf{c}^{p+1}$ for j $(1 \leq j \leq q_p + 1)$, where
   $\mathbf{c}^{ij} = [c_1^{ij}, c_2^{ij}, \cdots, c_{\nu_{11}}^{ij}, c_{\nu_{11}+1}^{ij}, \cdots, c_{\nu_1+\cdots+\nu_{i-1}+\nu_{i1}+\nu_{i2}+\cdots+\nu_{ij}}^{ij}, 0, \cdots, 0]^T \in K^n$,
   $\mathbf{c}^{p+1} = [c_1^{p+1}, c_2^{p+1}, \cdots, c_{\nu_{11}}^{p+1}, c_{\nu_{11}+1}^{p+1}, \cdots, c_n^{p+1}]^T \in K^n$
   and $\mathbf{e}_i = [0, \cdots, 0 \overset{i}{1}, 0, \cdots, 0]^T \in K^n$ (T denotes transposition).

The $F_{\alpha s}$,$F_{\beta s}$ and $F_{\gamma s}$ of the Quasi-reachable Standard System are represented by Figure 6.4, Figure 6.5 and Figure 6.6.

Theorem 6.11. Representation Theorem for equivalence classes.
For any finite-dimensional canonical 3-Commutative Linear Representation System, there exists a uniquely determined isomorphic Quasi-reachable Standard System with the vector index ν.

Proof. See (6-B.13) in Appendix 6-B.

Theorem 6.12. Theorem for existence criterion.
For a three-dimensional image $a \in F(N^3, Y)$, the following conditions are equivalent:

1) The image $a \in F(N^3, Y)$ is the behavior of a n-dimensional canonical 3-Commutative Linear Representation System
- 3-Commutative Linear Representation System.
2) There exist n linearly independent vectors and no more than n linearly independent vectors in a set $\{S_\alpha^i S_\beta^j S_\gamma^k a;\ i + j + k \leq n \text{ for } i, j, k \in N\}$.
3) The rank of the Hankel matrix H_a of the image a is n.

Proof. See (6-B.14) in Appendix 6-B.

Remark: Fliess [1974] has introduced the Hankel matrix of the non-commutative formal power series and shown that the recognizability of formal power series is equivalent to the finite rank of its Hankel matrix.

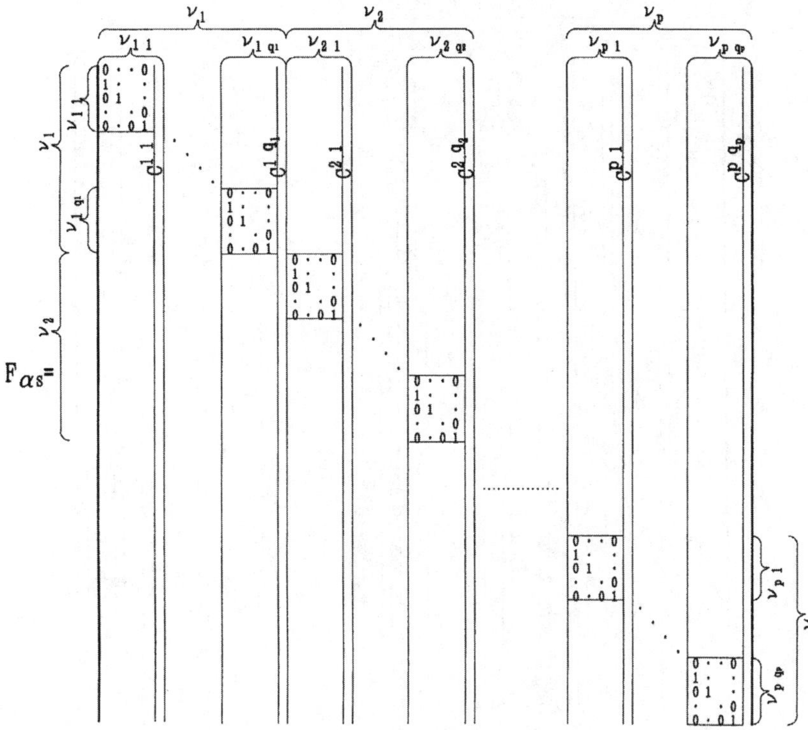

Fig. 6.4. $F_{\alpha s}$ in a Quasi-reachable Standard System

Theorem 6.13. The necessary and sufficient condition to ascertain that an image a is the behavior of the finite-dimensional 3-Commutative Linear Representation System is as follows:

The formal power series of a, that is, $\bar{a} = [\bar{a}_1, \bar{a}_2, \cdots, \bar{a}_t]^T \in (K[[z_\alpha^{-1}, z_\beta^{-1}, z_\gamma^{-1}]])^t$, is represented by the following rational function:

$$\bar{a}_s = z_\alpha z_\beta z_\gamma (\sum_{i=0}^{l-1} \sum_{j=0}^{m-1} \sum_{k=0}^{n-1} \lambda(i,j,k) z_\alpha^i z_\beta^j z_\gamma^k) / (q_\alpha(z_\alpha) q_\beta(z_\beta) q_\gamma(z_\gamma))$$

for $1 \leq s \leq t$, where $q_\alpha(z_\alpha), q_\beta(z_\beta)$ and $q_\gamma(z_\gamma)$ are the following monic polynomials:

$q_\alpha(z_\alpha) = z_\alpha^l + \alpha_{l-1} z_\alpha^{l-1} + \cdots + \alpha_1 z_\alpha + \alpha_0,$
$q_\beta(z_\beta) = z_\beta^m + \beta_{m-1} z_\beta^{m-1} + \cdots + \beta_1 z_\beta + \beta_0,$
$q_\gamma(z_\gamma) = z_\gamma^n + \gamma_{n-1} z_\gamma^{n-1} + \cdots + \gamma_1 z_\gamma + \gamma_0.$

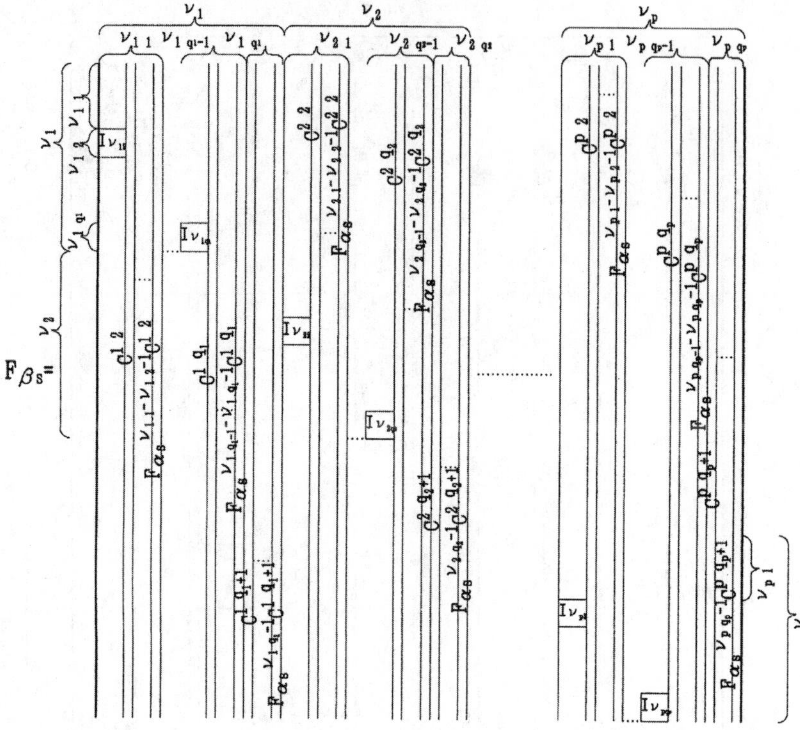

Fig. 6.5. $F_{\beta s}$ in a Quasi-reachable Standard System

Proof. See (6-B.15) in Appendix 6-B.

Remark 1: The equivalent condition that the commutative formal power series in one-variable becomes rational has already been presented in Gantmacher [1959] and R. E. Kalman, et al. [1969]. The two-variable case was presented in Chapter 3.

Remark 2: Fliess [1970, 1974] has pointed out that the recognizability of the three-variable commutative formal power series can be characterized by the same form of the rational function as presented in Theorem (6.13).

Theorem 6.14. Theorem for a realization procedure.

Let an image $a \in F(N^3, Y)$ satisfy the condition of Theorem (6.12). Then the Quasi-reachable Standard System realizing the image a can be obtained by the following procedure:

Step 1

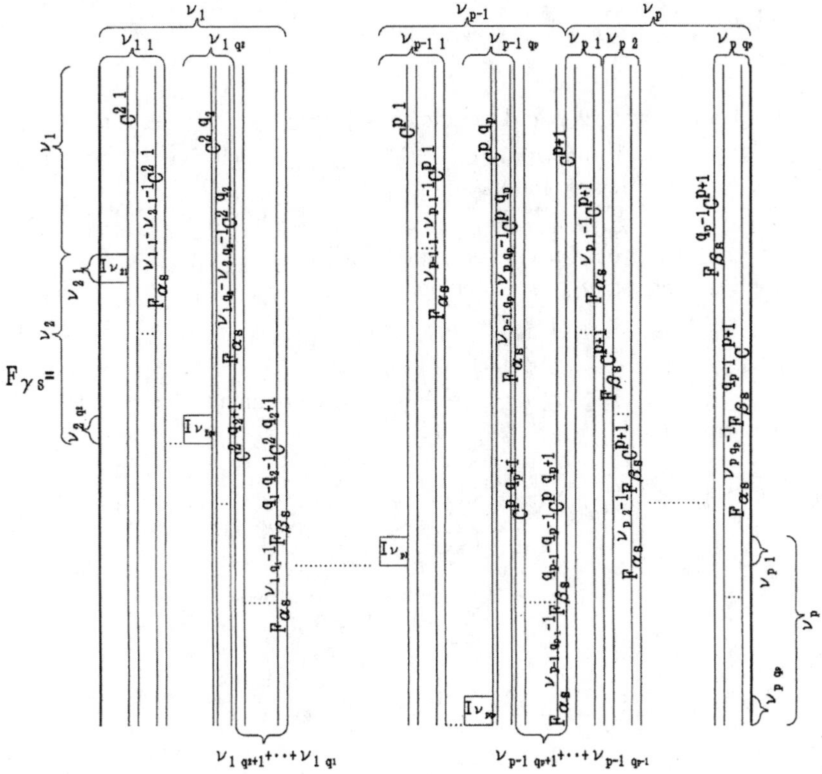

Fig. 6.6. $F_{\gamma s}$ in a Quasi-reachable Standard System

1-1) Find an integer ν_{11} and coefficients $\{c_i^{11} \in K;\ 1 \le i \le \nu_{11}\}$ such that $\{S_\alpha^{i-1}a;\ 1 \le i \le \nu_{11}\}$ are linearly independent and $\{S_\alpha^i a;\ 1 \le i \le \nu_{11}\}$ are linearly dependent. Namely, $S_\alpha^{\nu_{11}}a = \sum_{i=1}^{\nu_{11}} c_i^{11} S_\alpha^{i-1}a$ holds.

1-2) Find an integer ν_{12} and coefficients $\{c_i^{12} \in K;\ 1 \le i \le \nu_{12}\}$ such that $\{S_\beta^{j-1}S_\alpha^{i-1}a;\ 1 \le j \le 2,\ 1 \le i \le \nu_{1j}\}$ are linearly independent and $\{S_\alpha^{\nu_{12}}S_\beta a;\ 1 \le i \le \nu_{11}\}$ are linearly dependent. Namely, $S_\alpha^{\nu_{12}}S_\beta a = \sum_{j=1}^{2}\sum_{i=1}^{\nu_{1j}} c_{\nu_{1j}-1}^{12} S_\beta^{j-1}S_\alpha^{i-1}a$ holds.
This procedure is repeated until $1-q_1$).

1-q_1) Find an integer ν_{1q_1} and coefficients $\{c_{\nu_{11}+\cdots+\nu_{1j-1}+i}^{1q_1} \in K;\ 1 \le j \le q_1, 1 \le i \le \nu_{1j}\}$ and $\{c_{\nu_{11}+\cdots+\nu_{1j-1}+i}^{1q_1+1} \in K;\ 1 \le j \le q_1, 1 \le i \le \nu_{1j}\}$ such that $\{S_\beta^{j-1}S_\alpha^{i-1}a;\ 1 \le j \le q_1,\ 1 \le i \le \nu_{1j}\}$ are linearly independent, that is, the following equations hold:
$S_\alpha^{\nu_1 q_1}S_\beta^{q_1-1}a = \sum_{j=1}^{q_1}\sum_{i=1}^{\nu_{1j}} c_{\nu_{1j}+\cdots+\nu_{1j-1}+i}^{1q_1} S_\beta^{j-1}S_\alpha^{i-1}a$

and
$S_\beta^{q_1} a = \sum_{j=1}^{q_1} \sum_{i=1}^{\nu_{1j}} c^{1q_1+1}_{\nu_{1j}+\cdots+\nu_{1j-1}+i} S_\beta^{j-1} S_\alpha^{i-1} a$

Step 2

2-1) Find an integer ν_{21} and coefficients $\{c^{21}_{\nu_{11}+\cdots+\nu_{1j-1}+i} \in K;\ 1 \le j \le q_1, 1 \le i \le \nu_{1j}\}$ and $\{c^{21}_{\nu_1+i} \in K;\ 1 \le i \le \nu_{21}\}$ such that $\{S_\gamma S_\alpha^{i-1} a;\ 1 \le i \le \nu_{21}\}$ are linearly independent and
$S_\gamma S_\alpha^{\nu_{21}} a \;=\; \sum_{j=1}^{q_1} \sum_{i=1}^{\nu_{1j}} c^{21}_{\nu_{11}+\cdots+\nu_{1j-1}+i} S_\beta^{j-1} S_\alpha^{i-1} a \;+\; \sum_{i=1}^{\nu_{21}} c^{21}_{\nu_1+i} S_\gamma^{j-1}$ $S_\alpha^{i-1} a$. This procedure is continued until $2 - q_2$).

2-q_2) Find an integer ν_{2q_2} and coefficients $\{c^{2q_2}_{\nu_{k-1}+\nu_{k1}+\cdots+\nu_{kj-1}+i} \in K;\ 1 \le k \le 2,\ 1 \le j \le q_2, 1 \le i \le \nu_{1j}\}$ and $\{c^{2q_2+1}_{\nu_{k-1}+\nu_{k1}+\cdots+\nu_{kj-1}+i} \in K;\ 1 \le k \le 2,\ 1 \le j \le q_2, 1 \le i \le \nu_{1j}\}$ such that $\{S_\gamma^{k-1} S_\beta^{j-1} S_\alpha^{i-1} a;\ 1 \le k \le 2, 1 \le j \le q_k,\ 1 \le i \le \nu_{kj}\}$ are linearly independent Thus the following equations hold:

$$S_\gamma S_\beta^{q_2-1} S_\alpha^{\nu_{2q_2}} a = \sum_{k=1}^{2} \sum_{j=1}^{q_k} \sum_{i=1}^{\nu_{kj}} c^{2q_2}_{\nu_{k-1}+\nu_{k1}+\cdots+\nu_{kj-1}+i} S_\gamma^{k-1} S_\beta^{j-1} S_\alpha^{i-1} a$$
$$+ \sum_{i=1}^{\nu_{21}} c^{21}_{\nu_1+i} S_\gamma^{j-1} S_\alpha^{i-1} a$$

$$S_\gamma S_\beta^{q_2} a = \sum_{k=1}^{2} \sum_{j=1}^{q_k} \sum_{i=1}^{\nu_{kj}} c^{2q_2+1}_{\nu_{k-1}+\nu_{k1}+\cdots+\nu_{kj-1}+i} S_\gamma^{k-1} S_\beta^{j-1} S_\alpha^{i-1} a$$
$$+ \sum_{i=1}^{\nu_{21}} c^{21}_{\nu_1+i} S_\gamma^{j-1} S_\alpha^{i-1} a$$

These steps are repeated until Step p).

Step p.

p-1) Find an integer ν_{p1} and coefficients $\{c^{p1}_{\nu_1+\cdots\nu_{k-1}+\nu_{k1}+\cdots+\nu_{kj-1}+i} \in K;\ 1 \le k \le p-1,\ 1 \le j \le q_k, 1 \le i \le \nu_{kj}\}$ and $\{c^{p1}_{\nu_1+\cdots+\nu_{p-1}+i} \in K;\ 1 \le i \le \nu_{p1}\}$ such that $\{S_\gamma^{k-1} S_\beta^{j-1} S_\alpha^{i-1} a;\ 1 \le k \le p-1,\ 1 \le j \le q_k,\ 1 \le i \le \nu_{kj}\}$ and $\{S_\gamma^{p-1} S_\alpha^{i-1} a;\ 1 \le i \le \nu_{p1}\}$ are linearly independent, and the following equations hold:

$$S_\gamma^{p-1} S_\alpha^{\nu_{p1}} a = \sum_{k=1}^{p-1} \sum_{j=1}^{q_k} \sum_{i=1}^{\nu_{kj}} c^{p1}_{\nu_1+\cdots+\nu_{k-1}+\nu_{k1}+\cdots+\nu_{kj-1}+i} S_\beta^{k-1} S_\beta^{j-1} S_\alpha^{i-1} a$$
$$+ \sum_{i=1}^{\nu_{p1}} c^{p1}_{\nu_1+\cdots+\nu_{p-1}+i} S_\gamma^{p-1} S_\alpha^{i-1} a$$

This procedure is repeated until $p - q_p$).

p-q_p) Find an integer ν_{pq_p} and coefficients $\{c^{2q_2}_{\nu_1+\cdots\nu_{k-1}+\nu_{k1}+\cdots+\nu_{kj-1}+i} \in K;\ 1 \le k \le p,\ 1 \le j \le q_k,\ 1 \le i \le \nu_{kj}\}$ and $\{c^{p+1}_{\nu_1+\cdots+\nu_{k-1}+\nu_{k1}+\cdots+\nu_{kj-1}+i} \in K;\ 1 \le k \le p,\ 1 \le j \le q_k, 1 \le i \le \nu_{kj}\}$ such that $\{S^{k-1}_\gamma S^{j-1}_\beta S^{i-1}_\alpha a;\ 1 \le k \le p,\ 1 \le j \le q_k,\ 1 \le i \le \nu_{kj}\}$ are linearly independent. Then the following equations hold:

$$S^{p-1}_\gamma S^{q_p-1}_\beta S^{\nu_{pq_p}}_\alpha a = \sum_{k=1}^{p} \sum_{j=1}^{q_k} \sum_{i=1}^{\nu_{kj}} c^{pq_p}_{\nu_1+\cdots+\nu_{k-1}+\nu_{k1}+\cdots+\nu_{kj-1}+i} S^{k-1}_\gamma S^{j-1}_\beta S^{i-1}_\alpha a.$$

$$S^p_\gamma a = \sum_{k=1}^{p} \sum_{j=1}^{q_k} \sum_{i=1}^{\nu_{kj}} c^{2q_2+1}_{\nu_1+\cdots+\nu_{k-1}+\nu_{k1}+\cdots+\nu_{kj-1}+i} S^{k-1}_\gamma S^{j-1}_\beta S^{i-1}_\alpha a.$$

Set $n := \sum_{i=1}^{p} \sum_{j=1}^{q_i} \nu_{ij}$.

Step $p+1$
Let a state space be K^n, and let the initial state be $\mathbf{e}_1 \in K^n$.

Step $p+2$.
Let $F_{\alpha s}, F_{\beta s}, F_{\gamma s} \in K^{n \times n}$ be the same as in Definition (6.4).

Step $p+3$.
Let $h_s \in K^{t \times n}$ be
$h_s := [a(0,0,0), \cdots, a(\nu_{11}-1,0,0), a(0,1,0), \cdots, a(\nu_{12}-1,1,0), \cdots, a(\nu_{1q_1}-1, q_1-1, 0), a(0,0,1), \cdots, a(\nu_{pq_p}-1, q_p-1, p-1)]$.

Proof. See (6-B.16) in Appendix 6-B.

Example 6.15. Consider the $6 \times 6 \times 6$ periodic image depicted in Figure 6.7. Let K be $N/2N$ which is the quotient field modulo the prime number 2, and let the set Y of output values be K. We derive the Quasi-reachable Standard System which realizes the three-dimensional image $a \in F(N^3, K)$ in Figure 6.7 by using the realization procedure (6.14).

Step 1:
$1-1)$
The index $\nu_{11} = 3$. The coefficient $c^{11}_1 = 1$. Other coefficients are all 0.
$1-2)$
The index $\nu_{12} = 3$. The coefficient $c^{12}_1 = 1$. Other coefficients are all 0.
$1-q_1)$
The integer $q_1 = 3$. The index $\nu_{13} = 3$. The coefficient $c^{13}_1 = 1$ and $c^{14}_1 = 1$. Other coefficients are all 0.
Step 2:
$2-1)$
The index $\nu_{21} = 3$. The coefficient $c^{21}_1 = 1$. Other coefficients are all 0.
$2-2)$
The index $\nu_{22} = 3$. The coefficient $c^{22}_1 = 1$. Other coefficients are all 0.
$2-q_2)$
The integer $q_2 = 3$. The index $\nu_{23} = 3$. The coefficient $c^{23}_1 = 1$ and

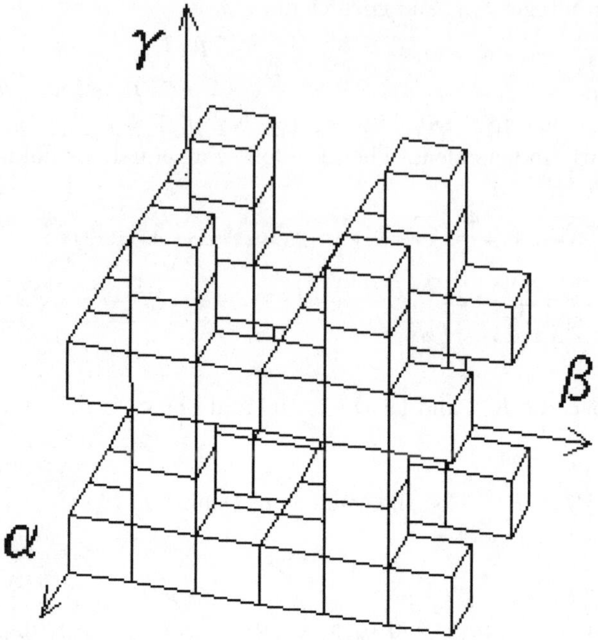

Fig. 6.7. The $6 \times 6 \times 6$ periodic image for Example (6.15)

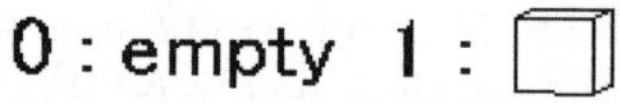

Fig. 6.8. The coding list for Example (6.15)

$c_1^{24} = 1$. Other coefficients are all 0.
Step 3:
$3-1)$
The index $\nu_{31} = 3$. The coefficient $c_1^{31} = 1$. Other coefficients are all 0.
$3-2)$
The index $\nu_{32} = 3$. The coefficient $c_1^{32} = 1$. Other coefficients are all 0.
$3-q_3)$
The integer $q_3 = 3$. The index $\nu_{33} = 3$. The coefficient $c_1^{33} = 1$ and $c_1^{34} = 1$. Other coefficients are all 0.

Then the dimension of the state space is $n = \sum_{i=1}^{3} \sum_{j=1}^{q_i} \nu_{ij} = 27$.
Step 4:
Let a state space be $K^n = K^{27}$, and let the initial state be $\mathbf{e}_1 \in K^{27}$.
Step 5:

Let
$F_{\alpha s} = I_3 \otimes I_3 \otimes F, \quad F_{\beta s} = I_3 \otimes F \otimes I_3, \quad F_{\gamma s} = F \otimes I_3 \otimes I_3,$
where

$$F = \begin{bmatrix} 0 & 0 & 1 \\ 1 & 0 & 0 \\ 0 & 1 & 0 \end{bmatrix}$$

Step 6:
$h_s = [1,1,1,0,0,1,0,0,1,0,0,0,0,0,1,0,0,0,0,0,0,0,0,1,0,0,0]$.

6.4 Partial Realization of Three-Dimensional Images

Now let us consider finite-sized three-dimensional images. Let $\underline{a}$ be a $(\underline{L}+1) \times (\underline{M}+1) \times (\underline{N}+1)$-sized image $\underline{a} \in F(\underline{\mathbf{L}} \times \underline{\mathbf{M}} \times \underline{\mathbf{N}}, K^t)$, where $\underline{L}$, $\underline{M}$, $\underline{N} \in N$.

A finite-dimensional 3-Commutative Linear Representation System $\sigma = ((X, F_\alpha, F_\beta, F_\gamma), x^0, h)$ is called a partial realization of $\underline{a}$ if $hF_\alpha^i F_\beta^j F_\gamma^k x^0 = \underline{a}(i,j,k)$ holds for any $0 \le i \le \underline{L}$, $0 \le j \le \underline{M}$ and $0 \le k \le \underline{N}$.

The Partial Realization Problem is stated as follows:
< For any given $\underline{a} \in F(\underline{\mathbf{L}} \times \underline{\mathbf{M}} \times \underline{\mathbf{N}}, K^t)$, find a partial realization σ of $\underline{a}$ such that the dimensions of state space X of σ is minimum. Such σ is called the minimal partial realization of $\underline{a}$. Moreover, show the minimal realizations are unique modulo isomorphisms.>

Proposition 6.16. There always exists the minimal partial realization of arbitrary $\underline{a} \in F(\underline{\mathbf{L}} \times \underline{\mathbf{M}} \times \underline{\mathbf{N}}, K^t)$, .

Proof. Set $\underline{a}(i,j,k) = 0$ for all $i > \underline{L}$, $j > \underline{M}$, $k > \underline{N}$. Then $\underline{a} \in F(N^3, Y)$. Theorem (6.12) implies that there exists a finite-dimensional partial realization of $\underline{a}$. Therefore, there always exists the minimal partial realization.

Minimal partial realizations are in general not unique modulo isomorphisms. Therefore, the notion of natural partial realization is introduced, and we show that natural partial realizations exist if and only if they are isomorphic.

Definition 6.17. For 3-Commutative Linear Representation System $\sigma = ((X, F_\alpha, F_\beta, F_\gamma), x^0, h)$ and some $l_1, m_1, n_1 \in N$, if $X = \ll \{F_\alpha^i F_\beta^j F_\gamma^k x^0;\ 0 \le i \le l_1,\ 0 \le j \le m_1,\ 0 \le k \le n_1\} \gg$, then σ is called (l_1, m_1, n_1)-quasi-reachable. Let l_2, m_2 and n_2 be some integer.

If $hF_\alpha^i F_\beta^j F_\gamma^k x = 0$ implies $x = 0$ for any $0 \le i \le l_2$, $0 \le j \le m_2$ and $0 \le k \le n_2$, then σ is called (l_2, m_2, n_2)-distinguishable.

For a given $\underline{a} \in F(\underline{\mathbf{L}} \times \underline{\mathbf{M}} \times \underline{\mathbf{N}}, K^t)$, if there exist l_1, m_1, n_1, l_2, m_2 and $n_2 \in N$ such that $l_1 + l_2 < \underline{L}$, $m_1 + m_2 < \underline{M}$, $n_1 + n_2 < \underline{N}$, and σ is (l_1, m_1, n_1)-quasi-reachable and (l_2, m_2, n_2)-distinguishable, then σ is called the natural partial realization of $\underline{a}$.

For a finite-sized image $\underline{a} \in F(\underline{\mathbf{L}} \times \underline{\mathbf{M}} \times \underline{\mathbf{N}}, K^t)$, the following matrix $H_{\underline{a}}$ is called a finite-sized Hankel-matrix of $\underline{a}$:

$$H_{\underline{a}}(l_1, m_1, n_1; L - l_1, M - m_1, N - n_1) =$$

$$\begin{array}{c} \\ (\tilde{i}, \tilde{j}, \tilde{k}) \end{array} \begin{pmatrix} & & (i,j,k) \\ & & \vdots \\ \cdots & \cdots & \underline{a}(\tilde{i}+i, \tilde{j}+j, \tilde{k}+k) \end{pmatrix}$$

where $0 \leq i \leq l_1$, $0 \leq j \leq m_1$, $0 \leq k \leq n_1$, $0 \leq \tilde{i} \leq \underline{L} - l_1$, $0 \leq \tilde{j} \leq \underline{M} - m_1$ and $0 \leq \tilde{k} \leq \underline{N} - n_1$.

Theorem 6.18. Let $H_{\underline{a}}(l_1, m_1, n_1; l_2, m_2, n_2)$ be the finite-sized Hankel-matrix of $\underline{a} \in F(\underline{\mathbf{L}} \times \underline{\mathbf{M}} \times \underline{\mathbf{N}}, K^t)$. Then there exists a natural partial realization of $\underline{a}$ if and only if the following condition holds:

$$\begin{aligned} &\text{rank } H_{\underline{a}}(l_1, m_1, n_1; \underline{L} - l_1, \underline{M} - m_1, \underline{N} - n_1) \\ &= \text{rank } H_{\underline{a}}(l_1 + 1, m_1, n_1; \underline{L} - l_1 - 1, \underline{M} - m_1, \underline{N} - n_1) \\ &= \text{rank } H_{\underline{a}}(l_1, m_1 + 1, n_1; \underline{L} - l_1, \underline{M} - m_1 - 1, \underline{N} - n_1) \\ &= \text{rank } H_{\underline{a}}(l_1, m_1, n_1 + 1; \underline{L} - l_1, \underline{M} - m_1, \underline{N} - n_1 - 1) \\ &= \text{rank } H_{\underline{a}}(l_1, m_1, n_1; \underline{L} - l_1 - 1, \underline{M} - m_1 - 1, \underline{N} - n_1 - 1) \\ &= \text{rank } H_{\underline{a}}(l_1 + 1, m_1 + 1, n_1 + 1; \underline{L} - l_1 - 1, \underline{M} - m_1 - 1, \underline{N} - n_1 - 1) \end{aligned}$$

Proof. See (6-C.7) in Appendix 6-C.

Theorem 6.19. There exists a natural partial realization of $\underline{a}$ for given finite-sized image $\underline{a} \in F(\underline{\mathbf{L}} \times \underline{\mathbf{M}} \times \underline{\mathbf{N}}, K^t)$ if and only if the minimal partial realizations of $\underline{a}$ are unique modulo isomorphisms.

Proof. See (6-C.9) in Appendix 6-C.

Define $\underline{S}_\alpha, \underline{S}_\beta$ and $\underline{S}_\gamma$ for some l, m and $n \in N$ as:
$\underline{S}_\alpha : F(\mathbf{l} \times \mathbf{m} \times \mathbf{n}, K^t) \to F((\mathbf{l} - \mathbf{1}) \times \mathbf{m} \times \mathbf{n}, K^t)$;
$\underline{a} \to \underline{S}_\alpha \underline{a}$ [; $(i, j, k) \mapsto \underline{a}(i + 1, j, k)$],
$\underline{S}_\beta : F(\mathbf{l} \times \mathbf{m} \times \mathbf{n}, K^t) \to F(\mathbf{l} \times (\mathbf{m} - \mathbf{1}) \times \mathbf{n}, K^t)$;
$\underline{a} \mapsto \underline{S}_\beta \underline{a}$ [; $(i, j, k) \mapsto \underline{a}(i, j + 1, k)$],
$\underline{S}_\gamma : F(\mathbf{l} \times \mathbf{m} \times \mathbf{n}, K^t) \to F(\mathbf{l} \times \mathbf{m} \times (\mathbf{n} - \mathbf{1}), K^t)$;
$\underline{a} \mapsto \underline{S}_\gamma \underline{a}$ [; $(i, j, k) \mapsto \underline{a}(i, j, k + 1)$] .

Theorem 6.20. There exists the natural partial realization of a given finite-sized image $\underline{a} \in F(\underline{\mathbf{L}} \times \underline{\mathbf{M}} \times \underline{\mathbf{N}}, K^t)$ if and only if the following whole algorithm can be completely executed:
Step 1.

$1-1)$

Find an integer ν_{11} such that column vectors $\{\underline{S}_{\alpha}^{i-1}\underline{a}\ ;\ 1 \le i \le \nu_{11}\}$ of $H_{\underline{a}}(\nu_{11}-1,0,0;\underline{L}-\nu_{11}+1,\underline{M},\underline{N})$ are linearly independent in the sense of $F((\mathbf{\underline{L}}-\nu_{11}+\mathbf{1})\times\mathbf{\underline{M}}\times\mathbf{\underline{N}},K^t)$ and column vectors $\{\underline{S}_{\alpha}^{i-1}\underline{a};\ 1 \le i \le \nu_{11}+1\}$ of $H_{\underline{a}}(\nu_{11},0,0;\underline{L}-\nu_{11},\underline{M},\underline{N})$ are linearly dependent in the sense of $F((\mathbf{\underline{L}}-\nu_{11})\times\mathbf{\underline{M}}\times\mathbf{\underline{N}},K^t)$. Find coefficients $\{c_i^{11} \in K;\ 1 \le i \le \nu_{11}\}$ such that the following relation holds:
$\underline{S}_{\alpha}^{\nu_{11}}\underline{a} = \sum_{i=1}^{\nu_{11}} c_i^{11}\underline{S}_{\alpha}^{i-1}\underline{a}$ in the sense of $F((\mathbf{\underline{L}}-\nu_{11})\times\mathbf{\underline{M}}\times\mathbf{\underline{N}},K^t)$.

$1-2)$

If column vectors $\{\underline{S}_{\alpha}^{i-1}\underline{a}; 1 \le i \le \nu_{11}\}$ of $H_{\underline{a}}(\nu_{11}-1,0,0;\underline{L}-\nu_{11},\underline{M}-1,\underline{N})$ are linearly dependent in the sense of $F((\mathbf{\underline{L}}-\nu_{11})\times(\mathbf{\underline{M}}-\mathbf{1})\times\mathbf{\underline{N}},K^t)$, then stop this algorithm. Otherwise, find an integer ν_{12} such that column vectors $\{\underline{S}_{\beta}^{j-1}\underline{S}_{\alpha}^{i-1}\underline{a};\ 1 \le j \le 2,\ 1 \le i \le \nu_{1j}\}$ of $H_{\underline{a}}(\nu_{11}-1,1,0;\underline{L}-\nu_{11}+1,\underline{M}-1,\underline{N})$ are linearly independent and column vectors $\{\underline{S}_{\beta}^{j-1}\underline{S}_{\alpha}^{i-1}\underline{a},\ \underline{S}_{\beta}\underline{S}_{\alpha}^{\nu_{12}}\underline{a};\ 1 \le j \le 2,\ 1 \le i \le \nu_{1j}\}$ of $H_{\underline{a}}(\nu_{11},1,0;\underline{L}-\nu_{11},\underline{M}-1,\underline{N})$ are linearly dependent in the sense of $F((\mathbf{\underline{L}}-\nu_{11})\times(\mathbf{\underline{M}}-\mathbf{1})\times\mathbf{\underline{N}},K^t)$.
Find coefficients $\{c_{\nu_{1j-1}+i}^{12} \in K;\ 1 \le j \le 2,\ 1 \le i \le \nu_{1j}\}$ such that the following relation holds: $\underline{S}_{\beta}\underline{S}_{\alpha}^{\nu_{12}}\underline{a} = \sum_{j=1}^{2}\sum_{i=1}^{\nu_{1j}} c_{\nu_{1j-1}+i}^{12}\underline{S}_{\beta}^{j-1}\underline{S}_{\alpha}^{i-1}\underline{a}$ in the sense of $F((\mathbf{\underline{L}}-\nu_{11})\times(\mathbf{\underline{M}}-\mathbf{1})\times\mathbf{\underline{N}},K^t)$. These procedures are repeated until the following $1-q_1)$ and $1-(q_1+1))$.

$1-q_1)$

If column vectors $\{\underline{S}_{\beta}^{j-1}\underline{S}_{\alpha}^{i-1}\underline{a};\ 1 \le j \le q_1,\ 1 \le i \le \nu_{1j}\}$ of $H_{\underline{a}}(\nu_{11}-1,q_1-2,0;\underline{L}-\nu_{11},\underline{M}-q_1+1,\underline{N})$ are linearly dependent in the sense of $F((\mathbf{\underline{L}}-\nu_{11})\times(\mathbf{\underline{M}}-\mathbf{q}_1+\mathbf{1})\times\mathbf{\underline{N}},K^t)$, then stop this algorithm. Otherwise, find an integer ν_{1q_1} such that column vectors $\{\underline{S}_{\beta}^{j-1}\underline{S}_{\alpha}^{i-1}\underline{a};\ 1 \le j \le q_1,\ 1 \le i \le \nu_{1j}\}$ of $H_{\underline{a}}(\nu_{11}-1,q_1-1,0;\underline{L}-\nu_{11}+1,\underline{M}-q_1+1,\underline{N})$ are linearly independent, column vectors $\{\underline{S}_{\beta}^{j-1}\underline{S}_{\alpha}^{i-1}\underline{a}\ ,\ \underline{S}_{\beta}^{q_1-1}\underline{S}_{\alpha}^{\nu_{1q_1}}\underline{a};\ 1 \le j \le q_1,\ 1 \le i \le \nu_{1j}\}$ of $H_{\underline{a}}(\nu_{11},q_1-1,0;\underline{L}-\nu_{11},\underline{M}-q_1+1,\underline{N})$ are linearly dependent in the sense of $F((\mathbf{\underline{L}}-\nu_{11})\times(\mathbf{\underline{M}}-\mathbf{q}_1+\mathbf{1})\times\mathbf{\underline{N}},K^t)$, and column vectors $\{\underline{S}_{\beta}^{j-1}\underline{S}_{\alpha}^{i-1}\underline{a}, \underline{S}_{\beta}^{q_1}\underline{a};\ 1 \le j \le q_1,\ 1 \le i \le \nu_{1j}\}$ of $H_{\underline{a}}(\nu_{11},q_1,0;\underline{L}-\nu_{11},\underline{M}-q_1,\underline{N})$ are linearly dependent in the sense of $F((\mathbf{\underline{L}}-\nu_{11})\times(\mathbf{\underline{M}}-\mathbf{q}_1)\times\mathbf{\underline{N}},K^t)$.
Find coefficients $\{c_{\nu_{11}+\cdots+\nu_{1j-1}+i}^{1q_1} \in K;\ 1 \le j \le q_1,\ 1 \le i \le \nu_{1j}\}$ such that the following relation holds:
$\underline{S}_{\beta}^{q_1-1}\underline{S}_{\alpha}^{\nu_{1q_1}}\underline{a} = \sum_{j=1}^{q_1}\sum_{i=1}^{\nu_{1j}} c_{\nu_{11}+\cdots+\nu_{1j-1}+i}^{1q_1}\underline{S}_{\beta}^{j-1}\underline{S}_{\alpha}^{i-1}\underline{a}$ in the sense of $F((\mathbf{\underline{L}}-\nu_{11})\times(\mathbf{\underline{M}}-\mathbf{q}_1)\times\mathbf{\underline{N}},K^t)$.

$1-(q_1+1)$)

If column vectors $\{\underline{S}_\beta^{j-1}\underline{S}_\alpha^{i-1}\underline{a};\ 1\le j\le q_1,\ 1\le i\le \nu_{1j}\}$ of $H_{\underline{a}}(\nu_{11}-1,q_1-1,0;\underline{L}-\nu_{11}+1,\underline{M}-q_1,\underline{N})$ are linearly dependent in the sense of $F((\underline{\mathbf{L}}-\nu_{11}+\mathbf{1})\times(\underline{\mathbf{M}}-\mathbf{q}_1)\times\underline{\mathbf{N}},K^t)$, then stop this algorithm. Otherwise, find coefficients $\{c^{1q_1+1}_{\nu_{11}+\cdots+\nu_{1j-1}+i}\in K;\ 1\le j\le q_1,\ 1\le i\le \nu_{1j}\}$ such that $\underline{S}_\beta^{q_1}\underline{a}=\sum_{j=1}^{q_1}\sum_{i=1}^{\nu_{1j}}c^{1q_1+1}_{\nu_{11}+\cdots+\nu_{1j-1}+i}\underline{S}_\beta^{j-1}\underline{S}_\alpha^{i-1}\underline{a}$ in the sense of $F((\underline{\mathbf{L}}-\nu_{11})\times(\underline{\mathbf{M}}-\mathbf{q}_1)\times\underline{\mathbf{N}},K^t)$.

Step 2

$2-1$)

If column vectors $\{\underline{S}_\beta^{j-1}\underline{S}_\alpha^{i-1}\underline{a};\ 1\le j\le q_1,\ 1\le i\le \nu_{1j}\}$ of $H_{\underline{a}}(\nu_{11}-1,q_1-1,0;\underline{L}-\nu_{11},\underline{M}-q_1,\underline{N}-1)$ are linearly dependent, then stop this algorithm. Otherwise, find an integer ν_{21} such that column vectors $\{\underline{S}_\beta^{j-1}\underline{S}_\alpha^{i-1}\underline{a},\underline{S}_\gamma\underline{S}_\alpha^{\bar{i}-1}\underline{a};\ 1\le j\le q_1,\ 1\le i\le \nu_{1j},\ 1\le \bar{i}\le \nu_{21}\}$ of $H_{\underline{a}}(\nu_{11},q_1-1,1;\underline{L}-\nu_{11},\underline{M}-q_1+1,\underline{N}-1)$ are linearly independent and column vectors $\{\underline{S}_\beta^{j-1}\underline{S}_\alpha^{i-1}\underline{a},\ \underline{S}_\gamma\underline{S}_\alpha^{\bar{i}}\underline{a};\ 1\le j\le q_1,\ 1\le i\le \nu_{1j},\ 1\le \bar{i}\le \nu_{21}\}$ of $H_{\underline{a}}(\nu_{11},q_1-1,1;\underline{L}-\nu_{11},\underline{M}-q_1+1,\underline{N}-1)$ are linearly dependent.

Find coefficients $\{c^{21}_{\nu_{11}+\cdots+\nu_{1j-1}+i}\in K;\ 1\le j\le q_1,\ 1\le i\le \nu_{1j}\}$ and $\{c^{21}_{\nu_1+i}\in K;\ 1\le i\le \nu_{21}\}$ such that the following relation holds: $\underline{S}_\gamma\underline{S}_\alpha^{\nu_{21}}\underline{a}=\sum_{j=1}^{q_1}\sum_{i=1}^{\nu_{1j}}c^{21}_{\nu_{11}+\cdots+\nu_{1j-1}+i}\underline{S}_\beta^{j-1}\underline{S}_\alpha^{i-1}\underline{a}+\sum_{i=1}^{\nu_{21}}c^{21}_{\nu_1+i}$ $\underline{S}_\gamma\underline{S}_\alpha^{i-1}\underline{a}$.

$2-2$)

If column vectors $\{\underline{S}_\beta^{j-1}\underline{S}_\alpha^{i-1}\underline{a},\ \underline{S}_\gamma\underline{S}_\alpha^{\bar{i}-1}\underline{a};\ 1\le j\le q_1,\ 1\le i\le \nu_{1j},\ 1\le \bar{i}\le \nu_{21}\}$ of $H_{\underline{a}}(\nu_{11}-1,q_1-1,1;\underline{L}-\nu_{11}+1,\underline{M}-q_1+1,\underline{N}-1)$ are linearly dependent, then stop this algorithm. Otherwise find an integer ν_{22} such that column vectors $\{\underline{S}_\beta^{j-1}\underline{S}_\alpha^{i-1}\underline{a},\ \underline{S}_\gamma\underline{S}_\beta^{\bar{j}-1}\underline{S}_\alpha^{\bar{i}-1}\underline{a};\ 1\le j\le q_1,\ 1\le i\le \nu_{1j},\ 1\le \bar{j}\le 2,\ 1\le \bar{i}\le \nu_{2\bar{j}}\}$ of $H_{\underline{a}}(\nu_{11}-1,q_1-1,1;\underline{L}-\nu_{11}+1,\underline{M}-q_1+1,\underline{N}-1)$ are linearly independent and column vectors $\{\underline{S}_\beta^{j-1}\underline{S}_\alpha^{i-1}\underline{a},$ $\underline{S}_\gamma\underline{S}_\beta^{\bar{j}-1}\underline{S}_\alpha^{\bar{i}-1}\underline{a},\ \underline{S}_\gamma\underline{S}_\beta\underline{S}_\alpha^{\nu_{22}}\underline{a};\ 1\le j\le q_1,\ 1\le i\le \nu_{1j},\ 1\le \bar{j}\le 2,\ 1\le \bar{i}\le \nu_{2\bar{j}}\}$ of $H_{\underline{a}}(\nu_{11},q_1,1;\underline{L}-\nu_{11},\underline{M}-q_1,\underline{N}-1)$ are linearly dependent. Find coefficients $\{c^{22}_{\nu_{11}+\cdots+\nu_{1j-1}+i}\in K;\ 1\le j\le q_1,\ 1\le i\le \nu_{1j}\}$ and $\{c^{22}_{\nu_1+\nu_{2j}+i}\in K;\ 0\le j\le 1,\ 1\le i\le \nu_{2j}\}$ such that the following relation holds:

$$\underline{S}_\gamma\underline{S}_\beta\underline{S}_\alpha^{\nu_{22}}\underline{a}=\sum_{j=1}^{q_1}\sum_{i=1}^{\nu_{1j}}c^{22}_{\nu_{11}+\cdots+\nu_{1j-1}+i}\underline{S}_\beta^{j-1}\underline{S}_\alpha^{i-1}\underline{a}+\sum_{j=0}^{1}\sum_{i=1}^{\nu_{2j}}c^{22}_{\nu_1+\nu_{2j}+i}\underline{S}_\gamma\underline{S}_\beta\underline{S}_\alpha^{i-1}\underline{a}.$$

This procedure is repeated until the following $2-q_2$) and $2-(q_2+1)$).

$2-q_2$)

If column vectors $\{\underline{S}_\gamma^{k-1}\underline{S}_\beta^{j-1}\underline{S}_\alpha^{i-1}\underline{a};\ 1\le k\le 2,\ 1\le j\le q_1,\ 1\le i\le \nu_{1j}\}$ of $H_{\underline{a}}(\nu_{11}-1,q_1-1,1;\underline{L}-\nu_{11}+1,\underline{M}-q_1+1,\underline{N}-1)$ are linearly dependent in the sense of $F((\underline{\mathbf{L}}-\nu_{11}+1)\times(\underline{\mathbf{M}}-\mathbf{q}_1+\mathbf{1})\times(\underline{\mathbf{N}}-\mathbf{1}),K^t)$,

then stop this algorithm. Otherwise find an integer ν_{2q_2} such that column vectors $\{\underline{S}_\gamma^{k-1}\underline{S}_\beta^{j-1}\underline{S}_\alpha^{i-1}\underline{a};\ 1 \le k \le 2, 1 \le j \le q_k,\ 1 \le i \le \nu_{kj}\}$ of
$H_{\underline{a}}(\nu_{11}, q_1-1, 1; \underline{L}-\nu_{11}, \underline{M}-q_1+1, \underline{N}-1)$ are linearly independent, column vectors $\{\underline{S}_\gamma^{k-1}\underline{S}_\beta^{j-1}\underline{S}_\alpha^{i-1}\underline{a},\ \underline{S}_\gamma\underline{S}_\beta^{q_2-1}\underline{S}_\alpha^{\nu_{2q_2}}\underline{a};\ 1 \le k \le 2,\ 1 \le j \le q_k,\ 1 \le i \le \nu_{kj}\}$ of $H_{\underline{a}}(\nu_{11}, q_1, 1; \underline{L}-\nu_{11}, \underline{M}-q_1, \underline{N}-1)$ are linearly dependent and column vectors $\{\underline{S}_\gamma^{k-1}\underline{S}_\beta^{j-1}\underline{S}_\alpha^{i-1}\underline{a},\ \underline{S}_\gamma\underline{S}_\beta^{q_2}\underline{a};\ 1 \le k \le 2,\ 1 \le j \le q_k,\ 1 \le i \le \nu_{kj}\}$ of $H_{\underline{a}}(\nu_{11}, q_1-1, 1; \underline{L}-\nu_{11}, \underline{M}-q_1+1, \underline{N}-1)$ are linearly dependent. Find coefficients $\{c^{2q_2}_{\nu_{k-1}+\nu_{k1}+\cdots+\nu_{kj-1}+i} \in K;\ 1 \le k \le 2,\ 1 \le j \le q_2,\ 1 \le i \le \nu_{1j}\}$ such that the following relations hold:
$\underline{S}_\gamma\underline{S}_\beta^{q_2-1}\underline{S}_\alpha^{\nu_{2q_2}}\underline{a} =$
$\sum_{k=1}^{2}\sum_{j=1}^{q_k}\sum_{i=1}^{\nu_{kj}} c^{2q_2}_{\nu_{k-1}+\nu_{k1}+\cdots+\nu_{kj-1}+i}\underline{S}_\gamma^{k-1}\underline{S}_\beta^{j-1}\underline{S}_\alpha^{i-1}\underline{a}.$

$2-(q_2+1)$)

If column vectors $\{\underline{S}_\gamma^{k-1}\underline{S}_\beta^{j-1}\underline{S}_\alpha^{i-1}\underline{a};\ 1 \le k \le 2,\ 1 \le j \le q_1,\ 1 \le i \le \nu_{1j}\}$ of $H_{\underline{a}}(\nu_{11}-1, q_1-1, 1; \underline{L}-\nu_{11}+1, \underline{M}-q_1, \underline{N}-1)$ are linearly dependent in the sense of $F((\mathbf{\underline{L}}-\nu_{11}+1)\times(\mathbf{\underline{M}}-q_1)\times(\mathbf{\underline{N}}-1), K^t)$, then stop this algorithm. Otherwise, find coefficients $\{c^{2q_2+1}_{\nu_{k-1}+\nu_{k1}+\cdots+\nu_{kj-1}+i} \in K;\ 1 \le k \le 2,\ 1 \le j \le q_2,\ 1 \le i \le \nu_{1j}\}$ such that the following relation holds:
$\underline{S}_\gamma\underline{S}_\beta^{q_2}\underline{a} = \sum_{k=1}^{2}\sum_{j=1}^{q_k}\sum_{i=1}^{\nu_{kj}} c^{2q_2+1}_{\nu_{k-1}+\nu_{k1}+\cdots+\nu_{kj-1}+i}\underline{S}_\gamma^{k-1}\underline{S}_\beta^{j-1}\underline{S}_\alpha^{i-1}\underline{a}$ in the sense of $F((\mathbf{\underline{L}}-\nu_{11})\times(\mathbf{\underline{M}}-\mathbf{q}_1)\times(\mathbf{\underline{N}}-\mathbf{1}), K^t)$.
These steps are repeated until step p.

Step p

$p-1$)

If column vectors $\{\underline{S}_\gamma^{k-1}\underline{S}_\beta^{j-1}\underline{S}_\alpha^{i-1}\underline{a};\ 1 \le k \le p-1,\ 1 \le j \le q_1,\ 1 \le i \le \nu_{kj}\}$ of $H_{\underline{a}}(\nu_{11}-1, q_1-1, p-1; \underline{L}-\nu_{11}+1, \underline{M}-q_1+1, \underline{N}-p)$ are linearly dependent, then stop this algorithm.
Otherwise, find an integer ν_{p1} such that column vectors $\{\underline{S}_\gamma^{k-1}\underline{S}_\beta^{j-1}\underline{S}_\alpha^{i-1}\underline{a};\ 1 \le k \le p-1,\ 1 \le j \le q_k,\ 1 \le i \le \nu_{kj}\}$ and $\{\underline{S}_\gamma^{p-1}\underline{S}_\alpha^{i-1}\underline{a};\ 1 \le i \le \nu_{p1}\}$ of $H_{\underline{a}}(\nu_{11}-1, q_1-1, p-1; \underline{L}-\nu_{11}+1, \underline{M}-q_1+1, \underline{N}-p+1)$ are linearly independent and column vectors $\{\underline{S}_\gamma^{k-1}\underline{S}_\beta^{j-1}\underline{S}_\alpha^{i-1}\underline{a},\ \underline{S}_\gamma^{p-1}\underline{S}_\alpha^{\bar{i}-1}\underline{a};\ 1 \le k \le p-1,\ 1 \le j \le q_k,\ 1 \le i \le \nu_{kj},\ 1 \le \bar{i} \le \nu_{p1}\}$ of $H_{\underline{a}}(\nu_{11}, q_1, p-1; \underline{L}-\nu_{11}, \underline{M}-q_1, \underline{N}-p+1)$ are linearly dependent. Find coefficients $\{c^{p1}_{\nu_1+\cdots+\nu_{k-1}+\nu_{k1}+\cdots+\nu_{kj-1}+i} \in K;\ 1 \le k \le p-1,\ 1 \le j \le q_k,\ 1 \le i \le \nu_{kj}\}$ and $\{c^{p1}_{\nu_1+\cdots+\nu_{p-1}+i} \in K;\ 1 \le i \le \nu_{p1}\}$ such that the following relation holds:
$\underline{S}_\gamma^{p-1}\underline{S}_\alpha^{\nu_{p1}}\underline{a} =$

$$\sum_{k=1}^{p-1}\sum_{j=1}^{q_k}\sum_{i=1}^{\nu_{kj}} c^{p1}_{\nu_1+\cdots+\nu_{k-1}+\nu_{k1}+\cdots+\nu_{kj-1}+i}\ \underline{S}_\gamma^{k-1}\underline{S}_\beta^{j-1}\underline{S}_\alpha^{i-1}\underline{a} + \sum_{i=1}^{\nu_{p1}} c^{p1}_{\nu_1+\cdots+\nu_{p-1}+i}\underline{S}_\gamma^{p-1}\underline{S}_\alpha^{i-1}\underline{a}$$

in the sense of $F((\mathbf{\underline{L}} - \nu_{11}) \times (\mathbf{\underline{M}} - \mathbf{q}_1) \times (\mathbf{\underline{N}} - \mathbf{p} + \mathbf{1}), K^t)$, where $\nu_i := \nu_{i1} + \cdots + \nu_{iq_i}$ for $1 \le i \le p$.

$p-2)$

If column vectors $\{\underline{S}_\gamma^{k-1} \underline{S}_\beta^{j-1} \underline{S}_\alpha^{i-1} \underline{a};\ 1 \le k \le p-1,\ 1 \le j \le q_1,\ 1 \le i \le \nu_{kj}\}$ of $H_{\underline{a}}(\nu_{11}, q_1 - 1, p-1; \underline{L} - \nu_{11}, \underline{M} - q_1 + 1, \underline{N} - p + 1)$ are linearly dependent, then stop this algorithm.

Otherwise, find an integer ν_{p2} such that column vectors $\{\underline{S}_\gamma^{k-1} \underline{S}_\beta^{j-1} \underline{S}_\alpha^{i-1} \underline{a},\ \underline{S}_\gamma^{p-1} \underline{S}_\beta^{\bar{j}-1} \underline{S}_\alpha^{\bar{i}-1} \underline{a};\ 1 \le k \le p-1,\ 1 \le j \le q_k,\ 1 \le i \le \nu_{kj},\ 1 \le \bar{j} \le 2, 1 \le \bar{i} \le \nu_{p1}\}$ of $H_{\underline{a}}(\nu_{11}, q_1 - 1, p-1; \underline{L} - \nu_{11}, \underline{M} - q_1 + 1, \underline{N} - p + 1)$ are linearly independent and column vectors $\{\underline{S}_\gamma^{k-1} \underline{S}_\beta^{j-1} \underline{S}_\alpha^{i-1} \underline{a},\ \underline{S}_\gamma^{p-1} \underline{S}_\beta^{\bar{j}-1} \underline{S}_\alpha^{\bar{i}-1} \underline{a},$
$\underline{S}_\gamma^{p-1} \underline{S}_\beta \underline{S}_\alpha^{\nu_{p2}} \underline{a};\ 1 \le k \le p-1, 1 \le j \le q_k,\ 1 \le i \le \nu_{kj},\ 1 \le \bar{j} \le 2,\ 1 \le \bar{i} \le \nu_{pj}\}$ of $H_{\underline{a}}(\nu_{11}, q_1, p-1; \underline{L} - \nu_{11}, \underline{M} - q_1, \underline{N} - p + 1)$ are linearly dependent. Find coefficients $\{c^{p2}_{\nu_1 + \cdots + \nu_{k-1} + \nu_{k1} + \cdots + \nu_{kj-1} + i} \in K;\ 1 \le k \le p-1,\ 1 \le j \le q_k,\ 1 \le i \le \nu_{kj}\}$ and $\{c^{p2}_{\nu_1 + \cdots + \nu_{p-1} + \nu_{pj} + i} \in K;\ 0 \le j \le 1,\ 1 \le i \le \nu_{pj}\}$ such that the following relation holds:

$$\underline{S}_\gamma^{p-1} \underline{S}_\beta \underline{S}_\alpha^{\nu_{p2}} \underline{a} = \sum_{k=1}^{p-1} \sum_{j=1}^{q_k} \sum_{i=1}^{\nu_{kj}} c^{p2}_{\nu_1 + \cdots + \nu_{k-1} + \nu_{k1} + \cdots + \nu_{kj-1} + i} \underline{S}_\gamma^{k-1} \underline{S}_\beta^{j-1} \underline{S}_\alpha^{i-1} \underline{a} + \sum_{j=0}^{1} \sum_{i=1}^{\nu_{pj}} c^{p2}_{\nu_1 + \cdots + \nu_{p-1} + \nu_{pj} + i} \underline{S}_\gamma^{p-1} \underline{S}_\alpha^{i-1} \underline{a}$$

in the sense of $F((\underline{L} - \nu_{11}) \times (\underline{M} - \mathbf{q}_1) \times (\underline{N} - \mathbf{p} + \mathbf{1}), K^t)$. This procedure is repeated until the following $p - q_p)$ and $p - (q_p + 1))$.

$p - q_p)$

If column vectors $\{\underline{S}_\gamma^{k-1} \underline{S}_\beta^{j-1} \underline{S}_\alpha^{i-1} \underline{a};\ 1 \le k \le p-1,\ 1 \le j \le q_1,\ 1 \le i \le \nu_{kj}\}$ of
$H_{\underline{a}}(\nu_{11} - 1, q_1 - 1, p-1; \underline{L} - \nu_{11}, \underline{M} - q_1 + 1, \underline{N} - p + 1)$ are linearly dependent in the sense of $F((\underline{L} - \nu_{11}) \times (\underline{M} - \mathbf{q}_1) \times (\underline{N} - \mathbf{p} + \mathbf{1}), K^t)$, then stop this algorithm.

Otherwise, find an integer ν_{pq_p} such that
$\{\underline{S}_\gamma^{k-1} \underline{S}_\beta^{j-1} \underline{S}_\alpha^{i-1} \underline{a};\ 1 \le k \le p,\ 1 \le j \le q_k,\ 1 \le i \le \nu_{kj}\}$ of
$H_{\underline{a}}(\nu_{11} - 1, q_1, p-1; \underline{L} - \nu_{11}, \underline{M} - q_1, \underline{N} - p + 1)$ are linearly independent, $\{\underline{S}_\gamma^{k-1} \underline{S}_\beta^{j-1} \underline{S}_\alpha^{i-1} \underline{a},\ \underline{S}_\gamma^{p-1} \underline{S}_\beta^{q_p-1} \underline{S}_\alpha^{\nu_{pq_p}} \underline{a};\ 1 \le k \le p,\ 1 \le j \le q_k,\ 1 \le i \le \nu_{kj}\}$ of $H_{\underline{a}}(\nu_{11} - 1, q_1 - 1, p-1; \underline{L} - \nu_{11} + 1, \underline{M} - q_1 + 1, \underline{N} - p + 1)$ are linearly dependent and $\{\underline{S}_\gamma^{k-1} \underline{S}_\beta^{j-1} \underline{S}_\alpha^{i-1} \underline{a},\ \underline{S}_\gamma^{p-1} \underline{S}_\alpha^{\nu_{pq_p}} \underline{a};\ 1 \le k \le p,\ 1 \le j \le q_k,\ 1 \le i \le \nu_{kj}\}$ of $H_{\underline{a}}(\nu_{11} - 1, q_1 - 1, p-1; \underline{L} - \nu_{11} + 1, \underline{M} - q_1 + 1, \underline{N} - p + 1)$ are linearly dependent. Find coefficients $\{c^{pq_p}_{\nu_1 + \cdots + \nu_{k-1} + \nu_{k1} + \cdots + \nu_{kj-1} + i} \in K;\ 1 \le k \le p,\ 1 \le j \le q_k,\ 1 \le i \le \nu_{kj}\}$ such that the following relation holds:

$$\underline{S}_\gamma^{p-1} \underline{S}_\beta^{q_p-1} \underline{S}_\alpha^{\nu_{pq_p}} \underline{a} = \sum_{k=1}^{p} \sum_{j=1}^{q_k} \sum_{i=1}^{\nu_{kj}} c^{pq_p}_{\nu_1 + \cdots + \nu_{k-1} + \nu_{k1} + \cdots + \nu_{kj-1} + i} \underline{S}_\gamma^{k-1} \underline{S}_\beta^{j-1} \underline{S}_\alpha^{i-1} \underline{a}$$

in the sense of $F((\underline{L} - \nu_{11} + \mathbf{1}) \times (\underline{M} - \mathbf{q}_1 + \mathbf{1}) \times (\underline{N} - \mathbf{p} + \mathbf{1}), K^t)$.

$p-q_p+1$) If column vectors $\{\underline{S}_\gamma^{k-1}\underline{S}_\beta^{j-1}\underline{S}_\alpha^{i-1}\underline{a};\ 1 \le k \le p,\ 1 \le j \le q_1,\ 1 \le i \le \nu_{kj}\}$ of $H_{\underline{a}}(\nu_{11}-1, q_1-1, p-1; \underline{L}-\nu_{11}+1, \underline{M}-q_1+1, \underline{N}-p)$ are linearly dependent, then stop this algorithm.
Otherwise, find coefficients $\{c^{p+1}_{\nu_1+\cdots+\nu_{k-1}+\nu_{k1}+\cdots+\nu_{kj-1}+i} \in K;\ 1 \le k \le p,\ 1 \le j \le q_k,\ 1 \le i \le \nu_{kj}\}$ such that the following relation holds:
$\underline{S}_\gamma^p\underline{a} = \sum_{k=1}^{p}\sum_{j=1}^{q_k}\sum_{i=1}^{\nu_{kj}} c^{p+1}_{\nu_1+\cdots+\nu_{k-1}+\nu_{k1}+\cdots+\nu_{kj-1}+i}\underline{S}_\gamma^{k-1}\underline{S}_\beta^{j-1}\underline{S}_\alpha^{i-1}\underline{a}$
in the sense of $F((\underline{L}-\nu_{11}) \times (\underline{M}-q_1) \times (\underline{N}-p), K^t)$.
Set $n := \sum_{i=1}^{p}\sum_{j=1}^{q_i} \nu_{ij}$.

Step p+1.
Let a state space be K^n and let the initial state be $\mathbf{e}_1$.

Step p+2.
Let $F_{\alpha s}, F_{\beta s}, F_{\gamma s} \in K^{n\times n}$ be the same as in Definition (6.10), where $\mathbf{c}^{ij} = [c_1^{ij}, c_2^{ij}, \cdots, c_{\nu_{11}}^{ij}, c_{\nu_{11}+1}^{ij}, \cdots, c^{ij}_{\nu_1+\cdots+\nu_{i-1}+\nu_{i1}+\cdots+\nu_{ij}}, 0, \cdots, 0]^T \in K^n$.

Step p+3.
Let $h_s \in K^{t\times n}$ be

$$h_s := [\underline{a}(0,0,0), \cdots, \underline{a}(\nu_{11}-1,0,0), \underline{a}(0,1,0), \cdots \underline{a}(\nu_{12}-1,1,0), \cdots,\ \underline{a}(\nu_{1q_1}-1, q_1-1, 0), \underline{a}(0,0,1), \cdots, \underline{a}(\nu_{pq_p}-1, q_p-1, p-1)].$$

Proof. Note that the following equation holds when the algorithm can be executed:

$$\begin{aligned}
&\text{rank } H_{\underline{a}}(\nu_{11}-1, q_1-1, p-1; L-\nu_{11}+1, M-q_1+1, N-p+1)\\
&= \text{rank } H_{\underline{a}}(\nu_{11}, q_1-1, p-1; L-\nu_{11}, M-q_1+1, N-p+1)\\
&= \text{rank } H_{\underline{a}}(\nu_{11}-1, q_1, p-1; L-\nu_{11}+1, M-q_1, N-p+1)\\
&= \text{rank } H_{\underline{a}}(\nu_{11}-1, q_1-1, p; L-\nu_{11}+1, M-q_1+1, N-p)\\
&= \text{rank } H_{\underline{a}}(\nu_{11}-1, q_1-1, p-1; L-\nu_{11}, M-q_1, N-p)\\
&= \text{rank } H_{\underline{a}}(\nu_{11}, q_1, p; L-\nu_{11}, M-q_1, N-p).
\end{aligned}$$

The detailed proof appears in (6-C.10) in Appendix 6-C.

Example 6.21. Consider the $6\times 6\times 1$ image depicted in Figure 6.9. Let K be $N/5N$ which is the quotient field modulo the prime number 5, and let the set Y of output values be K. The Quasi-reachable Standard System which realizes the three-dimensional image $\underline{a} \in F(\underline{\mathbf{L}} \times \underline{\mathbf{M}} \times \underline{\mathbf{N}}, K)$ is derived by using Theorem (6.20).

Step 1:

1 − 1) The index $\nu_{11} = 2$ as well as the coefficients $c_1^{11} = 4$ and $c_2^{11} = 0$ are obtained from the Hankel matrix $H_{\underline{a}}(2,0,0;1,3,3)$.

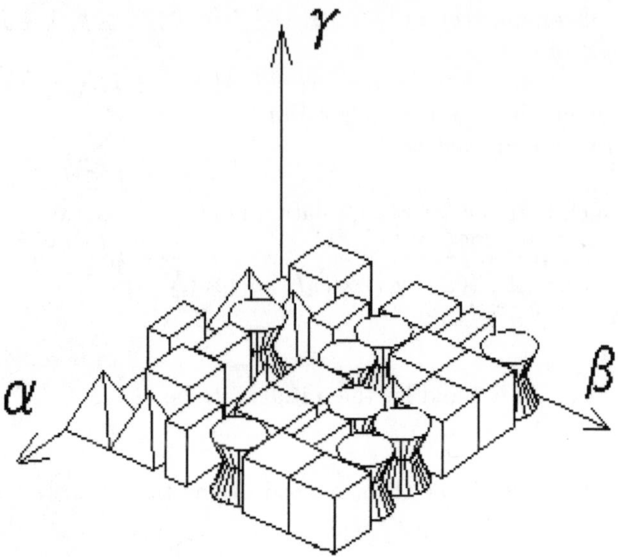

Fig. 6.9. The $6 \times 6 \times 1$ image for Example (6.21)

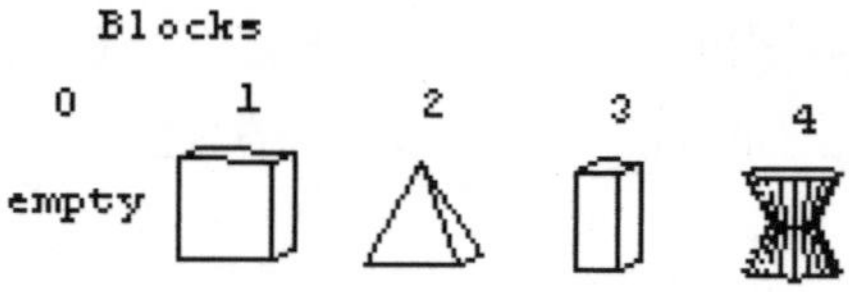

Fig. 6.10. The coding list for Example (6.21)

$1-2)$ The index $\nu_{12} = 1$ and the coefficients $c_1^{12} = 4$, $c_2^{12} = 2$ and $c_3^{12} = 3$ are obtained from the Hankel matrix $H_{\underline{a}}(2,1,0;1,2,3)$.

$1-(q_1+1))$ The integer $q_1 = 2$ along with the coefficients $c_1^{13} = 0$, $c_2^{13} = 3$ and $c_3^{13} = 4$ are obtained from the Hankel matrix $H_{\underline{a}}(1,1,0;2,2,3)$.

Step 2:

$2-1)$ The index $\nu_{21} = 0$ and the coefficients $c_1^{21} = 0$, $c_2^{21} = 0$ and $c_3^{21} = 1$ are obtained from the Hankel matrix $H_{\underline{a}}(2,1,1;1,2,2)$.

$2-2)$ The index $\nu_{22} = 0$ and the coefficients $c_1^{22} = 0$, $c_2^{22} = 3$ and $c_3^{22} = 4$ are obtained from the Hankel matrix $H_{\underline{a}}(2,2,1;1,1,2)$.

The Hankel matrices used in these first two steps can be seen at the end of this example.
The dimension of the state space is $n = 3$.

Step p:
The index p=1 and the coefficients $c_1^2 = 0,\ c_2^2 = 0,\ c_3^2 = 1$ have already been found in Step 2-1).

Step p+1:
Let the state space be $K^n = K^3$ and let the initial state be $\mathbf{e}_1 \in K^3$.

Step p+2:
Applying the coefficients from Step 1 and Step 2 to Definition (6.10), $F_{\alpha s}$, $F_{\beta s}$ and $F_{\gamma s}$ are determined as:

$$F_{\alpha s} = \begin{bmatrix} 0\,4\,4 \\ 1\,0\,2 \\ 0\,0\,3 \end{bmatrix}, \quad F_{\beta s} = \begin{bmatrix} 0\,4\,0 \\ 0\,2\,3 \\ 1\,3\,4 \end{bmatrix}, \quad F_{\gamma s} = \begin{bmatrix} 0\,4\,0 \\ 0\,2\,3 \\ 1\,3\,4 \end{bmatrix}.$$

Step p+3:
$h_s = [0, 2, 1]$.

$$H_{\underline{a}}(2,0,0;1,3,3) = \begin{bmatrix} 0\,2\,0 \\ 2\,0\,3 \\ 1\,2\,4 \\ 1\,2\,4 \\ 0\,3\,0 \\ 2\,4\,3 \\ 2\,4\,3 \\ 0\,3\,0 \\ 0\,3\,0 \\ 0\,3\,0 \\ 3\,0\,2 \\ 4\,3\,1 \\ 4\,3\,1 \\ 3\,0\,2 \\ 3\,0\,2 \\ 3\,0\,2 \\ 1\,4\,4 \\ 1\,4\,4 \\ 1\,4\,4 \\ 1\,4\,4 \end{bmatrix}, \quad H_{\underline{a}}(2,1,0;1,2,3) = \begin{bmatrix} 0\,2\,1\,2 \\ 2\,0\,2\,4 \\ 1\,2\,0\,3 \\ 1\,2\,0\,3 \\ 0\,3\,4\,3 \\ 2\,4\,3\,0 \\ 2\,4\,3\,0 \\ 0\,3\,1\,4 \\ 0\,3\,1\,4 \\ 0\,3\,1\,4 \\ 3\,0\,3\,1 \\ 4\,3\,0\,2 \\ 4\,3\,0\,2 \\ 3\,0\,4\,4 \\ 3\,0\,4\,4 \\ 3\,0\,4\,4 \\ 1\,4\,3\,1 \\ 1\,4\,3\,1 \\ 1\,4\,3\,1 \\ 1\,4\,3\,1 \end{bmatrix}$$

$$H_{\underline{a}}(2,2,0;1,1,3) = \begin{bmatrix} 0\,2\,1\,1 \\ 2\,0\,2\,2 \\ 1\,2\,0\,0 \\ 1\,2\,0\,0 \\ 0\,3\,4\,4 \\ 2\,4\,3\,3 \\ 2\,4\,3\,3 \\ 0\,3\,1\,1 \\ 0\,3\,1\,1 \\ 0\,3\,1\,1 \\ 3\,0\,3\,3 \\ 4\,3\,0\,0 \\ 4\,3\,0\,0 \\ 3\,0\,4\,4 \\ 3\,0\,4\,4 \\ 3\,0\,4\,4 \\ 1\,4\,3\,3 \\ 1\,4\,3\,3 \\ 1\,4\,3\,3 \\ 1\,4\,3\,3 \end{bmatrix}, \quad H_{\underline{a}}(2,1,1;1,2,2) = \begin{bmatrix} 0\,2\,1\,0 \\ 2\,0\,2\,3 \\ 1\,2\,0\,1 \\ 1\,2\,0\,1 \\ 0\,3\,4\,0 \\ 2\,4\,3\,4 \\ 2\,4\,3\,4 \\ 0\,3\,1\,3 \\ 0\,3\,1\,3 \\ 0\,3\,1\,3 \\ 3\,0\,3\,2 \\ 4\,3\,0\,4 \\ 4\,3\,0\,4 \\ 3\,0\,4\,1 \\ 3\,0\,4\,1 \\ 3\,0\,4\,1 \\ 1\,4\,3\,4 \\ 1\,4\,3\,4 \\ 1\,4\,3\,4 \\ 1\,4\,3\,4 \end{bmatrix},$$

$$H_{\underline{a}}(2,2,1;1,1,2) = \begin{bmatrix} 0\,2\,1\,0 \\ 2\,0\,2\,3 \\ 1\,2\,0\,1 \\ 1\,2\,0\,1 \\ 0\,3\,4\,0 \\ 2\,4\,3\,4 \\ 2\,4\,3\,4 \\ 0\,3\,1\,3 \\ 0\,3\,1\,3 \\ 0\,3\,1\,3 \\ 3\,0\,3\,2 \\ 4\,3\,0\,4 \\ 4\,3\,0\,4 \\ 3\,0\,4\,1 \\ 3\,0\,4\,1 \\ 3\,0\,4\,1 \\ 1\,4\,3\,4 \\ 1\,4\,3\,4 \\ 1\,4\,3\,4 \\ 1\,4\,3\,4 \end{bmatrix}$$

Example 6.22. Consider the $6 \times 6 \times 2$ image depicted in Figure 6.11. Let K be $N/3N$ which is the quotient field modulo the prime number 3, and let the set Y of output values be K.

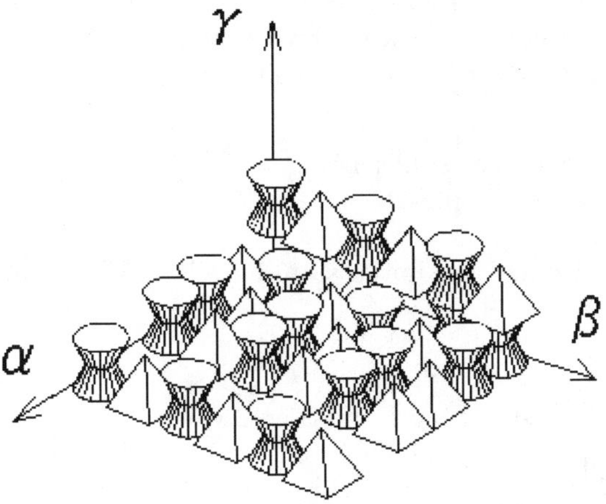

Fig. 6.11. The $6 \times 6 \times 2$ image for Example (6.22)

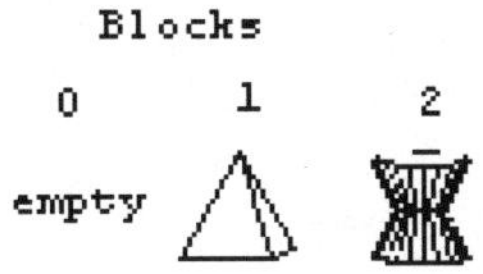

Fig. 6.12. The coding list for Example (6.22)

The Quasi-reachable Standard System which realizes the three-dimensional image $\underline{a} \in F(\underline{\mathbf{L}} \times \underline{\mathbf{M}} \times \underline{\mathbf{N}}, K)$ is derived in the following way.

Step 1:

1 − 1) The index $\nu_{11} = 3$ and the coefficients $c_1^{11} = 0$, $c_2^{11} = 1$ and $c_3^{11} = 2$ are found from the Hankel matrix $H_{\underline{a}}(3,0,0;0,3,3)$.

1 − 2) The index $\nu_{12} = 1$ and the coefficients $c_1^{12} = 0$, $c_2^{12} = 2$, $c_3^{12} = 0$ and $c_4^{12} = 0$ are found from the Hankel matrix $H_{\underline{a}}(3,1,0;0,2,3)$.

1 − $(q_1 + 1)$) The integer $q_1 = 2$ and the coefficients $c_1^{13} = 2$, $c_2^{13} = 3$, $c_3^{13} = 0$ and $c_4^{13} = 1$ are found from the Hankel matrix $H_{\underline{a}}(3,2,0;0,1,3)$.

Step 2:

2 − 1) The index $\nu_{21} = 0$ and the coefficients $c_1^{21} = 2$, $c_2^{21} = 0$, $c_3^{21} = 0$ and $c_4^{21} = 2$ are found from the Hankel matrix $H_{\underline{a}}(3,1,1;0,2,2)$.

2 − 2) The index $\nu_{22} = 0$ and the coefficients $c_1^{22} = 1$, $c_2^{22} = 0$, $c_3^{22} = 0$ and $c_4^{22} = 1$ are found from the Hankel matrix $H_{\underline{a}}(3,2,1;0,1,2)$.
Those Hankel matrices are shown at the end of this example.
The dimension of the state space is $n = 4$.

Step p:
The index p=1, and the coefficients $c_1^2 = 2$, $c_2^2 = 0$, $c_3^2 = 0$, $c_4^2 = 2$ have already been found in Step 2-1).
Step p+1:
Let the state space and the initial state be $K^n = K^4$ and $\mathbf{e}_1 \in K^4$, respectively.

Step p+2:
The coefficients found in Step 1 and Step 2 yield:

$$F_{\alpha s} = \begin{bmatrix} 0\,0\,0\,0 \\ 1\,0\,1\,2 \\ 0\,1\,2\,0 \\ 0\,0\,0\,0 \end{bmatrix}, \quad F_{\beta s} = \begin{bmatrix} 0\,0\,0\,2 \\ 0\,2\,0\,0 \\ 0\,0\,2\,0 \\ 1\,0\,0\,1 \end{bmatrix}, \quad F_{\gamma s} = \begin{bmatrix} 2\,0\,0\,1 \\ 0\,0\,0\,0 \\ 0\,0\,0\,0 \\ 2\,0\,0\,1 \end{bmatrix}$$

Step p+3:
$h_s = [0,2,1,2]$.

$$H_{\underline{a}}(3,0,0;0,3,3) = \begin{bmatrix} 1\,2\,2 \\ 2\,2\,2 \\ 2\,2\,2 \\ 0\,0\,0 \\ 2\,2\,2 \\ 2\,2\,2 \\ 0\,0\,0 \\ 0\,0\,0 \\ 0\,0\,0 \\ 1\,1\,1 \\ 2\,2\,2 \\ 2\,2\,2 \\ 0\,0\,0 \\ 0\,0\,0 \\ 0\,0\,0 \\ 1\,1\,1 \\ 1\,1\,1 \\ 0\,0\,0 \\ 3\,3\,3 \\ 3\,3\,3 \end{bmatrix}, \qquad H_{\underline{a}}(3,1,0;0,2,3) = \begin{bmatrix} 1\,2\,2\,2 \\ 2\,2\,2\,2 \\ 2\,2\,0\,0 \\ 0\,0\,0\,0 \\ 2\,2\,2\,2 \\ 2\,2\,0\,0 \\ 0\,0\,0\,0 \\ 0\,0\,1\,1 \\ 0\,0\,0\,0 \\ 1\,1\,3\,3 \\ 2\,2\,2\,2 \\ 2\,2\,0\,0 \\ 0\,0\,0\,0 \\ 0\,0\,1\,1 \\ 0\,0\,0\,0 \\ 1\,1\,3\,3 \\ 1\,1\,2\,2 \\ 0\,0\,0\,0 \\ 3\,3\,4\,4 \\ 3\,3\,4\,4 \end{bmatrix}$$

$$H_{\underline{a}}(3,2,0;0,1,3) = \begin{bmatrix} 1\,2\,2\,0 \\ 2\,2\,2\,0 \\ 2\,2\,0\,1 \\ 0\,0\,0\,0 \\ 2\,2\,2\,0 \\ 2\,2\,0\,1 \\ 0\,0\,0\,0 \\ 0\,0\,1\,2 \\ 0\,0\,0\,0 \\ 1\,1\,3\,4 \\ 2\,2\,2\,0 \\ 2\,2\,0\,1 \\ 0\,0\,0\,0 \\ 0\,0\,1\,2 \\ 0\,0\,0\,0 \\ 1\,1\,3\,4 \\ 1\,1\,2\,2 \\ 0\,0\,0\,0 \\ 3\,3\,4\,2 \\ 3\,3\,4\,2 \end{bmatrix}, \quad H_{\underline{a}}(3,1,1;0,2,2) = \begin{bmatrix} 1\,2\,2\,0\,0 \\ 2\,2\,2\,0\,0 \\ 2\,2\,0\,0\,0 \\ 0\,0\,0\,1\,1 \\ 2\,2\,2\,0\,0 \\ 2\,2\,0\,0\,0 \\ 0\,0\,0\,1\,1 \\ 0\,0\,1\,0\,0 \\ 0\,0\,0\,3\,3 \\ 1\,1\,3\,3\,3 \\ 2\,2\,2\,0\,0 \\ 2\,2\,0\,0\,0 \\ 0\,0\,0\,1\,1 \\ 0\,0\,1\,0\,0 \\ 0\,0\,0\,3\,3 \\ 1\,1\,3\,3\,3 \\ 1\,1\,2\,0\,0 \\ 0\,0\,0\,4\,4 \\ 3\,3\,4\,4\,4 \\ 3\,3\,4\,0\,0 \end{bmatrix}$$

$$H_{\underline{a}}(3,2,1;0,1,2) = \begin{bmatrix} 1\,2\,2\,0\,0 \\ 2\,2\,2\,0\,0 \\ 2\,2\,0\,0\,0 \\ 0\,0\,0\,1\,3 \\ 2\,2\,2\,0\,0 \\ 2\,2\,0\,0\,0 \\ 0\,0\,0\,1\,3 \\ 0\,0\,1\,0\,0 \\ 0\,0\,0\,3\,4 \\ 1\,1\,3\,3\,4 \\ 2\,2\,2\,0\,0 \\ 2\,2\,0\,0\,0 \\ 0\,0\,0\,1\,3 \\ 0\,0\,1\,0\,0 \\ 0\,0\,0\,3\,4 \\ 1\,1\,3\,3\,4 \\ 1\,1\,2\,0\,0 \\ 0\,0\,0\,4\,2 \\ 3\,3\,4\,4\,2 \\ 3\,3\,4\,0\,0 \end{bmatrix}$$

6.5 Historical Notes and Concluding Remarks

A new model for a three-dimensional infinite image is proposed, and the following results concerning the realization theory are established. The 3-Commutative Linear Representation System is regarded as a new model which describes any three-dimensional image. There exist at least two canonical 3-Commutative Linear Representation Systems which realize arbitrary three-dimensional image. Moreover, any two canonical 3-Commutative Linear Representaion Systems with the same behavior are isomorphic.

There have been various methods for treating three-dimensional images such as shape analysis or techniques whereby computer graphics display three-dimensional objects as realistic images on a two-dimensional screen [Nevatia, 1982; Serra, 1982]. There is a method of using oct-trees to encode 3-D objects [Jackins and Tanimoto, 1980]. However, the method presented in this chapter is quite different from them.

Our method depends on how characteristic rules are extracted preserving the connections of each pixel in all three directions for three-dimensional image. This realization theory provides a fundamental coding theory of three-dimensional images. Moreover, our method is a powerful and suitable tool for performing computer tasks such as describing geometrical images. Some examples are shown in Section 6.3.

According to the existence part of Theorem (6.3), the quotient canonical system Σ_q and the sub-canonical system Σ_s appearing in Corollary (6-A.20) of Appendix 6-A are new. The uniqueness part of Theorem (6.3) is also a new development in the sense of 3-Commutative Linear Representation Systems. Finite-dimensional 3-Commutative Linear Representation Systems, which can be implemented as computer programs or non-linear circuits, are investigated based on the proposed realization theory.

Several concrete examples provide to demonstrate the suitability of 3-Commutative Linear Representation Systems as the model to deal with geometrical images.

The main results of our investigation come in Section 6.3. First, the necessary and sufficient conditions for finite-dimensional 3-Commutative Linear Representation Systems to be canonical is derived. Second, the representation theorem for the equivalence classes of the finite-dimensional canonical 3-Commutative Linear Representation Systems is given. Third, we also provide the two necessary and sufficient conditions for an image to be regarded as the behavior of a finite-dimensional 3-Commutative Linear Representation System. One is the rank condition of the Hankel matrix. The other is the condition for rational function. These two conditions are new developments. Finally, the realization procedure is described to get the finite-dimensional canonical 3-Commutative Linear Representation System from the given image.

Section 6.4 deals with finite-sized three-dimensional images based on the ideal results discussed in the previous section. The notion of natural partial realization is introduced and it is seen that its existence is equivalent to the uniqueness of the minimal partial realization. Finally, the realization algorithm is presented by which to determine the finite-dimensional canonical 3-Commutative Linear Representation System from a given arbitrary finite-sized three-dimensional image.

In the field of 3-dimensional filtering, theories have been developed only in the area of rational function in three variables. See, for example, Kawakami [1991] and Zang and Burton [1994].

For the same reasons as in Section 3.5, we judge the new developments in view of approximations for three-dimensional images which are presented in this chapter are highly significant.

Appendix to Chapter 6

6-A Proof of the Realization Theory of Three-Dimensional Images

This appendix 6-A presents the proof of the realization Theorem (6.3) for three-dimensional images. In order to prove Theorem (6.3), we equivalently convert a 3-Commutative Linear Representation System into a sophisticated 3-Commutative Linear Representation System. Proof of the realization theorem for the sophisticated 3-Commutative Linear Representation System (6-A.19) implies proof of Theorem (6.3).

6-A.1 Linear State Structure: $\{\alpha, \beta, \gamma\}$-Actions

Definition 6-A.1. A collection represented by the following equations is written as $(X, F_\alpha, F_\beta, F_\gamma)$. It is called a $\{\alpha, \beta, \gamma\}$-action if X is a linear space over a field K and $F_\alpha, F_\beta, F_\gamma \in L(X)$ satisfy $F_\alpha F_\beta = F_\beta F_\alpha, F_\beta F_\gamma = F_\gamma F_\beta$ and $F_\gamma F_\alpha = F_\alpha F_\gamma$.

$$\begin{cases} x(i+1, j, k) = F_\alpha x(i, j, k) \\ x(i, j+1, k) = F_\beta x(i, j, k) \\ x(i, j, k+1) = F_\gamma x(i, j, k) \end{cases}$$

Let $(X_1, F_{\alpha 1}, F_{\beta 1}, F_{\gamma 1})$ and $(X_2, F_{\alpha 2}, F_{\beta 2}, F_{\gamma 2})$ be $\{\alpha, \beta, \gamma\}$-actions. Then the linear map $T : X_1 \to X_2$ is called a $\{\alpha, \beta, \gamma\}$-morphism $T : (X_1, F_{\alpha 1}, F_{\beta 1}, F_{\gamma 1}) \to (X_2, F_{\alpha 2}, F_{\beta 2}, F_{\gamma 2})$ if T satisfies $TF_{\alpha 1} = F_{\alpha 2}T$, $TF_{\beta 1} = F_{\beta 2}T$ and $TF_{\gamma 1} = F_{\gamma 2}T$.

Example 6-A.2. Let $K[z_\alpha, z_\beta, z_\gamma]$ be the set of any polynomials in three variables, z_α, z_β and z_γ. Let z_α,z_β and z_γ also be:

$z_\alpha : K[z_\alpha, z_\beta, z_\gamma] \to K[z_\alpha, z_\beta, z_\gamma]; \lambda \mapsto z_\alpha \lambda$,
$z_\beta : K[z_\alpha, z_\beta, z_\gamma] \to K[z_\alpha, z_\beta, z_\gamma]; \lambda \mapsto z_\beta \lambda$,
$z_\gamma : K[z_\alpha, z_\beta, z_\gamma] \to K[z_\alpha, z_\beta, z_\gamma]; \lambda \mapsto z_\gamma \lambda$.
Then $(K[z_\alpha, z_\beta, z_\gamma], z_\alpha, z_\beta, z_\gamma)$ is a $\{\alpha, \beta, \gamma\}$-action.

Example 6-A.3. In the set $F(N^3, Y)$ of any three-dimensional image, we define
$S_\alpha : F(N^3, Y) \to F(N^3, Y); a(i,j,k) \mapsto a(i+1,j,k)$,
$S_\beta : F(N^3, Y) \to F(N^3, Y); a(i,j,k) \mapsto a(i,j+1,k)$ and
$S_\gamma : F(N^3, Y) \to F(N^3, Y); a(i,j,k) \mapsto a(i,j,k+1)$.
Then $(F(N^3, Y), S_\alpha, S_\beta, S_\gamma)$ is a $\{\alpha, \beta, \gamma\}$-action.

Definition 6-A.4. For $\{\alpha, \beta, \gamma\}$-actions $(K[z_\alpha, z_\beta, z_\gamma], z_\alpha, z_\beta, z_\gamma)$ and $(F(N^3, Y), S_\alpha, S_\beta, S_\gamma)$ considered in Example (6-A.2) and (6-A.3), a $\{\alpha, \beta, \gamma\}$-morphism $A : (K[z_\alpha, z_\beta, z_\gamma], z_\alpha, z_\beta, z_\gamma) \to (F(N^3, Y), S_\alpha, S_\beta, S_\gamma)$ is called a linear input/output map. For a $\{\alpha, \beta, \gamma\}$-action $(X, F_\alpha, F_\beta, F_\gamma)$, a $\{\alpha, \beta, \gamma\}$-morphism $G : (K[z_\alpha, z_\beta, z_\gamma], z_\alpha, z_\beta, z_\gamma) \to (X, F_\alpha, F_\beta, F_\gamma)$ is called a linear input map, and a $\{\alpha, \beta, \gamma\}$-morphism $H : (X, F_\alpha, F_\beta, F_\gamma) \to (F(N^3, Y), S_\alpha, S_\beta, S_\gamma)$ is called a linear observation map.

Remark. A linear input/output map
$A : (K[z_\alpha, z_\beta, z_\gamma], z_\alpha, z_\beta, z_\gamma) \to (F(N^3, Y), S_\alpha, S_\beta, S_\gamma)$ is different from the map discussed in Perlman [1980].

Proposition 6-A.5. Let $(X, F_\alpha, F_\beta, F_\gamma)$ be a $\{\alpha, \beta, \gamma\}$-action. Then $G : (K[z_\alpha, z_\beta, z_\gamma], z_\alpha, z_\beta, z_\gamma) \to (X, F_\alpha, F_\beta, F_\gamma)$ is a linear input map if and only if $G(\lambda) = \sum_k \sum_j \sum_i \lambda(i,j,k) F_\alpha^i F_\beta^j F_\gamma^k G(\mathbf{1})$ for $\lambda = \sum_k \sum_j \sum_i \lambda(i,j,k) z_\alpha^i z_\beta^j z_\gamma^k \in K[z_\alpha, z_\beta, z_\gamma]$, where $\lambda(i,j,k) \in K$ and $\mathbf{1} \in K[z_\alpha, z_\beta, z_\gamma]$ is the unit element of multiplication.

Proof. First, we prove sufficiency. For any $\lambda = \sum_k \sum_j \sum_i \lambda(i,j,k) z_\alpha^i z_\beta^j z_\gamma^k \in K[z_\alpha, z_\beta, z_\gamma]$, $G(z_\alpha \lambda) = \sum_k \sum_j \sum_i \lambda F_\alpha^{i+1} F_\beta^j F_\gamma^k G(\mathbf{1}) = F_\alpha(\sum_k \sum_j \sum_i \lambda F_\alpha^i F_\beta^j F_\gamma^k G(\mathbf{1})) = F_\alpha G(\lambda)$.

Hence, $G z_\alpha = F_\alpha G$ holds. Similarly, $G z_\beta = F_\beta G$, $G z_\gamma = F_\gamma G$ hold. Therefore, G is a linear input map $G : (K[z_\alpha, z_\beta, z_\gamma], z_\alpha, z_\beta, z_\gamma) \to (X, F_\alpha, F_\beta, F_\gamma)$.

Next, we prove necessity. If G is a linear input map, then $G(z_\alpha^i z_\beta^j z_\gamma^k) = F_\alpha^i F_\beta^j F_\gamma^k G(\mathbf{1})$. Since the set $\{z_\alpha^i z_\beta^j z_\gamma^k; i,j,k \in N\}$ is basis in $K[z_\alpha, z_\beta, z_\gamma]$ [Bourbaki, 1974] , G can be extended uniquely to $K[z_\alpha, z_\beta, z_\gamma]$. This implies that G satisfies the condition.

We can introduce subspaces, quotient spaces and product spaces for $\{\alpha, \beta, \gamma\}$-actions in the usual sense, that is, in the same way these spaces are treated for linear space.

[Sub-$\{\alpha, \beta, \gamma\}$-actions]
Let $(X, F_\alpha, F_\beta, F_\gamma)$ be a $\{\alpha, \beta, \gamma\}$-action and $Z \subseteq X$ be an invariant sub-

space under F_α, F_β and F_γ. Then $(Z, F_\alpha, F_\beta, F_\gamma)$ is a sub-$\{\alpha, \beta, \gamma\}$-action of $(X, F_\alpha, F_\beta, F_\gamma)$.

[Quotient $\{\alpha, \beta, \gamma\}$-actions]
Let $(X, F_\alpha, F_\beta, F_\gamma)$ be a $\{\alpha, \beta, \gamma\}$-action and $Z \subseteq X$ be an invariant subspace under F_α, F_β and F_γ . Then $(X/Z, \dot{F}_\alpha, \dot{F}_\beta, \dot{F}_\gamma)$ is a quotient $\{\alpha, \beta, \gamma\}$-action, where $\dot{F}_\alpha : X/Z \to X/Z; [x] \mapsto [F_\alpha x]$, $\dot{F}_\beta : X/Z \to X/Z; [x] \mapsto [F_\beta x]$, and $\dot{F}_\gamma : X/Z \to X/Z; [x] \mapsto [F_\gamma x]$.

Corollary 6-A.6. Any $\{\alpha, \beta, \gamma\}$-morphism $T : (X_1, F_{\alpha 1}, F_{\beta 1}, F_{\gamma 1}) \to (X_2, F_{\alpha 2}, F_{\beta 2}, F_{\gamma 2})$ can be normally decomposed into $X_1 \xrightarrow{\pi} X_1/\ker T \xrightarrow{T^b} \operatorname{im} T \xrightarrow{j} X_2$, where π is the canonical surjection, T^b is the isomorphism associated with T and j is the canonical injection. Furthermore, π, T^b and j are all $\{\alpha, \beta, \gamma\}$-morphisms.

[Product $\{\alpha, \beta, \gamma\}$-actions]
Let $(X_1, F_{\alpha 1}, F_{\beta 1}, F_{\gamma 1})$ and $(X_2, F_{\alpha 2}, F_{\beta 2}, F_{\gamma 2})$ be $\{\alpha, \beta, \gamma\}$-actions. Define $F_{\alpha 1} \times F_{\alpha 2} : X_1 \times X_2 \to X_1 \times X_2; (x_1, x_2) \to (F_{\alpha 1} x_1, F_{\alpha 2} x_2)$ for the product space $X_1 \times X_2$. Similarly, define $F_{\beta 1} \times F_{\beta 2}$ and $F_{\gamma 1} \times F_{\gamma 2}$. Then $(X_1 \times X_2, F_{\alpha 1} \times F_{\alpha 2}, F_{\beta 1} \times F_{\beta 2}, F_{\gamma 1} \times F_{\gamma 2})$ is a product $\{\alpha, \beta, \gamma\}$-action.

6-A.2 Pointed $\{\alpha, \beta, \gamma\}$-Actions

In this section, we introduce pointed $\{\alpha, \beta, \gamma\}$-actions and discuss their reachability.

Definition 6-A.7. For a $\{\alpha, \beta, \gamma\}$-action $(X, F_\alpha, F_\beta, F_\gamma)$ and an initial state $x^0 \in X$, the collection $((X, F_\alpha, F_\beta, F_\gamma), x^0)$ is called a pointed $\{\alpha, \beta, \gamma\}$-action.

For the reachable set $R(x^0) = \{F_\alpha^i F_\beta^j F_\gamma^k x^0;\ i, j, k \in N\}$, if the linear hull $\ll R(x^0) \gg$ of $R(x^0)$ equals X, then $((X, F_\alpha, F_\beta, F_\gamma), x^0)$ is called quasi-reachable.

A pointed $\{\alpha, \beta, \gamma\}$-action$((X, F_\alpha, F_\beta, F_\gamma), x^0)$ is written by:

$$\begin{cases} x(i+1, j, k) = F_\alpha x(i, j, k) \\ x(i, j+1, k) = F_\beta x(i, j, k) \\ x(i, j, k+1) = F_\gamma x(i, j, k) \\ x(0, 0, 0) \quad = x^0 \end{cases}$$

For pointed $\{\alpha, \beta, \gamma\}$-actions $((X_1, F_{\alpha 1}, F_{\beta 1}, F_{\gamma 1}), x_1^0)$ and $((X_2, F_{\alpha 2}, F_{\beta 2}, F_{\gamma 2}), x_2^0)$, a $\{\alpha, \beta, \gamma\}$-morphism $T : (X_1, F_{\alpha 1}, F_{\beta 1}, F_{\gamma 1}) \to (X_2, F_{\alpha 2}, F_{\beta 2}, F_{\gamma 2})$ which satisfies $Tx_1^0 = x_2^0$ is called a pointed $\{\alpha, \beta, \gamma\}$-morphism $T : ((X_1, F_{\alpha 1}, F_{\beta 1}, F_{\gamma 1}), x_1^0) \to ((X_2, F_{\alpha 2}, F_{\beta 2}, F_{\gamma 2}), x_2^0)$.

Example 6-A.8. Let the $\{\alpha, \beta, \gamma\}$-action $(K[z_\alpha, z_\beta, z_\gamma], z_\alpha, z_\beta, z_\gamma)$ be the same as Example (6-A.2), and let $\mathbf{1}$ be the unit element of multiplication.

Then $((K[z_\alpha, z_\beta, z_\gamma], z_\alpha, z_\beta, z_\gamma), \mathbf{1})$ is a pointed $\{\alpha, \beta, \gamma\}$-action and quasi-reachable.

Let the $\{\alpha, \beta, \gamma\}$-action $(F(N^3, Y), S_\alpha, S_\beta, S_\gamma)$ be the same as Example (6-A.3), and let an image be $a \in F(N^3, Y)$.

Then $((F(N^3, Y), S_\alpha, S_\beta, S_\gamma), a)$ is a pointed $\{\alpha, \beta, \gamma\}$-action.

Proposition 6-A.9. There exists a unique pointed $\{\alpha, \beta, \gamma\}$-morphism $G : ((K[z_\alpha, z_\beta, z_\gamma], z_\alpha, z_\beta, z_\gamma), \mathbf{1}) \to ((X, F_\alpha, F_\beta, F_\gamma), x^0)$ for any pointed $\{\alpha, \beta, \gamma\}$-action $((X, F_\alpha, F_\beta, F_\gamma), x^0)$.

Proof. Set $G(\mathbf{1}) = x^0$ in Proposition(6-A.5). The fact that $\{z_\alpha^i z_\beta^j z_\gamma^k;\ i, j, k \in N\}$ is the basis in $K[z_\alpha, z_\beta, z_\gamma]$ results in G being unique.

Remark 1: According to Proposition (6-A.5) and (6-A.9), the linear input map $G : (K[z_\alpha, z_\beta, z_\gamma], z_\alpha, z_\beta, z_\gamma) \to (X, F_\alpha, F_\beta, F_\gamma)$ corresponds to the initial state $x^0 \in X$ uniquely and this correspondence is isomorphic.

Remark 2: If a pointed $\{\alpha, \beta, \gamma\}$-action $((X, F_\alpha, F_\beta, F_\gamma), x^0)$ in Proposition (6-A.9) is replaced with the $((F(N^3, Y), S_\alpha, S_\beta, S_\gamma), a)$ considered in Example (6-A.8), then a linear input/output map $A : (K[z_\alpha, z_\beta, z_\gamma], z_\alpha, z_\beta, z_\gamma) \to (F(N^3, Y), S_\alpha, S_\beta, S_\gamma)$ corresponds to an image $a \in F(N^3, Y)$ uniquely, and this correspondence is isomorphic. The definition of quasi-reachability and the formula for linear input maps presented in Proposition (6-A.5) lead to the following proposition:

Proposition 6-A.10. A pointed $\{\alpha, \beta, \gamma\}$-action $((X, F_\alpha, F_\beta, F_\gamma), x^0)$ is quasi-reachable if and only if the corresponding linear input map G is surjective.

6-A.3 $\{\alpha, \beta, \gamma\}$-Actions with a Readout Map

This section deals with $\{\alpha, \beta, \gamma\}$-actions with a readout map and its distinguishability.

Definition 6-A.11. For a $\{\alpha, \beta, \gamma\}$-action $(X, F_\alpha, F_\beta, F_\gamma)$ and a linear map $h : X \to Y$, a collection $((X, F_\alpha, F_\beta, F_\gamma), h)$ is called a $\{\alpha, \beta, \gamma\}$-action with a readout map. For any $i, j, k \in N$, if $hF_\alpha^i F_\beta^j F_\gamma^k x_1 = hF_\alpha^i F_\beta^j F_\gamma^k x_2$ implies $x_1 = x_2$, then $((X, F_\alpha, F_\beta, F_\gamma), h)$ is said to be distinguishable.

A $\{\alpha, \beta, \gamma\}$-action with a readout map $((X, F_\alpha, F_\beta, F_\gamma), h)$ is represented as:

$$\begin{cases} x(i+1, j, k) = F_\alpha x(i, j, k) \\ x(i, j+1, k) = F_\beta x(i, j, k) \\ x(i, j, k+1) = F_\gamma x(i, j, k) \\ \gamma(i, j, k) \quad\ = hx(i, j, k) \end{cases}$$

Let $((X_1, F_{\alpha 1}, F_{\beta 1}, F_{\gamma 1}), h_1)$ and $((X_2, F_{\alpha 2}, F_{\beta 2}, F_{\gamma 2}), h_2)$ be $\{\alpha, \beta, \gamma\}$-actions with a readout map. Then the $\{\alpha, \beta, \gamma\}$-morphism $T : (X_1, F_{\alpha 1}, F_{\beta 1}, F_{\gamma 1}) \to (X_2, F_{\alpha 2}, F_{\beta 2}, F_{\gamma 2})$ which satisfies $h_1 = h_2 T$ is called $\{\alpha, \beta, \gamma\}$-morphism with a readout map $T : ((X_1, F_{\alpha 1}, F_{\beta 1}, F_{\gamma 1}), h_1) \to ((X_2, F_{\alpha 2}, F_{\beta 2}, F_{\gamma 2}), h_2)$.

Lemma 6-A.12. For any linear map $\bar{a} : K[z_\alpha, z_\beta, z_\gamma] \to Y$, there exists a unique image $a \in F(N^3, Y)$ such that

$$\sum_k \sum_j \sum_i \lambda(i,j,k) a(i,j,k) = \bar{a}(\sum_k \sum_j \sum_i \lambda(i,j,k) z_\alpha^i z_\beta^j z_\gamma^k).$$

Proof. This lemma can be obtained easily.

Example 6-A.13. For the $\{\alpha, \beta, \gamma\}$-action $(K[z_\alpha, z_\beta, z_\gamma], z_\alpha, z_\beta, z_\gamma)$ and any image $a \in F(N^3, Y)$, $((K[z_\alpha, z_\beta, z_\gamma], z_\alpha, z_\beta, z_\gamma), a)$ is a $\{\alpha, \beta, \gamma\}$-actions with a readout map. See Lemma (6-A.12).

Define a linear map $0 : F(N^3, Y) \to Y; a \mapsto a(0,0,0)$. Then $((F(N^3, Y), S_\alpha, S_\beta, S_\gamma), 0)$ is a $\{\alpha, \beta, \gamma\}$-action with a readout map and it is distinguishable.

Proposition 6-A.14. For any $\{\alpha, \beta, \gamma\}$-action with a readout map $((X, F_\alpha, F_\beta, F_\gamma), h)$, there exists a unique linear observation map $H : (X, F_\alpha, F_\beta, F_\gamma) \to (F(N^3, Y), S_\alpha, S_\beta, S_\gamma)$ which satisfies $h = 0H$, where $Hx(i,j,k) = 0(S_\alpha^i S_\beta^j S_\gamma^k)(Hx) = 0 \cdot H(F_\alpha^i F_\beta^j F_\gamma^k)x = hF_\alpha^i F_\beta^j F_\gamma^k x$ holds for $x \in X,\ i,j,k \in N$.

Proof. Let $((X, F_\alpha, F_\beta, F_\gamma), h)$ be any $\{\alpha, \beta, \gamma\}$-action with a readout map. Defining $(Hx)(i,j,k) := hF_\alpha^i F_\beta^j F_\gamma^k x$ for any $x \in X, i,j,k \in N$ results in a linear observation map $H : (X, F_\alpha, F_\beta, F_\gamma) \to (F(N^3, Y),\ S_\alpha, S_\beta, S_\gamma)$ which satisfies $h = 0H$.

Next, we will prove uniqueness. Let H be a linear observation map $H : (X, F_\alpha, F_\beta, F_\gamma) \to (F(N^3, Y), S_\alpha, S_\beta, S_\gamma)$ which satisfies $h = 0H$. Then $(Hx)(i,j,k) = (S_\alpha^i S_\beta^j S_\gamma^k Hx)(0,0,0) = 0(S_\alpha^i S_\beta^j S_\gamma^k Hx) = 0(HF_\alpha^i F_\beta^j F_\gamma^k x) = hF_\alpha^i F_\beta^j F_\gamma^k x$ holds for any $x \in X,\ i,j,k \in N$. Hence, H is unique.

Remark 1: According to Proposition (6-A.14), the linear observation map $H : (X, F_\alpha, F_\beta, F_\gamma) \to (F(N^3, Y), S_\alpha, S_\beta, S_\gamma)$ corresponds to a linear map $h : X \to Y$ uniquely and this correspondence is isomorphic.

Remark 2: If $((X, F_\alpha, F_\beta, F_\gamma), h)$ in Proposition (6-A.14) is replaced with the $((F(N^3, Y), S_\alpha, S_\beta, S_\gamma), 0)$ considered in (6-A.13), the linear observation map $H : (K[z_\alpha, z_\beta, z_\gamma], z_\alpha, z_\beta, z_\gamma) \to (F(N^3, Y), S_\alpha, S_\beta, S_\gamma)$ is a linear input/output map.

The definition of distinguishability and Proposition (6-A.14) lead to the following proposition:

Proposition 6-A.15. A $\{\alpha, \beta, \gamma\}$-action with a readout map $((X, F_\alpha, F_\beta, F_\gamma), h)$ is distinguishable if and only if the corresponding linear observation map $H : (X, F_\alpha, F_\beta, F_\gamma) \to (F(N^3, Y), S_\alpha, S_\beta, S_\gamma)$ is injective.

6-A.4 3-Commutative Linear Representation System

In this section, the sophisticated 3-Commutative Linear Representation System is defined. It is shown that the sophisticated 3-Commutative Linear Representation System and the naive one, which is defined in Section 6.3, are the same.

Definition 6-A.16. A collection $\Sigma = ((X, F_\alpha, F_\beta, F_\gamma), G, H)$ is called a sophisticated 3-Commutative Linear Representation System if G is a linear input map of the form $G : (K[z_\alpha, z_\beta, z_\gamma], z_\alpha, z_\beta, z_\gamma) \to (X, F_\alpha, F_\beta, F_\gamma)$ and H is a linear observation map $H : (X, F_\alpha, F_\beta, F_\gamma) \to (F(N^3, Y), S_\alpha, S_\beta, S_\gamma)$.

The linear input/output map $A_\Sigma : (K[z_\alpha, z_\beta, z_\gamma], z_\alpha, z_\beta, z_\gamma) \to (F(N^3, Y), S_\alpha, S_\beta, S_\gamma)$ is called the behavior of Σ. For a linear input/output map A, if $A_\Sigma = A$, then the 3-Commutative Linear Representation System Σ is said to be a realization system of A.

A sophisticated 3-Commutative Linear Representation System $\Sigma = ((X, F_\alpha, F_\beta, F_\gamma), G, H)$ is called canonical if G is surjective and H is injective.

For $\Sigma_1 = ((X_1, F_{\alpha 1}, F_{\beta 1}, F_{\gamma 1}), G_1, H_1)$ and $\Sigma_2 = ((X_2, F_{\alpha 2}, F_{\beta 2}, F_{\gamma 2}), G_2, H_2)$, a $\{\alpha, \beta, \gamma\}$-morphism $T : (X_1, F_{\alpha 1}, F_{\beta 1}, F_{\gamma 1}) \to (X_2, F_{\alpha 2}, F_{\beta 2}, F_{\gamma 2})$ is called a sophisticated Linear Representation System morphism $T : \Sigma_1 \to \Sigma_2$. If T is bijective, then $T : \Sigma_1 \to \Sigma_2$ is called an isomorphism.

Example 6-A.17. Let the $\{\alpha, \beta, \gamma\}$-action $(K[z_\alpha, z_\beta, z_\gamma], z_\alpha, z_\beta, z_\gamma)$ be the same as the one in Example (6-A.2). Let the identity map on $K[z_\alpha, z_\beta, z_\gamma]$ denote I, and let a linear input/output map be $A : (K[z_\alpha, z_\beta, z_\gamma], z_\alpha, z_\beta, z_\gamma) \to (F(N^3, Y), S_\alpha, S_\beta,$
$S_\gamma)$. Then the collection $((K[z_\alpha, z_\beta, z_\gamma], z_\alpha, z_\beta, z_\gamma), I, A)$ is a sophisticated 3-Commutative Linear Representation System with the behavior A.

Let the $\{\alpha, \beta, \gamma\}$-action $(F(N^3, Y), S_\alpha, S_\beta, S_\gamma)$ be the same as in Example (6-A.3). Let a linear input/output map and identity map on $F(N^3, Y)$ denote A and I, respectively. Then the collection $((F(N^3, Y), S_\alpha, S_\beta, S_\gamma), A, I)$ is a sophisticated 3-Commutative Linear Representation System with the behavior A.

By virtue of Examples (6-A.9), (6-A.14) and (6-A.17), the relation between the sophisticated 3-Commutative Linear Representation System and the naive one will be investigated.

Proposition 6-A.18. Let $\Sigma = ((X, F_\alpha, F_\beta, F_\gamma), G, H)$ be any sophisticated 3-Commutative Linear Representation System. Then there exists a unique naive 3-Commutative Linear Representation System $\sigma = ((X, F_\alpha, F_\beta, F_\gamma),$

$x^0, h)$ which corresponds to the sophisticated one Σ and which satisfies the following equations:

$$\sum_k \sum_j \sum_i \lambda(i,j,k) F_\alpha^i F_\beta^j F_\gamma^k x^0 = G(\sum_k \sum_j \sum_i \lambda(i,j,k) z_\alpha^i z_\beta^j z_\gamma^k) \tag{a.1}$$

$$h F_\alpha^i F_\beta^j F_\gamma^k x = (Hx)(i,j,k) \text{for} x \in X, i,j,k \in N \tag{a.2}$$

This correspondence is isomorphic in the sense of category [Paregis, 1970].

Proof. Remark 1 of Proposition (6-A.9) and Remark 1 of Proposition (6-A.14) lead to this proposition.

6-A.5 Sophisticated 3-Commutative Linear Representation System

The Realization theorem (6.3) is proved in this section. According to Propositions (6-A.10), (6-A.15), (6-A.18) and Remark 2 of Proposition (6-A.9) or Remark 2 of Proposition (6-A.14), the realization theorem (6.3) is replaced with Theorem (6-A.19). Hence, the proof of Theorem (6-A.19) implies the proof of the Realization theorem (6.3).

Theorem 6-A.19. Sophisticated Realization Theorem of images.

(1) Existence: For any linear input/output map $A : (K[z_\alpha, z_\beta, z_\gamma], z_\alpha, z_\beta, z_\gamma) \to (F(N^3, Y), S_\alpha, S_\beta, S_\gamma)$, there exist at least two sophisticated canonical 3-Commutative Linear Representation Systems which realize A.
(2) Uniqueness: Let $\Sigma_1 = ((X_1, F_{\alpha 1}, F_{\beta 1}, F_{\gamma 1}), G_1, H_1)$ and $\Sigma_2 = ((X_2, F_{\alpha 2}, F_{\beta 2}, F_{\gamma 2}), G_2, H_2)$ be both sophisticated canonical 3-Commutative Linear Representation System which have the same behavior. Then there exists an isomorphism $T : \Sigma_1 \to \Sigma_2$.

Proof. Existence is proved by the Corollary (6-A.20) shown just below, while the Remark of the Lemma (6-A.24) leads to the uniqueness.

Corollary 6-A.20. For any linear input/output map $A : (K[z_\alpha, z_\beta, z_\gamma], z_\alpha, z_\beta, z_\gamma) \to (F(N^3, Y), S_\alpha, S_\beta, S_\gamma)$, the following sophisticated 3-Commutative Linear Representation Systems (1) and (2) are both canonical realization systems of A.

(1) $\Sigma_q = ((K[z_\alpha, z_\beta, z_\gamma]/\ker A, \dot{z_\alpha}, \dot{z_\beta}, \dot{z_\gamma}), \pi, A^i)$, where $\dot{z_\alpha} : K[z_\alpha, z_\beta, z_\gamma]/\ker A \to K[z_\alpha, z_\beta, z_\gamma]/\ker A; [\lambda] \mapsto [z_\alpha \lambda]$, $\dot{z_\beta} : K[z_\alpha, z_\beta, z_\gamma]/\ker A \to K[z_\alpha, z_\beta, z_\gamma]/\ker A; [\lambda] \mapsto [z_\beta \lambda]$ and $\dot{z_\gamma} : K[z_\alpha, z_\beta, z_\gamma]/\ker A \to K[z_\alpha,$

$z_\beta, z_\gamma]/\ker A$; $[\lambda] \mapsto [z_\gamma \lambda]$. π is the canonical surjection $\pi : K[z_\alpha, z_\beta, z_\gamma] \to K[z_\alpha, z_\beta, z_\gamma]/\ker A$. A^i is given by $A^i = jA^b$ for $A^b : K[z_\alpha, z_\beta, z_\gamma]/\ker A \to \mathrm{im}\ A$ being isomorphic with A. j is the canonical injection $j : \mathrm{im}\ A \to F(N^3, Y)$.

(2) $\Sigma_s = ((\mathrm{im}\ A, S_\alpha, S_\beta, S_\gamma), A^s, j)$, where $A^s = A^b\pi$.

Σ_q and Σ_s are called the quotient canonical realization and the subspace canonical realization, respectively.

Proof. Example (6-A.17), Corollary (6-A.6) and the definitions of canonicality and behavior lead to this corollary.

Next, in order to prove the uniqueness part of Theorem (6-A.19), the following morphism from $\Sigma_1 = ((X_1, F_{\alpha 1}, F_{\beta 1}, F_{\gamma 1}), G_1, H_1)$ to $\Sigma_2 = ((X_2, F_{\alpha 2}, F_{\beta 2}, F_{\gamma 2}), G_2, H_2)$ is defined the same way as in [Matsuo and Hasegawa, 2003], where Σ_1 and Σ_2 are both sophisticated 3-Commutative Linear Representation Systems. Mor $(\Sigma_1, \Sigma_2) := \{$a relation $T_{12} : X_1 \to X_2; GrT_{12}^{min} \subseteq GrT_{12} \subseteq GrT_{12}^{max}\}$, where GrT_{12}^{min} , GrT_{12} and GrT_{12}^{max} denote the graph of $T_{12}^{min} := G_2 G_1^{-1}$, T_{12} and $T_{12}^{max} := H_2^{-1} H_1$, respectively. The reason for introducing this morphism depends on the following lemma:

Lemma 6-A.21. $A_{\Sigma_1} = A_{\Sigma_2}$ if and only if Mor $(\Sigma_1, \Sigma_2) \neq \phi$.

Proof. This Lemma is proved the same way as in Matsuo and Hasegawa [2003].

Lemma 6-A.22. Let $A_{\Sigma_1} = A_{\Sigma_2}$ hold.

(1) If G_1 of Σ_1 is surjective, then dom $T_{12}^{min} = X_1$ holds, where dom T_{12}^{min} denotes the domain of T_{12}^{min}.
(2) If H_2 of Σ_2 is injective, then T_{12}^{max} is a partial function : $X_1 \to X_2$.

Proof. This Lemma is proved the same way as in Matsuo and Hasegawa [2003].

Lemma 6-A.23. Let $A_{\Sigma_1} = A_{\Sigma_2}$ hold. Then GrT_{12}^{max} is an invariant subproduct $\{\alpha, \beta, \gamma\}$-action of $(X_1, F_{\alpha 1}, F_{\beta 1}, F_{\gamma 1})$ and $(X_2, F_{\alpha 2}, F_{\beta 2}, F_{\gamma 2})$.

Proof. $GrT_{12}^{max} \in X_1 \times X_2$ is invariant under $F_{\alpha 1} \times F_{\alpha 2}, F_{\beta 1} \times F_{\beta 2}$ and $F_{\gamma 1} \times F_{\gamma 2}$. Therefore, $(GrT_{12}^{max}, F_{\alpha 1} \times F_{\alpha 2}, F_{\beta 1} \times F_{\beta 2}, F_{\gamma 1} \times F_{\gamma 2})$ is a $\{\alpha, \beta, \gamma\}$-action.

Lemma 6-A.24. Let $A_{\Sigma_1} = A_{\Sigma_2}$ hold. Let G_1 be surjective, and let H_2 be injective. Then $T_{12}^{min} = T_{12}^{max}$ holds. Moreover, set $T_{12} = T_{12}^{min}$. Then T_{12} is the linear representation system morphism $T_{12} : \Sigma_1 \to \Sigma_2$.

Proof. If G_1 is surjective and H_2 is injective, then Lemma (6-A.24) implies that $T_{12} \in$ Mor (Σ_1, Σ_2) is unique. Moreover, $T_{12} G_1 = G_2$ and $H_2 T_{12} = H_1$ hold. It follows from Lemma (6-A.23) that T_{12} is a $\{\alpha, \beta, \gamma\}$-morphism $T_{12} : (X_1, F_{\alpha 1}, F_{\beta 1}, F_{\gamma 1}) \to (X_2, F_{\alpha 2}, F_{\beta 2}, F_{\gamma 2})$.

Remark: The uniqueness part of the sophisticated realization Theorem (6-A.19) for images is proved by the canonicality of sophisticated canonical 3-Commutative Linear Representation Systems together with Lemma (6-A.24).

6-B Finite-Dimensional 3-Commutative Linear Representation Systems

This section provides the proofs for theorems, propositions and corollaries concerning finite-dimensionality set forth in Section 6.3.

6-B.1 Finite-Dimensional $\{\alpha, \beta, \gamma\}$-Actions

Appendix 6-A examined the $\{\alpha, \beta, \gamma\}$-actions. Here we will investigate the situation when state spaces are finite-dimensional. We will find that finite-dimensional $\{\alpha, \beta, \gamma\}$-actions can be represented by matrix-valued rational functions.

Definition 6-B.1. Let X be a linear space over the field K. Let a monoid morphism be $\phi : N^3 \to L(X)$, i.e., $\phi(0,0,0) = I$, where I is the identity map on X, and $\phi((l_1, m_1, n_1) + (l_2, m_2, n_2)) = \phi(l_1, m_1, n_1) \times \phi(l_2, m_2, n_2)$. Then the pair (X, ϕ) is called an N^3-module.

Note that N^3 is a monoid which has a unit element $(0,0,0)$ and operation $+ : N^3 \times N^3 \to N^3; ((l_1, m_1, n_1), (l_2, m_2, n_2)) \mapsto (l_1, m_1, n_1)(l_2, m_2, n_2) = (l_1 + l_2, m_1 + m_2, n_1 + n_2)$. Moreover, $L(X)$ is a monoid in terms of the composition of map.

Proposition 6-B.2. For any $\{\alpha, \beta, \gamma\}$-action $(X, F_\alpha, F_\beta, F_\gamma)$, there corresponds an N^3-module (X, ϕ) given by formula *) $\phi(i,j,k) = F_\alpha^i F_\beta^j F_\gamma^k$, and this correspondence is bijective.

Proof. Let $(X, F_\alpha, F_\beta, F_\gamma)$ be a given $\{\alpha, \beta, \gamma\}$-action. It is easily shown that ϕ satisfying the formula *) is a monoid morphism $\phi : N^3 \to L(X)$. The inverse correspondence is given by $F_\alpha := \phi(1,0,0)$, $F_\beta := \phi(0,1,0)$ and $F_\gamma := \phi(0,0,1)$. Thus, this correspondence is obviously bijective.

A $\{\alpha, \beta, \gamma\}$-action $(X, F_\alpha, F_\beta, F_\gamma)$ of which X is finite (n)-dimensional is called a finite (or n)-dimensional $\{\alpha, \beta, \gamma\}$-action.

Proposition 6-B.3. Let $(K^n, F_\alpha, F_\beta, F_\gamma)$ be a $\{\alpha, \beta, \gamma\}$-action and (K^n, ϕ) be the corresponding N^3-module. Then the formal power series $\bar{\phi}$ of the monoid morphism $\phi : N^3 \to L(X)$ is represented by the following matrix-valued rational function:
$\bar{\phi} = z_\alpha z_\beta z_\gamma [z_\alpha I - F_\alpha]^{-1} [z_\beta I - F_\beta]^{-1} [z_\gamma I - F_\gamma]^{-1} \in (K(z_\alpha, z_\beta, z_\gamma))^{n \times n}$.

Proof. ϕ can be represented as a matrix-valued formal power series: $\bar{\phi} = \sum_{k=0}^{\infty}\sum_{j=0}^{\infty}\sum_{i=0}^{\infty}\phi(i,j,k)z_\alpha^{-i}z_\beta^{-j}z_\gamma^{-k} = \sum_{k=0}^{\infty}\sum_{j=0}^{\infty}\sum_{i=0}^{\infty}F_\alpha^iF_\beta^jF_\gamma^kz_\alpha^{-i}z_\beta^{-j}z_\gamma^{-k} \in K^{n\times n}[[z_\alpha^{-1},z_\beta^{-1},z_\gamma^{-1}]]$.

Since the equation $K^{n\times n}[[z_\alpha^{-1},z_\beta^{-1},z_\gamma^{-1}]] = (K[[z_\alpha^{-1},z_\beta^{-1},z_\gamma^{-1}]])^{n\times n}$ holds and also $(K[[z_\alpha^{-1},z_\beta^{-1},z_\gamma^{-1}]])^{n\times n} \subseteq (K((z_\alpha^{-1},z_\beta^{-1},z_\gamma^{-1})))^{n\times n} = (K((z_\alpha,z_\beta,z_\gamma)))^{n\times n}$, then $\bar{\phi}[z_\alpha I - F_\alpha][z_\beta I - F_\beta][z_\gamma I - F_\gamma] = z_\alpha z_\beta z_\gamma$ holds.

Since $z_\alpha I - F_\alpha$, $z_\beta I - F_\beta$ and $z_\gamma I - F_\gamma$ are bijective in $(K((z_\alpha,z_\beta,z_\gamma)))^{n\times n}$, it follows that

$$\bar{\phi} = z_\alpha z_\beta z_\gamma[z_\alpha I - F_\alpha]^{-1}[z_\beta I - F_\beta]^{-1}[z_\gamma I - F_\gamma]^{-1}.$$

6-B.2 Finite-Dimensional Pointed $\{\alpha,\beta,\gamma\}$-Actions

Appendix 6-A stated that an initial object of any pointed $\{\alpha,\beta,\gamma\}$-action $((X,F_\alpha,F_\beta,F_\gamma),x^0)$ is $((K[z_\alpha,z_\beta,z_\gamma],z_\alpha,z_\beta,z_\gamma),\mathbf{1})$ and that the quasi-reachability of $((X,F_\alpha,F_\beta,F_\gamma),x^0)$ implies the surjectivity of the corresponding linear input map G. Therefore, equivalence classes of quasi-reachable pointed $\{\alpha,\beta,\gamma\}$-actions are characterized by ker G, namely, an ideal of $K[z_\alpha,z_\beta,z_\gamma]$.

This section provides a necessary and sufficient condition for pointed $\{\alpha,\beta,\gamma\}$-actions to be quasi-reachable.

The quasi-reachable standard form is defined, and we show that it is representative of pointed $\{\alpha,\beta,\gamma\}$-actions.

Let $((X,F_\alpha,F_\beta,F_\gamma),x^0)$ be a pointed $\{\alpha,\beta,\gamma\}$-action, and let G be the linear input map corresponding to an initial state x^0, namely, a $\{\alpha,\beta,\gamma\}$-morphism $G : (K[z_\alpha,z_\beta,z_\gamma],z_\alpha,z_\beta,z_\gamma) \to (X,F_\alpha,F_\beta,F_\gamma)$ which satisfies $G(\mathbf{1}) = x^0$. Let $P(\le l) := \{\lambda(i,j,k)z_\alpha^iz_\beta^jz_\gamma^k; \lambda(i,j,k) \in K,\ i+j+k \le l,\ i,j,k \in N\}$ for $l \in N$, and let J_l be the canonical injection $J_l : P(\le l) \to K[z_\alpha,z_\beta,z_\gamma]$. Moreover, let $G_l := GJ_l$. Then the following equation (*) holds for the range of G_l.

im $G_l = \ll \{F_\alpha^iF_\beta^jF_\gamma^kx^0;\ i,\ j,\ k \in N \text{ and } i+j+k \le l\} \gg$ ......(*)

This equation $(*)$ leads to the relation:

$$\text{im } G_0 \subseteq \text{im } G_1 \subseteq \cdots \subseteq \text{im } G_\infty = \text{im } G,$$

and results in the following lemma .

Lemma 6-B.4. If im G_{m-1} = im G_m for some integer $m \in N$, then im G_m = im G_{m+1}.

Lemma 6-B.5. im G_{n-1} = im G always holds for any pointed $\{\alpha,\beta,\gamma\}$-action $((K^n,F_\alpha,F_\beta,F_\gamma),x^0)$, and $((\text{im } G_{n-1},F_\alpha,F_\beta,F_\gamma),x^0)$ is a quasi-reachable pointed $\{\alpha,\beta,\gamma\}$-action.

Proof. This is a direct consequence of Lemma (6-B.4) and the definition of quasi-reachability.

Proposition 6-B.6. Let $((K^n, F_\alpha, F_\beta, F_\gamma), x^0)$ be a pointed $\{\alpha, \beta, \gamma\}$-action. Then $((K^n, F_\alpha, F_\beta, F_\gamma), x^0)$ is quasi-reachable if and only if rank $[x^0, F_\alpha x^0, \cdots, F_\alpha^{n-1}x^0, F_\beta x^0, F_\beta F_\alpha x^0, \cdots, F_\beta F_\alpha^{n-2}x^0, \cdots, F_\beta^{n-1}x^0, F_\gamma x^0, F_\gamma F_\alpha x^0, \cdots, F_\gamma F_\alpha^{n-1}x^0, \cdots, F_\gamma^{n-1}x^0] = n$ holds.

Proof. The necessary and sufficient condition for $((K^n, F_\alpha, F_\beta, F_\gamma), x^0)$ to be quasi-reachable is im $G = K^n$. This implies im $G_{n-1} = K^n$ by Lemma (6-B.5). Thus this proposition holds.

Definition 6-B.7. If a quasi-reachable pointed $\{\alpha, \beta, \gamma\}$-action $((K^n, F_\alpha, F_\beta, F_\gamma), x^0)$ is one of the Quasi-reachable Standard Systems described in Definition (6.10), then $((K^n, F_\alpha, F_\beta, F_\gamma), x^0)$ is called the quasi-reachable standard form with a vector index $\nu = (\nu_1, \nu_2, \nu_3, \cdots, \nu_p)$, where $\nu_i = (\nu_{i1}, \nu_{i2}, \cdots, \nu_{iq_i})$.

Theorem 6-B.8. For any quasi-reachable pointed $\{\alpha, \beta, \gamma\}$-action, there exists only one quasi-reachable standard form which is isomorphic to it.

Proof. The proof of this theorem is constructive. Let $((K^n, F_\alpha, F_\beta, F_\gamma), x^0)$ be any quasi-reachable pointed $\{\alpha, \beta, \gamma\}$ -action.

Step 1.

1-1) Determine an integer ν_{11} and coefficients $\{c_i^{11} \in K;\ 1 \le i \le \nu_{11}\}$ such that $\{F_\alpha^{i-1}a;\ 1 \le i \le \nu_{11}\}$ are linearly independent and $\{F_\alpha^i x^0;\ 1 \le i \le \nu_{11}\}$ are linearly dependent.
Thus, $F_\alpha^{\nu_{11}}x^0 = \sum_{i=1}^{\nu_{11}} c_i^{11} F_\alpha^{i-1}x^0$ holds.
Let $\mathbf{c}^{11} := [c_1^{11}, \cdots, c_{\nu_{11}}^{11}, 0, \cdots, 0]^T \in K^n$, and determine an integer ν_{12} and coefficients $\{c_i^{12} \in K;\ 1 \le i \le \nu_{12}\}$ such that $\{F_\beta^{j-1}F_\alpha^{i-1}x^0;\ 1 \le j \le 2,\ 1 \le i \le \nu_{1j}\}$ are linearly independent and $\{F_\beta F_\alpha^i x^0;\ 1 \le i \le \nu_{12}\}$ are linearly dependent.
Thus, $F_\alpha^{\nu_{12}} F_\beta x^0 = \sum_{j=1}^{2} \sum_{i=1}^{\nu_{1j}} c_{\nu_{1j}-1}^{12} F_\beta^{j-1} F_\alpha^{i-1} x^0$ holds.
Let $\mathbf{c}^{12} := [c_1^{12}, \cdots, c_{\nu_{11}}^{12}, c_{\nu_{11}+1}^{12}, \cdots, c_{\nu_{11}+\nu_{12}}^{12}, 0, \cdots, 0]^T \in K^n$.

This procedure is repeated until $1 - q_1$).

1-q_1) Determine an integer ν_{1q_1} , coefficients
$\{c_{\nu_{11}+\cdots+\nu_{1j-1}+i}^{1q_1} \in K;\ 1 \le j \le q_1,\ 1 \le i \le \nu_{1j}\}$ and
$\{c_{\nu_{11}+\cdots+\nu_{1j-1}+i}^{1q_1+1} \in K;\ 1 \le j \le q_1,\ 1 \le i \le \nu_{1j}\}$ such that
$\{F_\beta^{j-1}F_\alpha^{i-1}x^0;\ 1 \le j \le q_1,\ 1 \le i \le \nu_{1j}\}$ are linearly independent and the following equations hold:
$F_\alpha^{\nu_{1q_1}} F_\beta^{q_1} x^0 = \sum_{j=1}^{q_1} \sum_{i=1}^{\nu_{1j}} c_{\nu_{1j}+\cdots+\nu_{1j-1}+i}^{1q_1} F_\beta^{j-1} F_\alpha^{i-1} x^0$,
$F_\beta^{q_1} x^0 = \sum_{j=1}^{q_1} \sum_{i=1}^{\nu_{1j}} c_{\nu_{1j}+\cdots+\nu_{1j-1}+i}^{1q_1+1} F_\beta^{j-1} F_\alpha^{i-1} x^0$.
Let
$\mathbf{c}^{1q_1} := [c_1^{1q_1}, \cdots, c_{\nu_{11}}^{1q_1}, c_{\nu_{11}+1}^{1q_1}, \cdots, c_{\nu_{11}+\nu_{12}}^{1q_1}, \cdots, c_{\nu_1}^{1q_1}, 0, \cdots, 0]^T \in K^n$
and

$\mathbf{c}^{1q_1+1} := [c_1^{1q_1+1}, \cdots, c_{\nu_{11}}^{1q_1}, c_{\nu_{11}+1}^{1q_1+1}, \cdots, c_{\nu_{11}+\nu_{12}}^{1q_1+1}, \cdots, c_{\nu_1}^{1q_1+1}, 0, \cdots, 0]^T \in K^n$.

Step 2.

2-1) Determine an integer ν_{21}, coefficients $\{c_{\nu_{11}+\cdots+\nu_{1j-1}+i}^{21} \in K;\ 1 \leq j \leq q_1,\ 1 \leq i \leq \nu_{1j}\}$ and $\{c_{\nu_1+i}^{21} \in K;\ 1 \leq i \leq \nu_{21}\}$ such that $\{F_\gamma F_\alpha^{i-1}x^0;\ 1 \leq j \leq \nu_{21}\}$ are linearly independent and $F_\gamma F_\alpha^{\nu_{21}}x^0 = \sum_{j=1}^{q_1}\sum_{i=1}^{\nu_{1j}} c_{\nu_{11}+\cdots+\nu_{1j-1}+i}^{21} F_\beta^{j-1}F_\alpha^{i-1}x^0 + \sum_{i=1}^{\nu_{21}} c_{\nu_1+i}^{21} F_\gamma^{j-1}F_\alpha^{i-1}x^0$ holds.
Let $\mathbf{c}^{21} := [c_1^{21}, \cdots, c_{\nu_1}^{21}, \cdots, c_{\nu_1+\nu_{21}}^{21}, 0, \cdots, 0]^T \in K^n$.
This procedure is repeated until $2 - q_2$).

2-q_2) Determine an integer ν_{2q_2}, coefficients
$\{c_{\nu_{k-1}+\nu_{k1}+\cdots+\nu_{kj-1}+i}^{2q_2} \in K;\ 1 \leq k \leq 2,\ 1 \leq j \leq q_2,\ 1 \leq i \leq \nu_{1j}\}$ and
$\{c_{\nu_{k-1}+\nu_{k1}+\cdots+\nu_{kj-1}+i}^{2q_2+1} \in K;\ 1 \leq k \leq 2,\ 1 \leq j \leq q_2,\ 1 \leq i \leq \nu_{1j}\}$ such that $\{F_\gamma^{k-1}F_\beta^{j-1}F_\alpha^{i-1}x^0;\ 1 \leq k \leq 2,\ 1 \leq j \leq q_k,\ 1 \leq i \leq \nu_{kj}\}$ are linearly independent and the following equations hold:

$$F_\gamma F_\beta^{q_2-1}F_\alpha^{\nu_{2q_2}}x^0 = \sum_{k=1}^{2}\sum_{j=1}^{q_k}\sum_{i=1}^{\nu_{kj}} c_{\nu_{k-1}+\nu_{k1}+\cdots+\nu_{kj-1}+i}^{2q_2} F_\gamma^{k-1}F_\beta^{j-1}F_\alpha^{i-1}x^0 + \sum_{i=1}^{\nu_{21}} c_{\nu_1+i}^{21} F_\gamma^{j-1}F_\alpha^{i-1}x^0,$$

$$F_\gamma F_\beta^{q_2}x^0 = \sum_{k=1}^{2}\sum_{j=1}^{q_k}\sum_{i=1}^{\nu_{kj}} c_{\nu_{k-1}+\nu_{k1}+\cdots+\nu_{kj-1}+i}^{2q_2+1} F_\gamma^{k-1}F_\beta^{j-1}F_\alpha^{i-1}x^0 + \sum_{i=1}^{\nu_{21}} c_{\nu_1+i}^{21} F_\gamma^{j-1}F_\alpha^{i-1}x^0.$$

Let $\mathbf{c}^{2q_2} := [c_1^{2q_2}, \cdots, c_{\nu_1}^{2q_2}, \cdots, c_{\nu_1+\nu_2}^{2q_2}, 0, \cdots, 0]^T \in K^n$,
$\mathbf{c}^{2q_2+1} := [c_1^{2q_2+1}, \cdots, c_{\nu_1}^{2q_2+1}, \cdots, c_{\nu_1+\nu_2}^{2q_2+1}, 0, \cdots, 0]^T \in K^n$.

These steps are repeated until Step p.

Step p.

p-1) Determine an integer ν_{p1}, coefficients $\{c_{\nu_1+\cdots\nu_{k-1}+\nu_{k1}+\cdots+\nu_{kj-1}+i}^{p1} \in K;\ 1 \leq k \leq p-1,\ 1 \leq j \leq q_k,\ 1 \leq i \leq \nu_{kj}\}$ and $\{c_{\nu_1+\cdots+\nu_{p-1}+i}^{p1} \in K;\ 1 \leq i \leq \nu_{p1}\}$ such that $\{F_\gamma^{k-1}F_\beta^{j-1}F_\alpha^{i-1}x^0;\ 1 \leq k \leq p-1,\ 1 \leq j \leq q_k,\ 1 \leq i \leq \nu_{kj}\}$ and $\{F_\gamma^{p-1}F_\alpha^{i-1}x^0;\ 1 \leq i \leq \nu_{p1}\}$ are linearly independent, where

$$F_\gamma^{p-1}F_\alpha^{\nu_{p1}}x^0 = \sum_{k=1}^{p-1}\sum_{j=1}^{q_k}\sum_{i=1}^{\nu_{kj}} c_{\nu_1+\cdots+\nu_{k-1}+\nu_{k1}+\cdots+\nu_{kj-1}+i}^{p1} F_\beta^{k-1}F_\beta^{j-1}F_\alpha^{i-1}x^0 + \sum_{i=1}^{\nu_{p1}} c_{\nu_1+\cdots+\nu_{p-1}+i}^{p1} S_\gamma^{p-1}F_\alpha^{i-1}x^0.$$

Let $\mathbf{c}^{p1} := [c_1^{p1}, \cdots, c_{\nu_1}^{p1}, \cdots, c_{\nu_1+\cdots+\nu_{p-1}+\nu_{p1}}^{p1}, 0, \cdots, 0]^T \in K^n$.
This procedure is repeated until $p - q_p$).

p-q_p) Determine an integer ν_{pq_p}, coefficients
$\{c_{\nu_1+\cdots\nu_{k-1}+\nu_{k1}+\cdots+\nu_{kj-1}+i}^{2q_2} \in K;\ 1 \leq k \leq p,\ 1 \leq j \leq q_k,\ 1 \leq i \leq \nu_{kj}\}$
$\{c_{\nu_1+\cdots+\nu_{k-1}+\nu_{k1}+\cdots+\nu_{kj-1}+i}^{p+1} \in K;\ 1 \leq k \leq p,\ 1 \leq j \leq q_k,\ 1 \leq i \leq \nu_{kj}\}$
such that $\{F_\gamma^{k-1}F_\beta^{j-1}F_\alpha^{i-1}a;\ 1 \leq k \leq p,\ 1 \leq j \leq q_k,\ 1 \leq i \leq \nu_{kj}\}$ are linearly independent and the following equations hold:
$S_\gamma^{p-1}S_\beta^{q_p-1}S_\alpha^{\nu_{pq_p}}a$

$=\sum_{k=1}^{p}\sum_{j=1}^{q_k}\sum_{i=1}^{\nu_{kj}} c^{pq_p}_{\nu_1+\cdots+\nu_{k-1}+\nu_{k1}+\cdots+\nu_{kj-1}+i} F_\gamma^{k-1} F_\beta^{j-1} F_\alpha^{i-1} x^0,$

$S_\gamma^p a = \sum_{k=1}^{p}\sum_{j=1}^{q_k}\sum_{i=1}^{\nu_{kj}} c^{pq_p+1}_{\nu_1+\cdots+\nu_{k-1}+\nu_{k1}+\cdots+\nu_{kj-1}+i} F_\gamma^{k-1} F_\beta^{j-1} F_\alpha^{i-1} x^0.$

Let $\mathbf{c}^{pq_p} := [c_1^{pq_p}, \cdots, c_{\nu_1}^{pq_p}, \cdots, c_{\nu_1+\cdots+\nu_p}^{pq_p}]^T \in K^n$,

$\mathbf{c}^{pq_p+1} := [c_1^{pq_p+1}, \cdots, c_{\nu_1}^{pq_p+1}, \cdots, c_{\nu_1+\cdots+\nu_p}^{pq_p+1}]^T \in K^n.$

At this point, we set $n := \sum_{i=1}^{p}\sum_{j=1}^{q_i} \nu_{ij}$, where $\nu_i = \sum_{j=1}^{q_i} \nu_{ij}$.

Step $p+1$.
Let the state space be K^n and let the initial state be $\mathbf{e}_1 \in K^n$.

Step $p+2$.
Let $F_{\alpha s}, F_{\beta s}, F_{\gamma s} \in K^{n\times n}$ be the same as stated in Definition (6.10).

Step $p+3$.
Set $h_s := [a(0,0,0), \cdots, a(\nu_{11}-1,0,0), a(0,1,0), \cdots, a(\nu_{12}-1,1,0), \cdots, a(\nu_{1q_1}-1, q_1-1, 0), a(0,0,1), \cdots, a(\nu_{pq_p}-1, q_p-1, p-1)] \in K^{t\times n}$.

This procedure asserts that $0 \le \nu_{iq_i} \le \cdots \le \nu_{i2} \le \nu_{i1}$ and $0 \le \nu_{pj} \le \cdots \le \nu_{2j} \le \nu_{1j}$ for $1 \le i \le p$, $1 \le j \le q_p$. Moreover, its quasi-reachability leads to $n = \sum_{i=1}^{p}\sum_{j=1}^{q_i} \nu_{ij}$.

Let $\nu_m := \sum_{i=1}^{q_m} \nu_{mi}$, and let $T : K^n \to K^n$ be the linear map which satisfies $TF_\gamma^{k-1}F_\beta^{j-1}F_\alpha^{i-1}\mathbf{e}_1 = \mathbf{e}_{\nu_1+\cdots+\nu_{k-1}+\nu_{k1}+\cdots+\nu_{kj-1}+i}$ for any $1 \le k \le p$, $1 \le j \le q$, $1 \le i \le \nu_{kj}$. Then T is a regular matrix. Set $F_{\alpha s} := TF_\alpha T^{-1}$, $F_{\beta s} := TF_\beta T^{-1}$ and $F_{\gamma s} := TF_\gamma T^{-1}$. Then $F_{\alpha s}, F_{\beta s}$ and $F_{\gamma s} \in K^{n\times n}$ satisfy $F_{\alpha s}F_{\beta s} = F_{\beta s}F_{\alpha s}$, $F_{\beta s}F_{\gamma s} = F_{\gamma s}F_{\beta s}$ and $F_{\gamma s}F_{\alpha s} = F_{\alpha s}F_{\gamma s}$.

Hence, $((K^n, F_{\alpha s}, F_{\beta s}, F_{\gamma s}), \mathbf{e}_1)$ is a pointed $\{\alpha,\beta,\gamma\}$-action. Since the map $T : K^n \to K^n$ preserves the linear dependence and linear independence of each vector, $((K^n, F_{\alpha s}, F_{\beta s}, F_{\gamma s}), \mathbf{e}_1)$ is a quasi-reachable standard form with the vector index $\nu = (\nu_1, \nu_2, \nu_3, \cdots, \nu_p)$. Thus, T is bijective and a pointed $\{\alpha,\beta,\gamma\}$-morphism $T : ((K^n, F_\alpha, F_\beta, F_\gamma), x^0) \to ((K^n, F_{\alpha s}, F_{\beta s}, F_{\gamma s}), \mathbf{e}_1)$. Uniqueness is provided by the definition of the quasi-reachable standard form.

Proposition 6-B.9. The formal power series $X(z_\alpha^{-1}, z_\beta^{-1}, z_\gamma^{-1})$ of a pointed $\{\alpha,\beta,\gamma\}$-action $((K^n, F_\alpha, F_\beta, F_\gamma), x^0)$ is expressed by the vector-valued rational function:

$$X(z_\alpha^{-1}, z_\beta^{-1}, z_\gamma^{-1}) = z_\alpha z_\beta z_\gamma [z_\alpha I - F_\alpha]^{-1}[z_\beta I - F_\beta]^{-1}[z_\gamma I - F_\gamma]^{-1} x^0 \in (K(z_\alpha, z_\beta, z_\gamma))^n.$$

Proof. Let (K^n, ϕ) be the N^3-module corresponding to $(K^n, F_\alpha, F_\beta, F_\gamma)$. By Proposition (6-B.3), $\bar{\phi}$ is represented by

$$\bar{\phi} = z_\alpha z_\beta z_\gamma [z_\alpha I - F_\alpha]^{-1}[z_\beta I - F_\beta]^{-1}[z_\gamma I - F_\gamma]^{-1}$$

Therefore, $((K^n, F_\alpha, F_\beta, F_\gamma), x^0)$ can be expressed by

$$\bar{\phi}x^0 = z_\alpha z_\beta z_\gamma [z_\alpha I - F_\alpha]^{-1}[z_\beta I - F_\beta]^{-1}[z_\gamma I - F_\gamma]^{-1}x^0.$$

6-B.3 Finite-Dimensional $\{\alpha, \beta, \gamma\}$-Actions with a Readout Map

We noted in Appendix 6-A that a final object of the $\{\alpha, \beta, \gamma\}$-actions with a readout map is $((F(N^3, Y), S_\alpha, S_\beta, S_\gamma), 0)$, and that distinguishability of $((X, F_\alpha, F_\beta, F_\gamma), h)$ implies the injectivity of the corresponding linear observation map.

This section provides the necessary and sufficient conditions for a finite-dimensional $\{\alpha, \beta, \gamma\}$-action with a readout map to be distinguishable. Since the duality of the results in the preceding section covers the results in this section, proofs are omitted.

For a $\{\alpha, \beta, \gamma\}$-action with a readout map $((X, F_\alpha, F_\beta, F_\gamma), h)$, define $LO(m) :=\ll \{hF_\alpha^i F_\beta^j F_\gamma^k;\ i+j+k \le m,\ m \in N\} \gg$.

Then the sequence $LO(0) \subseteq LO(1) \subseteq \cdots \subseteq LO(m) \subseteq \cdots \subseteq LO(\infty)$ holds. Let H be the linear observation map corresponding to h, namely, the $\{\alpha, \beta, \gamma\}$-morphism $H : (X, F_\alpha, F_\beta, F_\gamma) \to (F(N^3, Y), S_\alpha, S_\beta, S_\gamma)$ satisfying $(Hx)(i, j, k) := hF_\alpha^i F_\beta^j F_\gamma^k x$ for any $x \in X,\ i, j, k \in N$. Let P_m be the canonical surjection $P_m : F(N^3, Y) \to F((N^3)_m, Y)$, where $F((N^3)_m, Y) := \{\underline{a} \in F((N^3)_m, Y);\ \underline{a} : (N^3)_m \to Y\}$ and $(N^3)_m := \{(\bar{i}, \bar{j}, \bar{k});\ \bar{i} + \bar{j} + \bar{k} \le m\}$ and let $H_m := P_m H$. Then ker $H_m = LO(m)^0$ holds, where $LO(m)^0 :=$ ker $H_m = \{x \in X; hx = 0 \text{ for } h \in LO(m)\}$. Moreover, ker $H = LO(\infty)^0$ holds.

Lemma 6-B.10. $LO(n-1) = LO(\infty)$ holds for the $\{\alpha, \beta, \gamma\}$-action with a readout map $((K^n, F_\alpha, F_\beta, F_\gamma), h)$.

Lemma 6-B.11. For a $\{\alpha, \beta, \gamma\}$ -action with the readout map $((K^n, F_\alpha, F_\beta, F_\gamma), h)$, (ker $H_{n-1}, F_\alpha, F_\beta, F_\gamma$) is a sub $\{\alpha, \beta, \gamma\}$-action of $(K^n, F_\alpha, F_\beta, F_\gamma)$ and $((K^n/\text{ker } H_{n-1}, \dot{F}_\alpha, \dot{F}_\beta, \dot{F}_\gamma), \dot{h})$ is a distinguishable $\{\alpha, \beta, \gamma\}$-action with a readout map, where $\dot{F}_\alpha, \dot{F}_\beta$ and $\dot{F}_\gamma$ are given as follows:
$\dot{F}_\alpha : K^n/\text{ker } H_{n-1} \to K^n/\text{ker } H_{n-1}; [x] \mapsto [F_\alpha x]$,
$\dot{F}_\beta : K^n/\text{ker } H_{n-1} \to K^n/\text{ker } H_{n-1}; [x] \mapsto [F_\beta x]$ and
$\dot{F}_\gamma : K^n/\text{ker } H_{n-1} \to K^n/\text{ker } H_{n-1}; [x] \mapsto [F_\gamma x]$.
$\dot{h}$ is defined as $h = \dot{h}\pi$ for the natural surjection $\pi : K^n \to K^n/\text{ker } H_{n-1}$.

Proposition 6-B.12. $\{\alpha, \beta, \gamma\}$-action with readout map $((K^n, F_\alpha, F_\beta, F_\gamma), h)$ is distinguishable if and only if rank $[h^T, (hF_\alpha)^T, \cdots, (hF_\alpha^{n-1})^T, (hF_\beta)^T, \cdots, (hF_\gamma)^T, \cdots, (hF_\gamma^{n-1})^T] = n$.

6-B.4 Finite-Dimensional 3-Commutative Linear Representation Systems

This section supplies the proof of Representation Theorem (6.11) for finite-dimensional canonical 3-Commutative Linear Representation Systems.

6-B.13. Proof of Representation Theorem (6.11).
Let $\sigma = ((K^n, F_\alpha, F_\beta, F_\gamma), x^0, h)$ be any finite-dimensional canonical 3-Commutative Linear Representation System.

Let $h_s := hT^{-1}$ for the quasi-reachable standard form $((K^n, F_{\alpha s}, F_{\beta s}, F_{\gamma s}), \mathbf{e}_1)$, where $T : ((K^n, F_\alpha, F_\beta, F_\gamma), x^0) \to ((K^n, F_{\alpha s}, F_{\beta s}, F_{\gamma s}), \mathbf{e}_1)$ is a pointed $\{\alpha, \beta, \gamma\}$-morphism and it is bijective, as it can be seen in the proof of Theorem (6-B.8). Then T is the linear representation system morphism $T : \sigma = ((K^n, F_\alpha, F_\beta, F_\gamma), x^0, h) \to \sigma_s = ((K^n, F_{\alpha s}, F_{\beta s}, F_{\gamma s}), \mathbf{e}_1, h_s)$, and it is bijective. Hence, σ_s is the unique Quasi-reachable Standard System. By Corollary (6.7), the behaviors of σ and σ_s are the same.

6-B.5 Existence Criterion for Finite-Dimensional 3-Commutative Linear Representation Systems

This section presents the proofs of Theorem (6.12) and Theorem (6.13).

6-B.14. Proof of Theorem (6.12)
Let A be the linear input/output map corresponding to a three-dimensional image $a \in F(N^3, Y)$. Obviously, im $A = \ll \{S_\alpha^i S_\beta^j S_\gamma^k a;\ i, j, k \in N\} \gg$. Let $A_l := AJ_l$, where J_l is the canonical injection $J_l :\ P(\leq l) \to K[z_\alpha, z_\beta, z_\gamma]$. The linear operator $A_{(l,m)} : P(\leq l) \to F((N^3)_m, Y)$ is defined by setting $A_{(l,m)} := P_m A J_l$, where P_m is the canonical surjection $P_m :\ F(N^3, Y) \to F((N^3)_m, Y)$. Then $A_{(l,m)}$ is represented by the partial Hankel matrix $H_{a(l,m)}$ from the Hankel matrix H_a, where $H_{a(l,m)} = [a(i + \bar{i}, j + \bar{j}, k + \bar{k})]$ for $i + j + k \leq l$ and $\bar{i} + \bar{j} + \bar{k} \leq m$.

First, we prove 1) $\Rightarrow$ 2).
Theorem (6.3) and Corollary (6-A.20) assert that im A is n-dimensional. If im $A_{n-1} \neq$ im A_n, then the dimension of im A_n is $n + 1$ or more by Lemma (6-B.4). Hence im $A_{n-1} =$ im $A_n = \cdots =$ im A. Thus, there exist n linearly independent vectors in $\{S_\alpha^i S_\beta^j S_\gamma^k a;\ i + j + k \leq n$ for $i, j, k \in N\}$, and n+1 or more linearly independent vectors are not included in it.

Second, we prove 2) $\Rightarrow$ 3).
im $A_{n-1} =$ im A_n implies im $A_{n-1} =$ im $A_n = \cdots =$ im A. Therefore, the dimension of im A_r is n for $r \geq n-1$. On the other hand, Corollary (6-A.20) and Lemma (6-B.4) indicate ker $P_s = 0$ for $s \geq n-1$. Thus, the dimension of im $P_s A J_r$ is n. Hence, the rank of partial Hankel matrix $H_{a(r,s)}$ corresponding to $P_s A J_r$ is n.

Finally, we prove 3) $\Rightarrow$ 1).
Since the rank of the Hankel matrix $H_{\underline{a}}$ is n, the range of the linear input/output map A (im A) corresponding to H_a is n-dimensional. im $A = \ll \{S_\alpha^i S_\beta^j S_\gamma^k a;\ i,j,k \in N\} \gg$ and Corollary (6-A.20) result in 1).

6-B.15. Proof of Theorem (6.13)
The necessary condition is obvious from Proposition (6-B.9) and

$$\begin{aligned}\bar{a_\sigma} &= \textstyle\sum_{k=0}^{\infty}\sum_{j=0}^{\infty}\sum_{i=0}^{\infty} a_\sigma(i,j,k) z_\alpha^{-i} z_\beta^{-j} z_\gamma^{-k} \\ &= \textstyle\sum_{k=0}^{\infty}\sum_{j=0}^{\infty}\sum_{i=0}^{\infty} hF_\alpha^i F_\beta^j F_\gamma^k x^0 z_\alpha^{-i} z_\beta^{-j} z_\gamma^{-k}.\end{aligned}$$

Let us prove the sufficient condition. This proof is given with $t=1$, since it is similar for all the other t.

Let the formal power series $\bar{a}$ of an image $a \in F(N^3, Y)$ be

$$\bar{a} = z_\alpha z_\beta z_\gamma \big(\textstyle\sum_{k=0}^{n-1}\sum_{j=0}^{m-1}\sum_{i=0}^{l-1} \lambda(i,j,k) z_\alpha^i z_\beta^j z_\gamma^k\big) / (q_\alpha(z_\alpha) q_\beta(z_\beta) q_\gamma(z_\gamma)),$$

where $q_\alpha(z_\alpha) = z_\alpha^l + \alpha_{l-1} z_\alpha^{l-1} + \cdots + \alpha_1 z_\alpha + \alpha_0$,

$$q_\beta(z_\beta) = z_\beta^m + \beta_{m-1} z_\beta + \cdots + \beta_1 z_\beta + \beta_0 \text{ and}$$

$$q_\gamma(z_\gamma) = z_\gamma^n + \gamma_{n-1} z_\gamma^{n-1} + \cdots + \gamma_1 z_\gamma + \gamma_0.$$

On the other hand, the formal power series $\bar{a}$ is represented by

$$\bar{a} = \textstyle\sum_{k=0}^{\infty}\sum_{j=0}^{\infty}\sum_{i=0}^{\infty} a(i,j,k) z_\alpha^{-i} z_\beta^{-j} z_\gamma^{-k}.$$

Hence, the following equation is obtained:

$$\begin{aligned}& z_\alpha z_\beta z_\gamma \big(\textstyle\sum_{k=0}^{n-1}\sum_{j=0}^{m-1}\sum_{i=0}^{l-1} \lambda(i,j,k) z_\alpha^i z_\beta^j z_\gamma^k\big) \\ &\quad = q_\alpha(z_\alpha) q_\beta(z_\beta) q_\gamma(z_\gamma) \big(\textstyle\sum_{k=0}^{\infty}\sum_{j=0}^{\infty}\sum_{i=0}^{\infty} a(i,j,k) z_\alpha^{-i} z_\beta^{-j} z_\gamma^{-k}\big). \qquad \cdots (**)\end{aligned}$$

Expanding and rearranging the equation (**), the coefficients of $z_\alpha^{-i} z_\beta^{-j} z_\gamma^{-k}$ on the right hand side of equation (**) are all zero for any non-negative integer i,j,k. Note that $q_\alpha(z_\alpha)$, $q_\beta(z_\beta)$ and $q_\gamma(z_\gamma)$ are polynomials in one-variable, z_α, z_β and z_γ, respectively. Then the coefficients of z_α^{-i} for any non-negative integer i in the equation $q_\alpha(z_\alpha)(\sum_{k=0}^{\infty}\sum_{j=0}^{\infty}\sum_{i=0}^{\infty} a(i,j,k) z_\alpha^{-i} z_\beta^{-j} z_\gamma^{-k})$ are all zero regardless of $q_\beta(z_\beta)$ and $q_\gamma(z_\gamma)$. Thus $q_\alpha(S_\alpha)a = 0$.

Similarly, $q_\beta(S_\beta)a = 0$ and $q_\gamma(S_\gamma)a = 0$ hold.

Let A be the linear input/output map corresponding to an image $a \in F(N^3, Y)$, and let $A_l := AJ_l$. Then im $A_{l+m+n} =$ im $A_{l+m+n+1}$ follows from $q_\alpha(S_\alpha)a = 0$, $q_\beta(S_\beta)a = 0$ and $q_\gamma(S_\gamma)a = 0$. Since

$$\text{im } A_{l+m+n} = \text{im } A_{l+m+n+1} = \cdots = \text{im } A = \ll \{S_\alpha^i S_\beta^j S_\gamma^k a;\ i,j,k \in N\} \gg$$

holds, the number of basis in im A is finite, and hence im A is finite-dimensional.

By Theorem (6.12), there exists a finite-dimensional 3-Commutative Linear Representation System which realizes an image $a \in F(N^3, Y)$.

6-B.6 Realization Procedure for Finite-Dimensional 3-Commutative Linear Representation Systems

This section proves Theorem for the realization procedure (6.14).

6-B.16 Proof of Theorem (6.14)
Let $R(a) := \{S_\alpha^i S_\beta^j S_\gamma^k a;\ i,j,k \in N\}$. Then $((\ll R(a) \gg, S_\alpha, S_\beta, S_\gamma), a, 0)$ is a canonical 3-Commutative Linear Representation System which realizes $a \in F(N^3, Y)$ by Examples (6-A.8) and (6-A.13), Proposition (6-A.18) and Corollary (6-A.20).

The linearly independent vectors

$$\{S_\gamma^{k-1} S_\beta^{j-1} S_\alpha^{i-1} a;\ 1 \le k \le p,\ 1 \le j \le q_k, 1 \le i \le \nu_{kj}\} \text{ satisfy}$$
$$\{S_\gamma^{k-1} S_\beta^{j-1} S_\alpha^{i-1} a;\ 1 \le k \le p,\ 1 \le j \le q_k,\ 1 \le i \le \nu_{kj}\} = R(a).$$

Let the linear map $T : \ll R(a) \gg \to K^n$ be
$TS_\gamma^{k-1} S_\beta^{j-1} S_\alpha^{i-1} a = \mathbf{e}_{\nu_1 + \cdots + \nu_{k-1} + \nu_{k1} + \cdots + \nu_{kj-1} + i}$
for any $1 \le k \le p,\ 1 \le j \le q,\ 1 \le i \le \nu_{kj}$.

Then, from the step 1 through the step p+2 of the realization procedure (6.14), we obtain $F_{\alpha s} T = TS_\alpha$, $F_{\beta s} T = TS_\beta$, $F_{\gamma s} T = TS_\gamma$, while the condition of the step p+3) leads to $h_s T = 0$.

Thus T is bijective, and it is the linear representation system morphism $T : ((\ll R(a) \gg, S_\alpha, S_\beta, S_\gamma), a, 0) \to \sigma_s = ((K^n, F_{\alpha s}, F_{\beta s}, F_{\gamma s}), \mathbf{e}_1, h_s)$.
By Corollary (6.7), the behavior of σ_s is a. The selection of $\{S_\gamma^{k-1} S_\beta^{j-1} S_\alpha^{i-1} a;$ $1 \le k \le p,\ 1 \le j \le q_k,\ 1 \le i \le \nu_{kj}\}$ and the determination of the map T affirm that σ_s is the Quasi-reachable Standard System.

6-C Partial Realization Theory

This section furnishes proofs for the theorems and the propositions concerning the partial realization problem set forth in Section 6.4.

The detailed notions and notations are the same as those in Appendix 6-A and 6-B.

6-C.1 Pointed $\{\alpha, \beta, \gamma\}$-Actions

Define $P(\le l_1, \le m_1, \le n_1)$ as
$P(\le l_1, \le m_1, \le n_1) := \{\sum_{k=0}^{n_1} \sum_{j=0}^{m_1} \sum_{i=0}^{l_1} \lambda(i,j,k) z_\alpha^i z_\beta^j z_\gamma^k \in K[z_\alpha, z_\beta, z_\gamma]\}$, and let $J_{(l_1, m_1, n_1)}$ be the canonical injection : $P(\le l_1, \le m_1, \le n_1) \to K[z_\alpha, z_\beta, z_\gamma]$.

Proposition 6-C.1. If a linear subspace S of $P(\le l_1, \le m_1, \le n_1)$ satisfies the following two conditions, then there uniquely exists an ideal $\bar{S} \cap$

$K[z_\alpha, z_\beta, z_\gamma]$ such that $\bar{S} \cap P(\leq l_1+1, \leq m_1+1, \leq n_1+1) = S$ and $P(\leq l_1+1, \leq m_1+1, \leq n_1+1)/S$ is isomorphic to $K[z_\alpha, z_\beta, z_\gamma]/\bar{S}$.

Moreover, the pointed $\{\alpha, \beta, \gamma\}$-action $((K[z_\alpha, z_\beta, z_\gamma]/\bar{S}, \dot{z}_\alpha, \dot{z}_\beta, \dot{z}_\gamma), \mathbf{1}+S)$ is (l_1, m_1, n_1)-quasi-reachable, where $\dot{z}_\alpha, \dot{z}_\beta$ and $\dot{z}_\gamma$ are given by $\dot{z}_\alpha(\lambda+\bar{S}) = z_\alpha\lambda+\bar{S}$, $\dot{z}_\beta(\lambda+\bar{S}) = z_\beta\lambda+\bar{S}$ and $\dot{z}_\gamma(\lambda+\bar{S}) = z_\gamma\lambda+\bar{S}$, respectively.

[**condition 1**]:

$\lambda \in P(\leq l_1, \leq m_1+1, \leq n_1+1) \cap S$ implies $z_\alpha\lambda \in S$,
$\lambda \in P(\leq l_1+1, \leq m_1, \leq n_1+1) \cap S$ implies $z_\beta\lambda \in S$ and
$\lambda \in P(\leq l_1+1, \leq m_1+1, \leq n_1) \cap S$ implies $z_\gamma\lambda \in S$.

[**condition 2**]:

1) There exist the coefficients $\{\lambda_1(i,j,k)\}$ such that a polynomial $z_\alpha^l z_\beta^{m_1+1} z_\gamma^{n_1+1} - \sum_{k=0}^{n_1}\sum_{j=0}^{m_1}\sum_{i=0}^{l_1} \lambda_1(i,j,k) z_\alpha^i z_\beta^j z_\gamma^k \in S$ for any $0 \leq l \leq l_1+1$.
2) There exists the coefficients $\{\lambda_2(i,j,k)\}$ such that a polynomial $z_\alpha^{l_1+1} z_\beta^{m} z_\gamma^{n_1+1} - \sum_{k=0}^{n_1}\sum_{j=0}^{m_1}\sum_{i=0}^{l_1} \lambda_2(i,j,k) z_\alpha^i z_\beta^j z_\gamma^k \in S$ for any $0 \leq m \leq m_1+1$.
3) There exists the coefficients $\{\lambda_3(i,j,k)\}$ such that a polynomial $z_\alpha^{l_1+1} z_\beta^{m_1+1} z_\gamma^{n} - \sum_{k=0}^{n_1}\sum_{j=0}^{m_1}\sum_{i=0}^{l_1} \lambda_3(i,j,k) z_\alpha^i z_\beta^j z_\gamma^k \in S$ for any $0 \leq n \leq n_1+1$.

Proof. Let $J_{(l_1,m_1,n_1;l_1+1,m_1+1,n_1+1)}$: $P(\leq l_1, \leq m_1, \leq n_1) \to P(\leq l_1+1, \leq m_1+1, \leq n_1+1)$ be a canonical injection, and let $\pi_S : P(\leq l_1+1, \leq m_1+1, \leq n_1+1) \to P(\leq l_1+1, \leq m_1+1, \leq n_1+1)/S$ be the canonical surjection. Then **condition 2** implies that the composition map $\pi_S J_{(l_1,m_1,n_1;l_1+1,m_1+1,n_1+1)}$ is surjective, while **condition 1** implies that $((P(\leq l_1+1, \leq m_1+1, \leq n_1+1)/S, \dot{z}_\alpha, \dot{z}_\beta, \dot{z}_\gamma), \mathbf{1}+S)$ is (l_1, m_1, n_1)-quasi-reachable by setting $\dot{z}_\alpha(\lambda_1+S) = z_\alpha\lambda_1+S$, $\dot{z}_\beta(\lambda_2+S) = z_\beta\lambda_2+S$ and $\dot{z}_\gamma(\lambda_3+S) = z_\gamma\lambda_3+S$ for any λ_1, λ_2 and $\lambda_3 \in K[z_\alpha, z_\beta, z_\gamma]$.

Let G be the linear input map corresponding to $((P(\leq l_1+1, \leq m_1+1, \leq n_1+1)/S, \dot{z}_\alpha, \dot{z}_\beta, \dot{z}_\gamma), \mathbf{1}+S)$.

Define $G_{(l_1,m_1,n_1)} := GJ_{(l_1,m_1,n_1)}$. Then $\ker G_{(l_1,m_1,n_1)} = S$ holds. Let $\bar{S} = \ker G$. Then $\bar{S} \cap P(\leq l_1+1, \leq m_1+1, \leq n_1+1) = S$ holds. Moreover, the surjectivity of G implies that $((P(\leq l_1+1, \leq m_1+1, \leq n_1+1)/S, \dot{z}_\alpha, \dot{z}_\beta, \dot{z}_\gamma), \mathbf{1}+S)$ is isomorphic to $((K[z_\alpha, z_\beta, z_\gamma]/\bar{S}, \dot{z}_\alpha, \dot{z}_\beta, \dot{z}_\gamma), \mathbf{1}+\bar{S})$ in the sense of pointed $\{\alpha, \beta, \gamma\}$-action. Therefore, $((K[z_\alpha, z_\beta, z_\gamma]/\bar{S}, \dot{z}_\alpha, \dot{z}_\beta, \dot{z}_\gamma), \mathbf{1}+\bar{S})$ is (l_1, m_1, n_1)- quasi-reachable. The uniqueness of $\bar{S}$ comes from the uniqueness of $\dot{z}_\alpha, \dot{z}_\beta, \dot{z}_\gamma$ and the input map G.

6-C.2 $\{\alpha, \beta, \gamma\}$-Actions with a Readout Map

Let $P_{(l,m,n)}$ be the canonical surjection $P_{(l,m,n)} : F(N^3, K^t) \to F(l \times m \times n, K^t); a \mapsto [; (i,j,k) \mapsto a(i,j,k)]$.

Define $\underline{S}_\alpha, \underline{S}_\beta$ and $\underline{S}_\gamma$ as:

$$\begin{aligned}
\underline{S}_\alpha &: F(\mathbf{l}\times\mathbf{m}\times\mathbf{n},K^t)\to F((\mathbf{l}-\mathbf{1})\times\mathbf{m}\times\mathbf{n},K^t);\\
&\quad a\mapsto \underline{S}_\alpha\underline{a}\ [;(i,j,k)\mapsto \underline{a}(i+1,j,k)],\\
\underline{S}_\beta &: F(\mathbf{l}\times\mathbf{m}\times\mathbf{n},K^t)\to F(\mathbf{l}\times(\mathbf{m}-\mathbf{1})\times\mathbf{n},K^t);\\
&\quad \underline{a}\mapsto \underline{S}_\beta\underline{a}\ [;(i,j,k)\mapsto \underline{a}(i,j+1,k)] \text{ and}\\
\underline{S}_\gamma &: F(\mathbf{l}\times\mathbf{m}\times\mathbf{n},K^t)\to F(\mathbf{l}\times\mathbf{m}\times(\mathbf{n}-\mathbf{1}),K^t);\\
&\quad \underline{a}\mapsto \underline{S}_\gamma\underline{a}\ [;(i,j,k)\mapsto \underline{a}(i,j,k+1)].
\end{aligned}$$

Proposition 6-C.2. If a subspace Z of $F((\mathbf{l}_2+\mathbf{1})\times(\mathbf{m}_2+\mathbf{1})\times(\mathbf{n}_2+\mathbf{1}),K^t)$ satisfies the following two conditions, then there exists uniquely a $\{\alpha,\beta,\gamma\}$-action $(X,S_\alpha,S_\beta,S_\gamma)$ such that the map $P_{(l_2,m_2,n_2)}|_X : X\to Z$ is isomorphic, where $P_{(l_2,m_2,n_2)}|_X$ is a restriction of the canonical surjection $P_{(l_2,m_2,n_2)}$ to X, and $P_{(l_2,m_2,n_2)} : F(N^3,K^t)\to F(\mathbf{l}_2\times\mathbf{m}_2\times\mathbf{n}_2,K^t)$. Furthermore, a $\{\alpha,\beta,\gamma\}$-action with the readout map $((X,S_\alpha,S_\beta,S_\gamma),0)$ is (l_2,m_2,n_2)-distinguishable.

[**condition 3**]:
The composition map
$\pi j : Z \xrightarrow{j} F(\mathbf{l}_2\times\mathbf{m}_2\times\mathbf{n}_2,K^t)\xrightarrow{\pi} F(\mathbf{l}_2\times\mathbf{m}_2\times\mathbf{n}_2,K^t)$ is injective,
where j is the canonical injection and π is the canonical surjection.

[**condition 4**]:
im $(\underline{S}_\alpha j)\subseteq$ im $(\pi_\alpha j)$ holds in the sense of $F((\mathbf{l}_2+\mathbf{1})\times\mathbf{m}_2\times\mathbf{n}_2,K^t)$,
im $(\underline{S}_\beta j)\subseteq$ im $(\pi_\beta j)$ holds in the sense of $F(\mathbf{l}_2\times(\mathbf{m}_2+\mathbf{1})\times\mathbf{n}_2,K^t)$ and
im $(\underline{S}_\gamma j)\subseteq$ im $(\pi_\gamma j)$ holds in the sense of $F(\mathbf{l}_2\times\mathbf{m}_2\times(\mathbf{n}_2+\mathbf{1}),K^t)$,
where π_α is the canonical surjection

$\pi_\alpha : F((\mathbf{l}_2+\mathbf{1})\times(\mathbf{m}_2+\mathbf{1})\times(\mathbf{n}_2+\mathbf{1}),K^t)\to F(\mathbf{l}_2\times(\mathbf{m}_2+\mathbf{1})\times(\mathbf{n}_2+\mathbf{1}),K^t)$,
and π_β is the canonical surjection

$\pi_\beta : F((\mathbf{l}_2+\mathbf{1})\times(\mathbf{m}_2+\mathbf{1})\times(\mathbf{n}_2+\mathbf{1}),K^t)\to F((\mathbf{l}_2+\mathbf{1})\times\mathbf{m}_2\times(\mathbf{n}_2+\mathbf{1}),K^t)$
and π_γ is the canonical surjection

$\pi_\gamma : F((\mathbf{l}_2+\mathbf{1})\times(\mathbf{m}_2+\mathbf{1})\times(\mathbf{n}_2+\mathbf{1}),K^t)\to F((\mathbf{l}_2+\mathbf{1})\times(\mathbf{m}_2+\mathbf{1})\times\mathbf{n}_2,K^t)$.

Proof. Set $F_\alpha(z)=(\pi_\alpha j)^{-1}(\underline{S}_\alpha j)(z)$, $F_\beta(z)=(\pi_\beta j)^{-1}(\underline{S}_\beta j)(z)$ and $F_\gamma(z)=(\pi_\gamma j)^{-1}(\underline{S}_\gamma j)(z)$ for any $z\in Z$. Then by **condition 3** and **4**, $((Z,F_\alpha,F_\beta,F_\gamma),\underline{0})$ is a $\{\alpha,\beta,\gamma\}$-action with a readout map, where $\underline{0}{:}Z\to K^t; a\mapsto a(0,0,0)$ is a map. The injectivity of πj implies that $((Z,F_\alpha,F_\beta,F_\gamma),\underline{0})$ is (l_2,m_2,n_2)-distinguishable. It follows that the linear observation map H corresponding to $((Z,F_\alpha,F_\beta,F_\gamma),\underline{0})$ is injective. Set $X:=$ im H. Then the map $H^{-1}:X\to Z$ is clearly the restriction of the map $P_{(l_2\times m_2\times n_2)}$: $F(N^3,K^t)\to F(\mathbf{l}_2\times\mathbf{m}_2\times\mathbf{n}_2,K^t)$ to X. The equation $\underline{0}=0H$ implies that $((X, S_\alpha, S_\beta, S_\gamma), \underline{0})$ is isomorphic to $((Z, F_\alpha, F_\beta, F_\gamma), 0)$ in the sense of $\{\alpha, \beta, \gamma\}$-action with a readout map. Therefore, $((X, S_\alpha, S_\beta, S_\gamma), 0)$ is (l_2, m_2, n_2)-distinguishable. The uniqueness of X is guaranteed by the uniqueness of $F_\alpha, F_\beta, F_\gamma$ and H.

6-C.3 Partial Realization Problem

We can consider a partial linear input/output map $A_{(l_1,m_1,n_1;l_1+1,m_1+1,n_1+1)}: P(\leq l_1, \leq m_1, \leq n_1) \rightarrow F((\underline{\mathbf{L}} - \mathbf{l}_1) \times (\underline{\mathbf{M}} - \mathbf{m}_1) \times (\underline{\mathbf{N}} - \mathbf{n}_1), K^t)$ for $\underline{a} \in F(\underline{\mathbf{L}} \times \underline{\mathbf{M}} \times \underline{\mathbf{N}}, K^t)$ in the same way as the linear input/output map $A : (K[z_\alpha, z_\beta, z_\gamma], z_\alpha, z_\beta, z_\gamma) \rightarrow (F(N^3, Y), S_\alpha, S_\beta, S_\gamma)$ for $a \in F(N^3, K^t)$ presented in Appendix 6-A.1.

Lemma 6-C.3. Let $A_{(l_1,m_1,n_1;\underline{L}-l_1,\underline{M}-m_1,\underline{N}-n_1)}$ be the partial linear input/output map corresponding to $\underline{a} \in F(\underline{\mathbf{L}} \times \underline{\mathbf{M}} \times \underline{\mathbf{N}}, K^t)$. Then the following diagrams commute:

1)

$$\begin{array}{ccc} P(\leq l_1, \leq m_1, \leq n_1) & \xrightarrow{A_{(l_1,m_1,n_1\underline{L}-l_1,\underline{M}-m_1,\underline{N}-n_1)}} & F((\underline{\mathbf{L}}-\mathbf{l}_1)\times(\underline{\mathbf{M}}-\mathbf{m}_1)\times(\underline{\mathbf{N}}-\mathbf{n}_1),K^t) \\ \underline{i}\downarrow & & \downarrow\pi \\ P(\leq l_1+i, \leq m_1+j, \leq n_1+k) & \xrightarrow{A_{(l_1+i,m_1+j,n_1+k,\underline{L}-l_1-i,\underline{M}-m_1-j,\underline{N}-n_1-k)}} & F((\underline{\mathbf{L}}-\mathbf{l}_1-\mathbf{i})\times(\underline{\mathbf{M}}-\mathbf{m}_1-\mathbf{j})\times(\underline{\mathbf{N}}-\mathbf{n}_1-\mathbf{k}),K^t) \end{array}$$

where $\underline{i}$ is canonical injection and π is canonical surjection.

2)

$$\begin{array}{ccc} P(\leq l_1, \leq m_1, \leq n_1) & \xrightarrow{A_{(l_1,m_1,n_1;\underline{L}-l_1,\underline{M}-m_1,\underline{N}-n_1)}} & F((\underline{\mathbf{L}}-\mathbf{l}_1)\times(\underline{\mathbf{M}}-\mathbf{m}_1)\times(\underline{\mathbf{N}}-\mathbf{n}_1),K^t) \\ z_\alpha^i z_\beta^j z_\gamma^k\downarrow & & \downarrow\underline{S}_\alpha^i\underline{S}_\beta^j\underline{S}_\gamma^k \\ P(\leq l_1+i, \leq m_1+j, \leq n_1+k) & \xrightarrow{A_{(l_1+i,m_1+j,n_1+k;\underline{L}-l_1-i,\underline{M}-m_1-j,\underline{N}-n_1-k)}} & F((\underline{\mathbf{L}}-\mathbf{l}_1-\mathbf{i})\times(\underline{\mathbf{M}}-\mathbf{m}_1-\mathbf{j})\times(\underline{\mathbf{N}}-\mathbf{n}_1-\mathbf{k}),K^t) \end{array}$$

Proof. Direct calculation results in this lemma.

Proposition 6-C.4. Let $A_{(l_1,m_1,n_1;\underline{L}-l_1,\underline{M}-m_1,\underline{N}-n_1)}$ be the partial linear input/output map corresponding to $\underline{a} \in F(\mathbf{L} \times \mathbf{M} \times \mathbf{N}, K^t)$, and let l_2, m_2 and n_2 be any integers such that $0 \leq l_2 \leq l_1 < \underline{L}, 0 \leq m_2 \leq m_1 < \underline{M}$ and $0 \leq n_2 \leq n_1 < \underline{N}$. If
(a) im $A_{(l_2+1,m_2,n_2;\underline{L}-l_2-1,\underline{M}-m_2,\underline{N}-n_2)} =$ im $A_{(l_2,m_2,n_2;\underline{L}-l_2-1,\underline{M}-m_2,\underline{N}-n_2)}$
(b) im $A_{(l_2,m_2+1,n_2;\underline{L}-l_2,\underline{M}-m_2-1,\underline{N}-n_2)} =$ im $A_{(l_2,m_2,n_2;\underline{L}-l_2,\underline{M}-m_2-1,\underline{N}-n_2)}$
(c) im $A_{(l_2,m_2,n_2+1;\underline{L}-l_2,\underline{M}-m_2,\underline{N}-n_2-1)} =$ im $A_{(l_2,m_2,n_2;\underline{L}-l_2,\underline{M}-m_2,\underline{N}-n_2-1)}$

hold, then
im $A_{(l_1+1,m_1+1,n_1+1;\underline{L}-l_1-1,\underline{M}-m_1-1,\underline{N}-n_1-1)}$
$=$ im $A_{(l_1+1,m_1,n_1;\underline{L}-l_1-1,\underline{M}-m_1-1,\underline{N}-n_1-1)}$
$=$ im $A_{(l_1,m_1+1,n_1;\underline{L}-l_1-1,\underline{M}-m_1-1,\underline{N}-n_1-1)}$
$=$ im $A_{(l_1,m_1,n_1+1;\underline{L}-l_1-1,\underline{M}-m_1-1,\underline{N}-n_1-1)}$
$=$ im $A_{(l_1,m_1,n_1;\underline{L}-l_1-1,\underline{M}-m_1-1,\underline{N}-n_1-1)}$
hold.

Proof. Direct calculation yields this proposition.

Proposition 6-C.5. Let $A_{(l_1,m_1,n_1;\underline{L}-l_1,\underline{M}-m_1,\underline{N}-n_1)}$ be the partial linear input/output map corresponding to $\underline{a} \in F(\mathbf{L} \times \mathbf{M} \times \mathbf{N}, K^t)$, and let l_2, m_2 and n_2 be any integers such that $0 \leq l_2 \leq l_1 - 1 < \underline{L}$, $0 \leq m_2 \leq m_1 - 1 < \underline{M}$ and $0 \leq n_2 \leq n_1 - 1 < \underline{N}$.
If
(a) ker $A_{(l_1-1,m_1,n_1;\underline{L}-l_1+1,\underline{M}-m_1,\underline{N}-n_1)} =$ ker $A_{(l_1-1,m_1,n_1;\underline{L}-l_1,\underline{M}-m_1,\underline{N}-n_1)}$
(b) ker $A_{(l_1,m_1-1,n_1;\underline{L}-l_1,\underline{M}-m_1+1,\underline{N}-n_1)} =$ ker $A_{(l_1,m_1-1,n_1;\underline{L}-l_1,\underline{M}-m_1,\underline{N}-n_1)}$
(c) ker $A_{(l_1,m_1,n_1-1;\underline{L}-l_1,\underline{M}-m_1,\underline{N}-n_1+1)} =$ ker $A_{(l_1,m_1,n_1-1;\underline{L}-l_1,\underline{M}-m_1,\underline{N}-n_1)}$

hold, then
ker $A_{(l_2,m_2,n_2;\underline{L}-l_2,\underline{M}-m_2,\underline{N}-n_2)}$
$=$ ker $A_{(l_2,m_2,n_2;\underline{L}-l_2-1,\underline{M}-m_2,\underline{N}-n_2)}$
$=$ ker $A_{(l_2,m_2,n_2;\underline{L}-l_2,\underline{M}-m_2-1,\underline{N}-n_2)}$
$=$ ker $A_{(l_2,m_2,n_2;\underline{L}-l_2,\underline{M}-m_2,\underline{N}-n_2-1)}$
$=$ ker $A_{(l_2,m_2,n_2;\underline{L}-l_2-1,\underline{M}-m_2-1,\underline{N}-n_2-1)}$
hold.

Proof. This proposition is also proved by direct calculation.

Lemma 6-C.6. For a partial linear input/output map $A_{(\ ,\ ,\ ;\ ,\ ,\ )}$ corresponding to $\underline{a} \in F(\mathbf{L}\times\mathbf{M}\times\mathbf{N}, K^t)$ and 3-Commutative Linear Representation System $\sigma = ((X, F_\alpha, F_\beta, F_\gamma), x^0, h)$, the following statements hold:

1) σ is a partial realization of $\underline{a}$ if and only if the following diagram commutes for any l_1, m_1 and n_1 such that $0 \leq l_1 < \underline{L}$, $0 \leq m_1 < \underline{M}$ and $0 \leq n_1 < \underline{N}$.
2) σ is a natural partial realization of $\underline{a}$ if and only if the following diagram commutes. Moreover, $G_{(l_1,m_1,n_1)}$ is surjective and $H_{(\underline{L}-l_1-1,\underline{M}-m_1-1,\underline{L}-n_1-1)}$ is injective for any l_1, m_1 and n_1 such that

$0 \le l_1 < \underline{L}$, $0 \le m_1 < \underline{M}$ and $0 \le n_1 < \underline{N}$. Note that $G_{(l_1,m_1,n_1)} := GJ_{(l_1,m_1,n_1)}$, $H_{(l_1,m_1,n_1)} := P_{(l_1,m_1,n_1)}H$, where G is the linear input map corresponding to x^0 and H is the linear observation map corresponding to h. Furthermore, $A_{(l_1,m_1,n_1;l_2,m_2,n_2)} := H_{(l_2,m_2,n_2)}G_{(l_1,m_1,n_1)}$.

Proof. The definitions of partial realization and natural partial realization lead to this lemma.

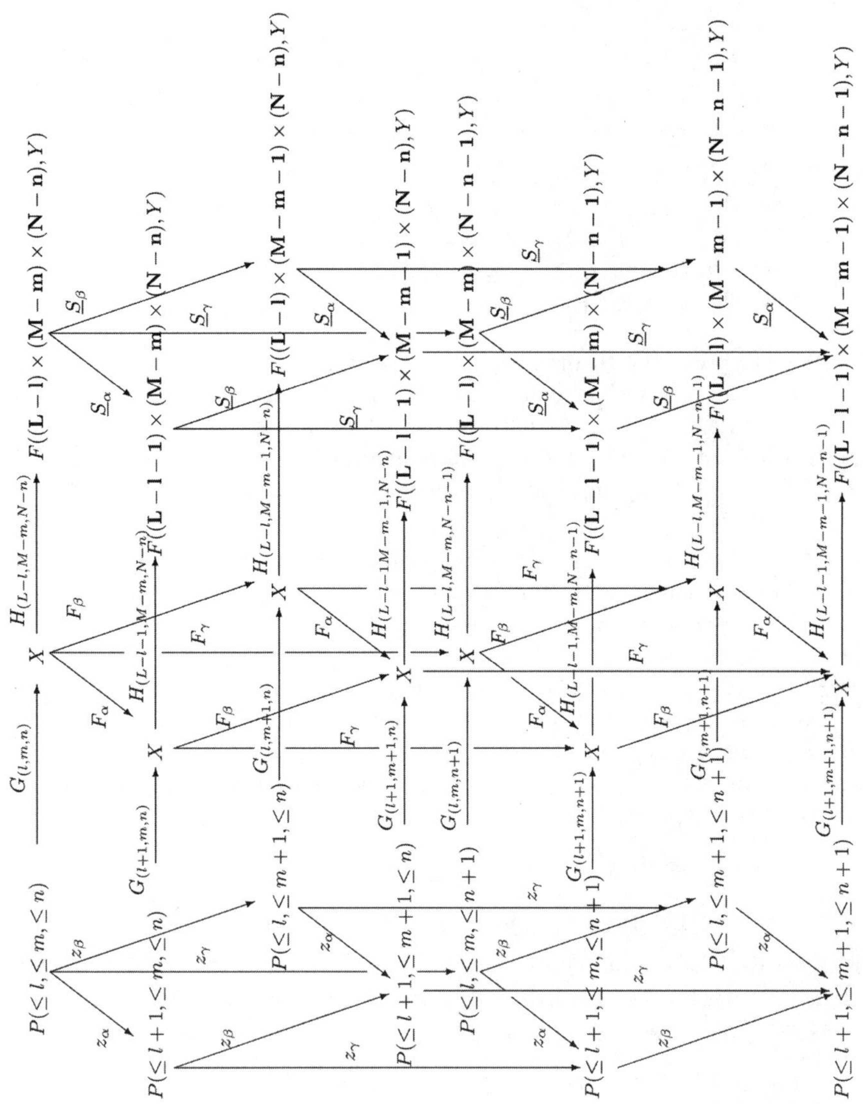

6-C.7. Proof of Theorem (6.18)
This theorem is proved by rewriting the conditions of the partial Hankel matrices in Theorem (6.18) to the partial linear input/output map $A_{(, , ; , ,)}$ corresponding to $\underline{a} \in F(\underline{\mathbf{L}} \times \underline{\mathbf{M}} \times \underline{\mathbf{N}}, K^t)$.

The conditions of Hankel matrices can be equivalently transformed into the following equations (1) and (2) by using Proposition (6-C.4) and (6-C.5).

(1) $\operatorname{im} A_{(l_1,m_1,n_1;\underline{L}-l_1-1,\underline{M}-m_1-1,\underline{N}-n_1-1)}$
$= \operatorname{im} A_{(l_1+1,m_1,n_1;\underline{L}-l_1-1,\underline{M}-m_1-1,\underline{N}-n_1-1)}$
$= \operatorname{im} A_{(l_1,m_1+1,n_1;\underline{L}-l_1-1,\underline{M}-m_1-1,\underline{N}-n_1-1)}$
$= \operatorname{im} A_{(l_1,m_1,n_1+1;\underline{L}-l_1-1,\underline{M}-m_1-1,\underline{N}-n_1-1)}$
$= \operatorname{im} A_{(l_1+1,m_1+1,n_1+1;\underline{L}-l_1-1,\underline{M}-m_1-1,\underline{N}-n_1-1)}$

(2) $\ker A_{(l_1,m_1,n_1;\underline{L}-l_1,\underline{M}-m_1,\underline{N}-n_1)}$
$= \ker A_{(l_1,m_1,n_1;\underline{L}-l_1-1,\underline{M}-m_1,\underline{N}-n_1)}$
$= \ker A_{(l_1,m_1,n_1;\underline{L}-l_1,\underline{M}-m_1-1,\underline{N}-n_1)}$
$= \ker A_{(l_1,m_1,n_1;\underline{L}-l_1,\underline{M}-m_1,\underline{N}-n_1-1)}$
$= \ker A_{(l_1,m_1,n_1;\underline{L}-l_1-1,\underline{M}-m_1-1,\underline{N}-n_1-1)}$

We will prove the theorem by using (1) and (2).

First, we prove that equations (1) and (2) are necessary. Let $\sigma = ((X, F_\alpha, F_\beta, F_\gamma), x^0, h)$ be a natural partial realization of $\underline{a} \in F(\underline{\mathbf{L}} \times \underline{\mathbf{M}} \times \underline{\mathbf{N}}, K^t)$. Then σ is (l_1, m_1, n_1)-quasi-reachable, and (l_2, m_2, n_2)-distinguishable for some l_1, m_1, n_1, l_2, m_2 and n_2 such that $l_1 + l_2 < \underline{L}$, $m_1 + m_2 < \underline{M}$ and $n_1 + n_2 < \underline{N}$.
Let G be the linear input map corresponding to x^0, and let H be the linear observation map corresponding to h. let $l_1 \le \bar{l_1}$, $m_1 \le \bar{m_1}$, $n_1 \le \bar{n_1}$, $l_2 \le \bar{l_2}$, $m_2 \le \bar{m_2}$ and $n_2 \le \bar{n_2}$. Then $G_{(\bar{l_1},\bar{m_1},\bar{n_1})} := GJ_{(\bar{l_1},\bar{m_1},\bar{n_1})}$ is onto, and $H_{(\bar{l_2},\bar{m_2},\bar{n_2})} := P_{(\bar{l_2},\bar{m_2},\bar{n_2})}H$ is one-to-one. Therefore, $A_{(\bar{l_1},\bar{m_1},\bar{n_1};\bar{l_2},\bar{m_2},\bar{n_2})} := H_{(\bar{l_2},\bar{m_2},\bar{n_2})}G_{(\bar{l_1},\bar{m_1},\bar{n_1})}$ satisfies conditions (1) and (2).

Next, let us prove that equations (1) and (2) are sufficient.
Set $S := \ker A_{(l_1+1,m_1+1,n_1+1;\underline{L}-l_1-1,\underline{M}-m_1-1,\underline{N}-n_1-1)}$ and $Z := \operatorname{im} A_{(l_1,m_1,n_1;\underline{L}-l_1,\underline{M}-m_1,\underline{N}-n_1)}$. Then equation (2) implies that the composition map

$$\pi j : Z \xrightarrow{j} F((\underline{\mathbf{L}} - \mathbf{l}_1) \times (\underline{\mathbf{M}} - \mathbf{m}_1) \times (\underline{\mathbf{N}} - \mathbf{n}_1), K^t) \xrightarrow{\pi}$$
$$F((\underline{\mathbf{L}} - \mathbf{l}_1 - \mathbf{1}) \times (\underline{\mathbf{M}} - \mathbf{m}_1 - \mathbf{1}) \times (\underline{\mathbf{N}} - \mathbf{n}_1 - \mathbf{1}), K^t)$$

is injective, where π and j are the same as presented in Proposition (6-C.2). Hence Z satisfies the **condition 3** in Proposition (6-C.2). Equation (1) implies that there exist λ_1, λ_2 and $\lambda_3 \in P(\le l_1, \le m_1, \le n_1)$ such that

$A_{(l_1+1,m_1+1,n_1+1;\underline{L}-l_1-1,\underline{M}-m_1-1,\underline{N}-n_1-1)}(z_\alpha^l z_\beta^{m_1+1} z_\gamma^{n_1+1})$
$= A_{(l_1,m_1,n_1;\underline{L}-l_1-1,\underline{M}-m_1-1,\underline{N}-n_1-1)}(\lambda_1)$,

$A_{(l_1+1,m_1+1,n_1+1;\underline{L}-l_1-1,\underline{M}-m_1-1,\underline{N}-n_1-1)}(z_\alpha^{l_1+1} z_\beta^{m} z_\gamma^{n_1+1})$

$= A_{(l_1,m_1,n_1;\underline{L}-l_1-1,\underline{M}-m_1-1,\underline{N}-n_1-1)}(\lambda_2)$ and

$A_{(l_1+1,m_1+1,n_1+1;\underline{L}-l_1-1,\underline{M}-m_1-1,\underline{N}-n_1-1)}(z_\alpha^{l_1+1} z_\beta^{m_1+1} z_\gamma^{n})$
$= A_{(l_1,m_1,n_1;\underline{L}-l_1-1,\underline{M}-m_1-1,\underline{N}-n_1-1)}(\lambda_3)$

for any integers l, m and n ($0 \le l \le l_1+1$, $0 \le m \le m_1+1$ and $0 \le n \le n_1+1$).

1) in Lemma (6-C.3) results in

$A_{(l_1+1,m_1+1,n_1+1;\underline{L}-l_1-1,\underline{M}-m_1-1,\underline{N}-n_1-1)}(z_\alpha^{l} z_\beta^{m_1+1} z_\gamma^{n_1+1} - \bar{i}\lambda_1) = 0,$

$A_{(l_1+1,m_1+1,n_1+1;\underline{L}-l_1-1,\underline{M}-m_1-1,\underline{N}-n_1-1)}(z_\alpha^{l_1+1} z_\beta^{m} z_\gamma^{n_1+1} - \bar{i}\lambda_2) = 0$ and

$A_{(l_1+1,m_1+1,n_1+1;\underline{L}-l_1-1,\underline{M}-m_1-1,\underline{N}-n_1-1)}(z_\alpha^{l_1+1} z_\beta^{m_1+1} z_\gamma^{n} - \bar{i}\lambda_3) = 0.$

Hence, $z_\alpha^{l} z_\beta^{m_1+1} z_\gamma^{n_1+1} - \bar{i}\lambda_1$, $z_\alpha^{l_1+1} z_\beta^{m} z_\gamma^{n_1+1} - \bar{i}\lambda_2$ and $z_\alpha^{l_1+1} z_\beta^{m_1+1} z_\gamma^{n} - \bar{i}\lambda_3 \in S$. These imply that S satisfies **condition 2** in Proposition (6-C.1).

Let $\bar{j}$ be the canonical injection $\bar{j} : A_{(l_1,m_1,n_1;\underline{L}-l_1-1,\underline{M}-m_1-1,\underline{N}-n_1-1)} \to F((\underline{\mathbf{L}} - \mathbf{l}_1 - \mathbf{1}) \times (\underline{\mathbf{M}} - \mathbf{m}_1 - \mathbf{1}) \times (\underline{\mathbf{N}} - \mathbf{n}_1 - \mathbf{1}), K^t)$. Let π_α, π_β and π_γ be the same as presented in the proof of Proposition (6-C.2). Then $B := \bar{j}^{-1}\pi_\alpha\pi_\beta\pi_\gamma : Z \to \text{im } \mathrm{A}_{(l_1,m_1,n_1;\underline{L}-l_1-1,\underline{M}-m_1-1,\underline{N}-n_1-1)}$ is bijective by

using equation (1). Let us consider the bijective linear map $A^b := A^b_{(l_1+1,m_1+1,n_1+1;\underline{L}-l_1-1,\underline{M}-m_1-1,\underline{N}-n_1-1)} : P(\le l_1+1, \le m_1+1, \le n_1+1)/S \to F((\underline{\mathbf{L}} - \mathbf{l}_1 - \mathbf{1}) \times (\underline{\mathbf{M}} - \mathbf{m}_1 - \mathbf{1}) \times (\underline{\mathbf{N}} - \mathbf{n}_1 - \mathbf{1}), K^t)$ associated with

$A_{(l_1+1,m_1+1,n_1+1;\underline{L}-l_1-1,\underline{M}-m_1-1,\underline{N}-n_1-1)} : P(\le l_1+1, \le m_1+1, \le n_1+1) \to F((\underline{\mathbf{L}} - \mathbf{l}_1 - 1) \times (\underline{\mathbf{M}} - \mathbf{m}_1 - 1) \times (\underline{\mathbf{N}} - \mathbf{n}_1 - 1), K^t)$.

Equation (1) implies that $B^{-1}A^b$ is a bijective linear map $B^{-1}A^b : P(\le l_1+1, \le m_1+1, \le n_1+1)/S \to Z$. The injectivity of $B^{-1}A^b$ leads to $A_{(l_1,m_1+1,n_1+1;\underline{L}-l_1,\underline{M}-m_1-1,\underline{N}-n1-1)}(\lambda) = 0$ for any $\lambda \in P(\le l_1, \le m_1+1, \le n_1+1) \cap S$. Hence,

$A_{(l_1+1,m_1+1,n_1+1;\underline{L}-l_1-1,\underline{M}-m_1-1,\underline{N}-n_1-1)}(z_\alpha\lambda)$
$= \underline{S}_\alpha A_{(l_1,m_1+1,n_1+1;\underline{L}-l_1,\underline{M}-m_1-1,\underline{N}-n_1-1)}(\lambda) = 0$

holds by using 2) in Lemma (6-C.3). This implies that $z_\alpha\lambda \in S$. In the same manner, we find $z_\beta\lambda \in S$ for any $\lambda \in P(\le l_1+1, \le m_1, \le n_1+1) \cap S$ and $z_\gamma\lambda \in S$ for any $\lambda \in P(\le l_1+1, \le m_1+1, \le n_1) \cap S$. Therefore, S satisfies **condition 1** in Proposition (6-C.1). Then Proposition (6-C.1) implies that a pointed $\{\alpha,\beta,\gamma\}$-action with a readout map $((K[z_\alpha, z_\beta, z_\gamma]/\bar{S}, \dot{z}_\alpha, \dot{z}_\beta, \dot{z}_\gamma), \mathbf{1} + \bar{S})$ is (l_1, m_1, n_1)-quasi-reachable. Let $\underline{j}_\alpha$ be the canonical injection: im $A_{(l_1,m_1,n_1;\underline{L}-l_1-1,\underline{M}-m_1,\underline{N}-n_1)} \to F((\underline{\mathbf{L}} - \mathbf{l}_1 - \mathbf{1}) \times (\underline{\mathbf{M}} - \mathbf{m}_1) \times (\underline{\mathbf{N}} - \mathbf{n}_1), K^t)$. Then equation (2) implies that the map $B_\alpha := \bar{j}_\alpha^{-1}\pi_\alpha j : Z \to \text{im } A_{(l_1,m_1,n_1;\underline{L}-l_1-1,\underline{M}-m_1,\underline{N}-n_1)}$ is bijective, while equation (1) implies that there exists $x \in \text{im } A_{(l_1,m_1,n_1;\underline{L}-l_1-1,\underline{M}-m_1,\underline{N}-n_1)}$ such that $\bar{j}_\alpha(x) = \underline{S}_\alpha j(z)$ for any $z \in Z$. Moreover, by the surjectivity of

B_α, there exists $z' \in Z$ such that $B_\alpha(z') = x$. Hence, $\underline{S}_\alpha j(z) = \bar{j}_\alpha(x) = \bar{j}_\alpha B_\alpha(z') = \pi_\alpha j(z')$, which implies im $(\underline{S}_\alpha j) \subseteq$ im $(\pi_\alpha j)$.

In the same manner, we obtain im $(\underline{S}_\beta j) \subseteq$ im $(\pi_\beta j)$ and im $(\underline{S}_\gamma j) \subseteq$ im $(\pi_\gamma j)$. It follows that Z satisfies **condition 4** in Proposition (6-C.2) and the $\{\alpha, \beta, \gamma\}$-action with readout map $((Z, F_\alpha, F_\beta, F_\gamma), \underline{0})$ is $(\underline{L} - l_1, \underline{M} - m_1, \underline{N} - n_1)$-distinguishable. Since we can also prove that $B^{-1}A^b$ is the $\{\alpha, \beta, \gamma\}$-morphism : $(P(\leq l_1 + 1, \leq m_1 + 1, \leq n_1 + 1)/S, \dot{z}_\alpha, \dot{z}_\beta, \dot{z}_\gamma) \to (Z, F_\alpha, F_\beta, F_\gamma)$, the 3-Commutative Linear Representation System $\sigma_1 = ((P(\leq l_1 + 1, \leq m_1 + 1, \leq n_1 + 1)/S, \dot{z}_\alpha, \dot{z}_\beta, \dot{z}_\gamma), \mathbf{1} + S, \underline{0}B^{-1}A^b)$ is isomorphic to the 3-Commutative Linear Representation System $\sigma_2 = ((Z, F_\alpha, F_\beta, F_\gamma), B^{-1}A^b(\mathbf{1} + S), 0)$. It follows that σ_1 and σ_2 are natural partial realizations of $\underline{a} \in F(\underline{\mathbf{L}} \times \underline{\mathbf{M}} \times \underline{\mathbf{N}}, K^t)$.

Lemma 6-C.8. Two canonical 3-Commutative Linear Representation Systems are isomorphic if and only if their behaviors are the same.

Proof. Theorem (6.3) and Corollary (6.7) lead to this lemma.

6-C.9. Proof of Theorem (6.19)
Let $A_{(\ ,\ ,\ ;\ ,\ ,\ )}$ be the partial linear input/output map corresponding to $\underline{a} \in F(\underline{\mathbf{L}} \times \underline{\mathbf{M}} \times \underline{\mathbf{N}}, K^t)$.

In order to prove necessity, we assume the existence of the natural partial realization of $\underline{a}$.

Let Theorem (6.19) hold for the integers l_1, $\bar{l_1}$, m_1, $\bar{m_1}$, n_1 and $\bar{n_1}$ such that $l_1 \neq \bar{l_1}$, $m_1 \neq \bar{m_1}$ and $n_1 \neq \bar{n_1}$. Namely,

(1) im $A_{(\dot{l_1}, \dot{m_1}, \dot{n_1}; \underline{L} - \dot{l_1} - 1, \underline{M} - \dot{m_1} - 1, \underline{N} - \dot{n_1} - 1)}$
$= \text{im } A_{(\dot{l_1} + 1, \dot{m_1}, \dot{n_1}; \underline{L} - \dot{l_1} - 1, \underline{M} - \dot{m_1} - 1, \underline{N} - \dot{n_1} - 1)}$
$= \text{im } A_{(\dot{l_1}, \dot{m_1} + 1, \dot{n_1}; \underline{L} - \dot{l_1} - 1, \underline{M} - \dot{m_1} - 1, \underline{N} - \dot{n_1} - 1)}$
$= \text{im } A_{(\dot{l_1}, \dot{m_1}, \dot{n_1} + 1; \underline{L} - \dot{i_1} - 1, \underline{M} - \dot{m_1} - 1, \underline{N} - \dot{n_1} - 1)}$

(2) ker $A_{(\dot{l_1}, \dot{m_1}, \dot{n_1}; \underline{L} - \dot{l_1}, \underline{M} - \dot{m_1}, \underline{N} - \dot{n_1})}$
$= \ker A_{(\dot{l_1}, \dot{m_1}, \dot{n_1}; \underline{L} - \dot{l_1} - 1, \underline{M} - \dot{m_1}, \underline{N} - \dot{n_1})}$
$= \ker A_{(\dot{l_1}, \dot{m_1}, \dot{n_1}; \underline{L} - \dot{l_1}, \underline{M} - \dot{m_1} - 1, \underline{N} - \dot{n_1})}$
$= \ker A_{(\dot{l_1}, \dot{m_1}, \dot{n_1}; \underline{L} - \dot{l_1}, \underline{M} - \dot{m_1}, \underline{N} - \dot{n_1} - 1)}$
$= \ker A_{(\dot{l_1}, \dot{m_1}, \dot{n_1}; \underline{L} - \dot{l_1} - 1, \underline{M} - \dot{m_1} - 1, \underline{N} - \dot{n_1} - 1)}$

hold for $\dot{l_1} = l_1$, $\dot{m_1} = m_1$, $\dot{n_1} = n_1$ and $\dot{l_1} = \bar{l_1}$, $\dot{m_1} = \bar{m_1}$, $\dot{n_1} = \bar{n_1}$.

Then Propositions (6-C.4) and (6-C.5) imply that the dimension of
$Z = \text{im } A_{(l_1, m_1, n_1; \underline{L} - l_1, \underline{M} - m_1, \underline{N} - n_1)}$ equals the dimension of
$\bar{Z} = \text{im } A_{(\bar{l_1}, \bar{m_1}, \bar{n_1}; \underline{L} - \bar{l_1}, \underline{M} - \bar{m_1}, \underline{N} - \bar{n_1})}$.

Let σ and $\bar{\sigma}$ be the natural partial realizations of $\underline{a}$ whose state space is Z, and $\bar{Z}$, respectively, where σ and $\bar{\sigma}$ are obtained by the same procedure as in (6-C.7). Then clearly σ is isomorphic to $\bar{\sigma}$. Thus, by virtue of Lemma (6-C.8), the behavior of σ is equivalent to that of $\bar{\sigma}$. This implies that the behavior of the natural partial realization is always the same regardless of different integers l_1, m_1, n_1 and $\bar{l_1}, \bar{m_1}, \bar{n_1}$. Therefore, according to Lemma (6-C.8), the natural partial realization of a is a unique modulo isomorphism.

Sufficiency is proved by contraposition. Assume that there does not exist a natural partial realization of $\underline{a} \in F(\underline{\mathbf{L}} \times \underline{\mathbf{M}} \times \underline{\mathbf{N}}, K^t)$. Then the minimum-dimensional partial realization σ of $\underline{a}$ is (l_1, m_1, n_1)-quasi-reachable and (l_2, m_2, n_2)-distinguishable for $l_1 + l_2 < \underline{L}$, $m_1 + m_2 < \underline{M}$ and $n_1 + n_2 < \underline{N}$. Moreover, there exists a state $x \in \sigma$ such that x can be reachable by $z_\alpha^{l_1} z_\beta^m z_\gamma^n$ for some $m, n \in N$. The remaining data of $F((\underline{\mathbf{L}} - \mathbf{l}_1) \times (\underline{\mathbf{M}} - \mathbf{m}) \times (\underline{\mathbf{N}} - \mathbf{n}), K^t)$ can not determine a new state $F_\alpha x$. Therefore, the transition matrix F_α can not be determined uniquely in terms of (l_2, m_2, n_2)- distinguishability. This implies that the minimum dimensional realization of $\underline{a}$ is not unique.

6-C.10. Proof of Theorem (6.20)
Consider the natural partial realization $\sigma_2 = ((Z, F_\alpha, F_\beta, F_\gamma), B^{-1}A^b(\mathbf{1} + S), 0)$ of $\underline{a} \in F(\underline{\mathbf{L}} \times \underline{\mathbf{M}} \times \underline{\mathbf{N}}, K^t)$ presented in (6-C.7).

The Quasi-reachable Standard System $\sigma_s = ((K^n, F_{\alpha s}, F_{\beta s}, F_{\gamma s}), \mathbf{e}_1, h_s)$ is derived from σ_2 as in Theorem (6.14).

7 Structures of 3-Commutative Linear Representation Systems

In this chapter, we discuss the simple state structure of 3-Commutative Linear Representation Systems, which is an extension of the linear system theory and 2-Commutative Linear Representation Systems. Based on the results, the effective encoding of three-dimensional images is discussed as an extension of the two-dimensional case.

The results of Chapter 6 can be summarized as follows:

First, we established a foundation for new realization problems of three-dimensional images. It was shown that there exist canonical, that is, quasi-reachable and distinguishable, 3-Commutative Linear Representation Systems which realize, that is, which faithfully describe arbitrary three-dimensional images, and they are isomorphic to each other.

Next, finite-dimensional 3-Commutative Linear Representation Systems were investigated with the following main results:

The representation of an isomorphic class of finite-dimensional 3-Commutative Linear Representation Systems is characterized by the Quasi-reachable Standard Systems. Necessary and sufficient conditions for three-dimensional images to be behaviors of finite-dimensional 3-Commutative Linear Representation Systems was provided by the rank condition of the Hankel matrix and the rationality of formal power series in three variables. A procedure to obtain the Quasi-reachable Standard System from a given three-dimensional image was also presented.

Finally, we dealt with partial realization problems, with the following results:

There always exists the minimal dimensional 3-Commutative Linear Representation System which realizes any given finite-sized three-dimensional image. The rank condition of the finite-sized Hankel matrix provides the necessary and sufficient condition for the unique existence of the minimal dimensional 3-Commutative Linear Representation System which realizes the given three-dimensional image. It is also found that our algorithm provides the Quasi-reachable Standard System which has partially the same behavior for the given finite-sized three-dimensional image.

This chapter investigates and clarifies the structure of finite-dimensional 3-Commutative Linear Representation Systems based on the results in Chapter 6. This matter is a new development. The structure problem can be stated as follows:

Find a 3-Commutative Linear Representation System in the class of finite-dimensional 3-Commutative Linear Representation Systems with the same behavior but which has a simpler state space and state transition.

7.1 Structure Theory of 3-Commutative Linear Representation Systems

In this section, we propose a new structure theory which is an extension of the structure theory of two-dimensional images to three-dimensional ones. A new 3-Commutative Linear Representation System called the Invariant Standard System is introduced. It is characterized by a simple state space and transition matrices. By virtue of the obtained structure theory, the transition matrix is composed of fewer parameters than in the other 3-Commutative Linear Representation Systems. We will also posit that there uniquely exists an Invariant Standard System in the isomorphic class of canonical 3-Commutative Linear Representation System with the same behavior.

Note that the structure theory of linear systems provides that the state space of a system is divided into the direct sum of sub-state spaces and its transition matrix is block-diagonalized by the direct sum decomposition. Similarly, the structure theory of three-dimensional images contains these features of the structure theory of linear systems.

Definition 7.1. A canonical n dimensional 3-Commutative Linear Representation System $\sigma_I = ((K^n, F_{\alpha I}, F_{\beta I}, F_{\gamma I}), \mathbf{e}_1, h_I)$ shown in Figure 7.1 is called an Invariant Standard System with an index $\nu = (\nu_1, \nu_2, \cdots, \nu_p)$, where $n = \sum_{i=1}^{p} \nu_i$, $\nu_i = \sum_{j=1}^{q_i} \nu_{ij}$. A $\{\alpha, \beta, \gamma\}$-action $(K^n, F_{\alpha I}, F_{\beta I}, F_{\gamma I})$ in the Invariant Standard System is represented as follows:
$\mathbf{B}_{ij} \in K^{\nu_{ij}}$ of $\mathbf{B}_i \in K^{\nu_i}$ which is the submatrix of $F_{\beta I} = [\mathbf{B}_1, \mathbf{B}_2, \cdots, \mathbf{B}_p]$ is represented as $\mathbf{B}_{ij} := [\mathbf{b}_{ij}^1, \cdots, F_{\alpha I}^{\ i\ell_j - 1}\mathbf{b}_{ij}^1, \mathbf{b}_{ij}^2, \cdots, F_{\alpha I}^{\ i\ell_{j+1} - 1}\mathbf{b}_{ij}^2, \cdots, \mathbf{b}_{ij}^{q_i - j + 1}, \cdots, F_{\alpha I}^{\ i\ell_{q_i} - 1}\mathbf{b}_{ij}^{q_i - j + 1}]$.
$\mathbf{b}_{ij}^1 \in K^n$ is given by
$\mathbf{b}_{ij}^1 = [\mathbf{b}_{ij}^{11}, \mathbf{b}_{ij}^{12}, \mathbf{b}_{ij}^{13}, \cdots, \mathbf{b}_{ij}^{1i-1}, \mathbf{b}_{ij}^{1i}, \mathbf{0}]^T$ for i, j $(1 \le i \le p,\ 1 \le j \le q_i)$,
where $\mathbf{b}_{ij}^{1\hat{i}} := [\mathbf{b}_{ij}^{1\hat{i}1}, \mathbf{b}_{ij}^{1\hat{i}2}, \cdots, \mathbf{b}_{ij}^{1\hat{i}j}, \mathbf{0}]^T \in K^{\nu_{\hat{i}}}$ for $\hat{i}$ $(1 \le \hat{i} \le i - 1)$,
$\mathbf{b}_{ij}^{1\hat{i}\hat{j}} := [\mathbf{0}, \mathbf{b}_{ij}^{1\hat{i}\hat{j}}, \mathbf{0}]^T \in K^{\nu_{\hat{i}\hat{j}}}$, $\mathbf{b}_{ij}^{1\hat{i}\hat{j}} \in K^{\nu_{ij} - \nu_{ij+1}}$ for $\hat{j}$ $(1 \le \hat{j} \le j)$.

In particular, $\mathbf{b}_{ij}^{1i} := [\mathbf{b}_{ij}^{1i1}, \mathbf{b}_{ij}^{1i2}, \cdots, \mathbf{b}_{ij}^{1ij}, \mathbf{e}_1^s, \mathbf{0}]^T \in K^{\nu_i}$,
$\mathbf{b}_{ij}^{1i\hat{j}} := [\mathbf{0}, \mathbf{b}_{ij}^{1i\hat{j}}, \mathbf{0}]^T \in K^{\nu_{i\hat{j}}}$, $\mathbf{b}_{ij}^{1i\hat{j}} \in K^{\nu_{ij} - \nu_{ij+1}}$ for $\hat{j}$ $(1 \le \hat{j} \le j)$,
$\mathbf{b}_{ij}^{1ij} := [\mathbf{0}, \mathbf{b}_{ij}^{1ij}, \mathbf{0}]^T \in K^{\nu_{ij}}$, $\mathbf{e}_1^s \in K^{\nu_{ij+1}}$.

$\mathbf{D}_{ij} \in K^{\nu_{ij}}$ of $\mathbf{D}_i \in K^{\nu_i}$ which is the submatrix of $F_{\gamma I} = [\mathbf{D}_1, \mathbf{D}_2, \cdots, \mathbf{D}_p]$ is expressed as $\mathbf{D}_{ij} := [\mathbf{d}_{ij}^1, \cdots, F_{\alpha I}^{\,{}_i\ell_j - 1}\mathbf{d}_{ij}^1, \mathbf{d}_{ij}^2, \cdots, F_{\alpha I}^{\,{}_i\ell_{j+1}-1}\mathbf{d}_{ij}^2, \cdots, \mathbf{d}_{ij}^{q_i-j+1}, \cdots, F_{\alpha I}^{\,{}_i\ell_{q_i}-1}\mathbf{d}_{ij}^{q_i-j+1}]$.

$\mathbf{d}_{i1}^1 \in K^n$ is given by

$\mathbf{d}_{i1}^1 := [\mathbf{d}_{i1}^{11}, \mathbf{d}_{i1}^{12}, \cdots, \mathbf{d}_{i1}^{1i}, \mathbf{e}_1^s, \mathbf{0}]$, $\mathbf{d}_{i1}^{1\hat{i}} = [\mathbf{d}_{i1}^{1\hat{i}1}, \underset{\smile}{\cdots,} \mathbf{d}_{i1}^{1\hat{i}j}, \mathbf{0}] \in K^{\nu_{\hat{i}}}$ for $\hat{i},\ \hat{j}\ (1 \leq \hat{i} \leq i,\ 1 \leq \hat{j} \leq j)$, where $\mathbf{d}_{i1}^{1\hat{i}\hat{j}} = [\mathbf{0}, \widetilde{\mathbf{d}_{i1}^{1\hat{i}\hat{j}}}, \mathbf{0}] \in K^{\nu_{\hat{i}\hat{j}}}$, $\widetilde{\mathbf{d}_{i1}^{1\hat{i}\hat{j}}} \in K^{\nu_{i1}-\nu_{i+11}}$, $\mathbf{e}_1^s \in K^{\nu_{i+1}}$.

Through out this chapter, for the linear space K^n and its subspace K^ℓ, $\mathbf{e}_1 \in K^n$, $\mathbf{e}_i^s \in K^\ell$ and $\mathbf{e}_1^s \in K^\ell$ are defined as:

$$\mathbf{e}_1 = [\overbrace{1, 0, \cdots, 0}^{n}]^T,\ \mathbf{e}_1^s = [\overbrace{1, 0, \cdots, 0}^{\ell}]^T,\ \mathbf{e}_i^s = [\overbrace{0, \cdots, 0, \underset{i}{1}, 0, \cdots, 0}^{\ell}]^T.$$

Theorem 7.2. There uniquely exists an Invariant Standard System $((K^n, F_{\alpha I}, F_{\beta I}, F_{\gamma I}), \mathbf{e}_1, h_I)$ which is isormorphic to any n-dimensional canonical 3-Commutative Linear Representation System $((K^n, F_\alpha, F_\beta, F_\gamma), x^0, h)$.

Proof. The proof is presented in (7-A.20) in Appendix 7-A.

Remark 1: The Invariant Standard System $((K^n, F_{\alpha I}, F_{\beta I}, F_{\gamma I}), \mathbf{e}_1, h_I)$ with an index $\nu = (\nu_1, \nu_2, \cdots, \nu_p)$ has the following properties:

(1) $K^n = K^{\nu_1} \oplus K^{\nu_2} \oplus \cdots \oplus K^{\nu_p}$, $\bar{B}_{ij} \in L(K^{\nu_j}, K^{\nu_i})$.
(2) For $i, j (1 \leq i \leq p, 1 \leq j \leq q_i)$, the minimal polynomial $\phi_{ij}(\lambda)$ of F_{ij} is represented as $\phi_{ij}(\lambda) = \chi_{ij+1}(\lambda) \cdots \chi_{iq_i}(\lambda)$, where $\phi_{ij}(\lambda)$ can be divided by the minimal polynomial $\phi_{ij+1}(\lambda)$ of F_{ij+1} in $F_{\alpha I}$. Moreover, $\phi_{ij}(\lambda)/\phi_{ij+1}(\lambda) = \chi_{ij}(\lambda)$ holds for $i,\ j\ (1 \leq i \leq p,\ 1 \leq j \leq q_i - 1)$. In particular, the minimal polynomial of $F_{\alpha I}$ is $\phi_{11}(\lambda)$. α^{ij} appeared in Figure 7.1 is represented as $\alpha^{ij} = [\alpha_1^{ij}, \alpha_2^{ij}, \cdots, \alpha_{{}_i\ell_j-1}^{ij}, \alpha_{{}_i\ell_j}^{ij}]^T$ for $\chi_{ij}(\lambda) = \lambda^{\,{}_i\ell_j} - \alpha_{{}_i\ell_j}^{ij}\lambda^{\,{}_i\ell_j-1} - \alpha_{{}_i\ell_j-1}^{ij}\lambda^{\,{}_i\ell_j-2} - \cdots - \alpha_2^{ij}\lambda - \alpha_1^{ij}$.
(3) The $\{\alpha, \beta, \gamma\}$-action $(K^n, F_{\alpha I}, F_{\beta I}, F_{\gamma I})$ in the Invariant Standard System σ_I is determined by the minimal parameters. The number of parameters is $\nu_{11} + \nu_1 + 2 \times \nu_2 + \cdots + i \times \nu_i + \cdots + p \times \nu_p + \nu_{11} + \nu_{21} + \cdots + \nu_{p1}$. See Proposition (7-A.18).

Remark 2: The structure Theorem (7.2) is a direct extension of the structure theory [(8.1) in Kalman et al., 1969] for linear systems to n-dimensional 3-Commutative Linear Representation Systems.

Remark 3: The commutativity of two matrices was discussed in [Gantmacher, 1959], while the commutativity of three matrices is discussed for the first time.

$$F_{ij} = \left[\begin{array}{cccc|c|c|c|ccccc|c|c|cccc|cc}
\multicolumn{4}{c|}{\leftarrow \; {}_i\ell_j \; \rightarrow} & & \multicolumn{6}{c|}{\leftarrow \; {}_i\ell_{j+1} \; \rightarrow} & & \cdot\;\cdot & \multicolumn{4}{c}{\leftarrow \; {}_i\ell_{q_i} \; \rightarrow} & & \\
0 & \cdot & 0 & 0 & & & & & & & & & & & & & & & \\
1 & 0 & \cdot & 0 & & & & & & & & & & & & & & & \\
0 & \cdot & & & \alpha^{ij} & & & & & & & & & & & & & & \\
\cdot & 0 & 1 & 0 & & & & & & & & & & & & & & & \\
0 & \cdot & 0 & 1 & & & & & & & & & & & & & & & \\
\hline
& & 0 & 0 & & 1 & 0 & \cdot & 0 & 0 & & & & & & & & & \\
& & 0 & 0 & & 0 & 1 & 0 & 0 & 0 & \alpha^{ij+1} & & & & & & & & \\
& & \cdot & 0 & & 0 & 0 & \cdot & & & & & & & & & & & \\
& & 0 & 0 & & \cdot & \cdot & 0 & 1 & 0 & & & & & & & & & \\
& & 0 & 0 & & 0 & 0 & \cdot & 0 & 1 & & & & & & & & & \\
\hline
& & & & & & & & & & 1 & 0 & \cdot\;\cdot\;0\;0 & & & & & & \\
& & & & & & & & & & 0 & 1 & & & & & & & \\
& & & & & & & & & & 0 & \cdot & 1 & & & & & & \\
& & & & & & & & & & \cdot & & 0\;\cdot & & & & & & \\
& & & & & & & & & & 0 & & \cdot\;0\;1\;0 & & & & & & \\
& & & & & & & & & & 0 & & \cdot\;0\;0\;1 & & & & & & \\
\hline
& & & & & & & & & & & & 0\;1 & 0 & 0 & 0 & & & \\
& & & & & & & & & & & & 0 & 1 & 0 & 0 & \alpha^{iq_i} & & \\
& & & & & & & & & & & & \cdot & 0 & 1 & 0 & & & \\
& & & & & & & & & & & & 0 & 0 & 0 & 1 & & &
\end{array}\right] \in K^{\nu_{ij} \times \nu_{ij}}$$

$$F_i = \begin{bmatrix} F_{i1} & \mathbf{0} & \cdots & \cdots & \mathbf{0} \\ \mathbf{0} & F_{i2} & \mathbf{0} & \ddots & \vdots \\ \vdots & \mathbf{0} & \ddots & \ddots & \vdots \\ \vdots & \ddots & \ddots & \ddots & \mathbf{0} \\ \mathbf{0} & \cdots & \cdots & \mathbf{0} & F_{iq_i} \end{bmatrix} \in K^{\nu_i \times \nu_i}, \quad F_{\alpha I} = \begin{bmatrix} F_1 & \mathbf{0} & \cdots & \cdots & \mathbf{0} \\ \mathbf{0} & F_2 & \mathbf{0} & \ddots & \vdots \\ \vdots & \mathbf{0} & \ddots & \ddots & \vdots \\ \vdots & \ddots & \ddots & \ddots & \mathbf{0} \\ \mathbf{0} & \cdots & \cdots & \mathbf{0} & F_p \end{bmatrix} \in K^{n \times n}$$

$F_{\beta I} = [\mathbf{B}_1, \mathbf{B}_2, \mathbf{B}_3, \cdots, \mathbf{B}_p]$, $F_{\gamma I} = [\mathbf{D}_1, \mathbf{D}_2, \mathbf{D}_3, \cdots, \mathbf{D}_p]$,

where $\mathbf{B}_i$ and $\mathbf{D}_i \in K^{n \times \nu_i}$ are expressed as:

$\mathbf{B}_i = [\mathbf{B}_{i1}, \mathbf{B}_{i2}, \mathbf{B}_{i3}, \cdots, \mathbf{B}_{iq_i}]$, $\mathbf{D}_i = [\mathbf{D}_{i1}, \mathbf{D}_{i2}, \mathbf{D}_{i3}, \cdots, \mathbf{D}_{iq_i}]$,

$\mathbf{B}_{ij} = [\mathbf{b}_{ij}^1 \cdots F_{\alpha I}^{\,{}_i\ell_j - 1} \mathbf{b}_{ij}^1 \mathbf{b}_{ij}^2 \cdots F_{\alpha I}^{\,{}_i\ell_{j+1} - 1} \mathbf{b}_{ij}^2 \cdots \mathbf{b}_{ij}^{q_i - j + 1} \cdots F_{\alpha I}^{\,{}_i\ell_{q_j} - 1} \mathbf{b}_{ij}^{q_i - j + 1}]$,

$\mathbf{D}_{ij} = [\mathbf{d}_{ij}^1 \cdots F_{\alpha I}^{\,{}_i\ell_j - 1} \mathbf{d}_{ij}^1 \mathbf{d}_{ij}^2 \cdots F_{\alpha I}^{\,{}_i\ell_{j+1} - 1} \mathbf{d}_{ij}^2 \cdots \mathbf{d}_{ij}^{q_i - j + 1} \cdots F_{\alpha I}^{\,{}_i\ell_{q_j} - 1} \mathbf{d}_{ij}^{q_i - j + 1}]$.

Furthermore, $\mathbf{b}_{ij}^{\hat{j}}$ and $\mathbf{d}_{ij}^{\hat{j}}$ $(1 \le i \le p)$ are expressed as:

$\mathbf{b}_{ij}^{\hat{j}} = \chi_{ij}(F_{\alpha I})\chi_{ij+1}(F_{\alpha I}) \cdots \chi_{ij+\hat{j}-2}(F_{\alpha I})\mathbf{b}_{ij}^1$,

$\mathbf{d}_{ij}^{\hat{j}} = \chi_{ij}(F_{\alpha I})\chi_{ij+1}(F_{\alpha I}) \cdots \chi_{ij+\hat{j}-2}(F_{\alpha I})\mathbf{d}_{ij}^1$ for $\hat{j}$ $(2 \le \hat{j} \le q_i)$,

$\phi_{ij}(F_{\alpha I})\mathbf{b}_{ij}^1 = \chi_{ij}(F_{\alpha I})\chi_{ij+1}(F_{\alpha I}) \cdots \chi_{iq_i}(F_{\alpha I})\mathbf{b}_{ij}^1 = \mathbf{0}$ and

$\phi_{ij}(F_{\alpha I})\mathbf{d}_{ij}^1 = \chi_{ij}(F_{\alpha I})\chi_{ij+1}(F_{\alpha I}) \cdots \chi_{iq_i}(F_{\alpha I})\mathbf{d}_{ij}^1 = \mathbf{0}$ for j $(1 \le j \le q_i)$.

Fig. 7.1. Invariant Standard System $\sigma_I = ((K^n, F_{\alpha I}, F_{\beta I}, F_{\gamma I}), \mathbf{e}_1, h_I)$ with an index $\nu = (\nu_1, \nu_2, \cdots, \nu_p)$.

7.2 Structure Theory and a Coding Theory of Three-Dimensional Images

In the previous section, we presented a structure theory of 3-Commutative Linear Representation Systems which contains a state structure having the fewest parameters. Therefore, we can now discuss an effective encoding of three-dimensional images. The encoding of three-dimensional images is regarded as encoding of channel. Note that an encoding of information source corresponds to the partial realization algorithm discussed in Chapter 6. We have found the algorithm to obtain the Quasi-reachable Standard System $\sigma_s = ((K^n, F_{\alpha s}, F_{\beta s}, F_{\gamma s}), \mathbf{e}_1, h_s)$ which has the same partial behavior of the given finite-sized three-dimensional image.

Let $\phi_{ij}(\lambda)$ be the minimal polynomial of the matrix F_{ij} in Figure 7.1, where

$$\begin{aligned}
&\phi_{ij}(\lambda) = \lambda^{\nu_{ij}} - c^{ij}_{\nu_1+\cdots+\nu_{i-1}+\nu_{i1}+\nu_{i2}+\cdots+\nu_{ij}}\\
&\times\lambda^{\nu_{ij}-1} - c^{ij}_{\nu_1+\cdots+\nu_{i-1}+\nu_{i1}+\nu_{i2}+\cdots+\nu_{ij}-1}\\
&\times\lambda^{\nu_{ij}-2} - \cdots - c^{ij}_{\nu_1+\cdots+\nu_{i-1}+\nu_{i1}+\nu_{i2}+\cdots+\nu_{ij-1}+2}\\
&\times\lambda - c^{ij}_{\nu_1+\cdots+\nu_{i-1}+\nu_{i1}+\nu_{i2}+\cdots+\nu_{ij-1}+1}.
\end{aligned}$$

For the $\mathbf{c}^{ij} := [c^{ij}_1, c^{ij}_2, \cdots, c^{ij}_{\nu_1+\cdots+\nu_{i-1}+\nu_{i1}+\nu_{i2}+\cdots+\nu_{ij}}, \mathbf{0}]^T$, we also use the following notations for the subsequent discussions:

$$\begin{aligned}
&\mathbf{c}^{ij} := [C^{ij}_1, C^{ij}_2, \cdots, C^{ij}_{i-1}, C^{ij}_{i1}, C^{ij}_{i2}, \cdots, C^{ij}_{ij}, \mathbf{0}]^T,\\
&C^{ij}_{\bar{i}} := [c^{ij}_{\nu_1+\cdots+\nu_{\bar{i}-1}+1}, \cdots, c^{ij}_{\nu_1+\cdots+\nu_{\bar{i}-1}+\nu_{\bar{i}}}]^T \text{ and}\\
&C^{ij}_{i\bar{j}} := [c^{ij}_{\nu_1+\cdots+\nu_{i-1}+\nu_{i1}+\cdots+\nu_{i\bar{j}-1}+1}, \cdots, c^{ij}_{\nu_1+\cdots+\nu_{i-1}+\nu_{i1}+\cdots+\nu_{i\bar{j}}}]^T
\end{aligned}$$

for $\bar{i}$ $(1 \le \bar{i} \le i-1)$ and $\bar{j}$ $(1 \le \bar{j} \le j)$.

7.3. Procedure for the effective encoding
Let $\sigma_s = ((K^n, F_{\alpha s}, F_{\beta s}, F_{\gamma s}), \mathbf{e}_1, h_s)$ be Quasi-reachable Standard System. Then Invariant Standard System $\sigma_I = ((K^n, F_{\alpha I}, F_{\beta I}, F_{\gamma I}), \mathbf{e}_1, h_I)$ which is isomorphic to σ_s is obtained by the following procedure.

Let T_s be an isomorphism $T_s : ((K^n, F_{\alpha s}, F_{\beta s}, F_{\gamma s}), \mathbf{e}_1, h_s) \to ((K^n, F_{\alpha I}, F_{\beta I}, F_{\gamma I}), \mathbf{e}_1, h_I)$, where $T_s := [\mathbf{T}_1, \mathbf{T}_2, \cdots, \mathbf{T}_p] \in K^{n\times n}$, $\mathbf{T}_i = [\mathbf{T}_{i1}, \mathbf{T}_{i2}, \cdots, \mathbf{T}_{iq_i}] \in K^{n\times\nu_i}$, $\mathbf{T}_{ij} = [\mathbf{t}^1_{ij} \cdots, F^{\nu_{ij}-1}_{\alpha I}\mathbf{t}^1_{ij}] \in K^{n\times\nu_{ij}}$ and $\mathbf{t}^1_{ij} = F^{j-1}_{\beta I} F^{i-1}_{\gamma I}\mathbf{e}_1$.

① $F_{\alpha I}$ can be found in this way:
Since $\phi_{ij}(\lambda) =$

$$\begin{aligned}
&\lambda^{\nu_{ij}} - c^{ij}_{\nu_1+\cdots+\nu_{i-1}+\nu_{i1}+\nu_{i2}+\cdots+\nu_{ij}}\\
&\times\lambda^{\nu_{ij}-1} - c^{ij}_{\nu_1+\cdots+\nu_{i-1}+\nu_{i1}+\nu_{i2}+\cdots+\nu_{ij}-1}\\
&\times\lambda^{\nu_{ij}-2} - \cdots - c^{ij}_{\nu_1+\cdots+\nu_{i-1}+\nu_{i1}+\nu_{i2}+\cdots+\nu_{ij-1}+2}\\
&\times\lambda - c^{ij}_{\nu_1+\cdots+\nu_{i-1}+\nu_{i1}+\nu_{i2}+\cdots+\nu_{ij-1}+1},
\end{aligned}$$

The following equations hold for any i $(1 \le i \le p)$:
$\phi_{iq_i}(\lambda) = \chi_{iq_i}(\lambda)$, $\phi_{iq_i-1}(\lambda) = \chi_{iq_i-1}(\lambda)\chi_{iq_i}, \cdots, \phi_{ij}(\lambda) = \chi_{ij}(\lambda)\chi_{ij+1}(\lambda)\cdots\chi_{iq_i}(\lambda), \cdots$, and $\phi_{i1}(\lambda) = \chi_{i1}(\lambda)\chi_{i2}(\lambda)\cdots\chi_{iq_i}(\lambda)$.

Then $\chi_{ij}(\lambda) = \phi_{ij}(\lambda)/\phi_{ij+1}(\lambda)(1 \le i \le p, 1 \le j \le q_i - 1)$ is calculated in turn. Hence, the companion form F_{ij} associated with the polynomial $\chi_{ij}(\lambda)$ leads the transition matrix $F_{\alpha I}$.

②

②-1 Consider $\mathbf{B}_1 \in K^{n\times\nu_1}$ of $F_{\beta I}$, where $\mathbf{B}_1 = \left[\mathbf{B}_{11}, \mathbf{B}_{12}, \cdots, \mathbf{B}_{1q_1}\right]$ and $\mathbf{B}_{1j} := [\mathbf{b}_{1j}^1, \cdots, F_{\alpha I}^{\ {}_1\ell_j-1}\mathbf{b}_{1j}^1, \mathbf{b}_{1j}^2, \cdots, F_{\alpha I}^{\ {}_1\ell_{j+1}-1}\mathbf{b}_{1j}^2, \cdots, \mathbf{b}_{1j}^{q_1-j+1} \cdots, F_{\alpha I}^{\ {}_1\ell_{q_1}-1}\mathbf{b}_{1j}^{q_1-j+1}]$. $\mathbf{b}_{11}^1$ is expressed as:
$\mathbf{b}_{11}^1 = [\mathbf{b}_{11}^{111}, \mathbf{e}_1^s, \mathbf{0}]^T$, where $\mathbf{b}_{11}^{111} = [\widetilde{\mathbf{b}_{11}^{111}}, \mathbf{0}]^T \in K^{\nu_{11}}$, $\widetilde{\mathbf{b}_{11}^{111}} \in K^{\nu_{11}-\nu_{12}}$ and $\mathbf{e}_1^s \in K^{\nu_{12}}$, ${}_1\ell_1 = \nu_{11} - \nu_{12}$.
$\widetilde{\mathbf{b}_{11}^{111}}$ is given as follows:
$\widetilde{\mathbf{b}_{11}^{111}} = \{[\phi_{12}(F_{11})\mathbf{e}_1^s, \cdots, \phi_{12}(F_{11})\mathbf{e}^s_{\ {}_1\ell_1}]^T[\phi_{12}(F_{11})\mathbf{e}_1^s, \cdots, \phi_{12}(F_{11})\mathbf{e}^s_{\ {}_1\ell_1}]\}^{-1} \times [\phi_{12}(F_{11})\mathbf{e}_1^s, \cdots, \phi_{12}(F_{11})\mathbf{e}^s_{\ {}_1\ell_1}]^T\beta_{11}^{111}$, where $\beta_{11}^{111} = [\mathbf{e}_1^s, \cdots, F_{11}^{\nu_{11}-1}\mathbf{e}_1^s]C_{11}^{12}$. $\mathbf{b}_{1j}^1$ can be expressed as follows:

$\mathbf{b}_{1j}^1 = [\mathbf{b}_{1j}^{111}, \mathbf{b}_{1j}^{112}, \cdots, \mathbf{b}_{1j}^{11j}, \mathbf{e}_1^s, \mathbf{0}]^T, \mathbf{b}_{1j}^{11\hat{\jmath}} = [\mathbf{0}, \widetilde{\mathbf{b}_{1j}^{11\hat{\jmath}}}, \mathbf{0}]^T \in K^{\nu_{1\hat{\jmath}}}$ $(1 \le \hat{\jmath} \le j)$.
$\widetilde{\mathbf{b}_{1j}^{11\hat{\jmath}}} \in K^{\nu_{1j}-\nu_{1j+1}}$ is given as follows:

$$\begin{aligned}\widetilde{\mathbf{b}_{1j}^{11\hat{\jmath}}} = &\{[\phi_{1j+1}(F_{1\hat{\jmath}})\mathbf{e}^s_{\nu_{1\hat{\jmath}}-\nu_{1j}+1} \cdots, \phi_{1j+1}(F_{1\hat{\jmath}})\mathbf{e}^s_{\nu_{1\hat{\jmath}}-\nu_{1j+1}}]^T\\ &\times [\phi_{1j+1}(F_{1\hat{\jmath}})\mathbf{e}^s_{\nu_{1\hat{\jmath}}-\nu_{1j}+1}, \cdots, \phi_{1j+1}(F_{1\hat{\jmath}})\mathbf{e}^s_{\nu_{1\hat{\jmath}}-\nu_{1j+1}}]\}^{-1}\\ &\times [\phi_{1j+1}(F_{1\hat{\jmath}})\mathbf{e}^s_{\nu_{1\hat{\jmath}}-\nu_{1j}+1}, \cdots, \phi_{1j+1}(F_{1\hat{\jmath}})\mathbf{e}^s_{\nu_{1\hat{\jmath}}-\nu_{1j+1}}]^T\\ &\times \{\beta_{1j}^{11\hat{\jmath}} - \phi_{1j+1}(F_{1\hat{\jmath}})(\mathbf{tb})_{11}^{11\hat{\jmath}}\},\end{aligned}$$

where $$\begin{aligned}\beta_{1j}^1 &:= [\mathbf{T}_{11},\ \mathbf{T}_{12},\ \cdots,\ \mathbf{T}_{1j}][C_{11}^{1j+1}, C_{12}^{1j+1}, \cdots, C_{1j}^{1j+1}]^T\\ &= [\beta_{1j}^{111}, \cdots, \beta_{1j}^{11j}, \mathbf{0}],\ \beta_{1j}^{111} \in K^{\nu_{11}}, \cdots, \beta_{1j}^{11j} \in K^{\nu_{1j}}.\end{aligned}$$

$\mathbf{b}_{1q_1}^1$ can be found by the equation $\mathbf{b}_{1q_1}^1 = T_s\mathbf{c}^{1q_1+1} - (\mathbf{tb})_{1q_1}^1$, where $(\mathbf{tb})_{1q_1}^1 = F_{\beta I}^{q_1-1}(\mathbf{b}_{11}^1 - \mathbf{e}_{\nu_{11}+1}) + F_{\beta I}^{q_1-2}(\mathbf{b}_{12}^1 - \mathbf{e}_{\nu_{11}+\nu_{12}+1}) + \cdots + F_{\beta I}(\mathbf{b}_{1q_1-1}^1 - \mathbf{e}_{\nu_{11}+\cdots+\nu_{1q_1-1}+1})$.

The number of parameter of $\mathbf{b}_{1j}^1$ is $j \times {}_1\ell_j$ $(= j \times (\nu_{1j} - \nu_{1j+1}))$ for j $(1 \le j \le q_1)$. $\mathbf{b}_{1q_1}^1 \in K^n$ of $F_{\beta I}$ is determined by $q_1 \times {}_1\ell_{q_1}$ parameters, which is the minimal parameters, where ${}_1\ell_{q_1} = \nu_{1q_1}$.

②-2 Next, consider $\mathbf{T}_1 \in K^{n\times\nu_1}$ of T_s. Since $\mathbf{t}_{1j}^1$ holds the following equation: $\mathbf{t}_{1j}^1 = (\mathbf{tb})_{1j-1}^1 + \mathbf{b}_{1j-1}^1$ for j $(2 \le j \le q_1)$, in particular, $\mathbf{t}_{11}^1 = \mathbf{e}_1$, where $(\mathbf{tb})_{1j-1}^1 = F_{\beta I}(\mathbf{tb})_{1j-2}^1 + F_{\beta I}(\mathbf{b}_{1j-2}^1 - \mathbf{e}_{\nu_{11}+\cdots+\nu_{1j-2}+1})$ for j $(3 \le j \le q_1)$ and $(\mathbf{tb})_{11}^1 = 0$,

$\mathbf{t}_{1j}^1$ is expressed as:
$\mathbf{t}_{1j}^1 = [\mathbf{t}_{1j}^{111}, \mathbf{t}_{1j}^{112}, \cdots, \mathbf{t}_{1j}^{11j}, \mathbf{e}_1^s, \mathbf{0}]^T$, where $\mathbf{t}_{1j}^{11\hat{j}} \in K^{\nu_{1\hat{j}}}$ for $\hat{j}$ $(1 \leq \hat{j} \leq j)$.

③

③-1 Let us consider $\mathbf{D}_1 \in K^{n \times \nu_1}$ of $F_{\gamma I}$, where
$\mathbf{D}_1 = \left[\mathbf{D}_{11}, \mathbf{D}_{12}, \cdots, \mathbf{D}_{1q_1}\right]$
and $\mathbf{D}_{1j} := [\mathbf{d}_{1j}^1, \cdots, F_{\alpha I}^{\ 1\ell_j - 1}\mathbf{d}_{1j}^1, \mathbf{d}_{1j}^2, \cdots, F_{\alpha I}^{\ 1\ell_{j+1}-1}\mathbf{d}_{1j}^2, \cdots,$
$\mathbf{d}_{1j}^{q_1-j+1}, \cdots, F_{\alpha I}^{\ 1\ell_{q_1}-1}\mathbf{d}_{1j}^{q_1-j+1}]$.
$\mathbf{d}_{11}^1 \in K^n$ can be expressed by $\mathbf{d}_{11}^1 = [\mathbf{d}_{11}^{11}, \mathbf{e}_1^s, \mathbf{0}, \cdots, \mathbf{0}]^T \in K^n$,
where $\mathbf{d}_{11}^{11} = [\mathbf{d}_{11}^{111}, \mathbf{0}]^T \in K^{\nu_1}$, $\mathbf{d}_{11}^{111} := [\widetilde{\mathbf{d}_{11}^{111}}, \mathbf{0}]^T \in K^{\nu_{11}}$, $\widetilde{\mathbf{d}_{11}^{111}} \in K^{\nu_{11}-\nu_{21}}$, $\mathbf{e}_1^s \in K^{\nu_2}$.
$\widetilde{\mathbf{d}_{11}^{111}}$ is given as:
$\widetilde{\mathbf{d}_{11}^{111}} = \{[\phi_{21}(F_{11})\mathbf{e}_1^s, \cdots, \phi_{21}(F_{11})\mathbf{e}_{\nu_{11}-\nu_{21}}^s]^T[\phi_{21}(F_{11})\mathbf{e}_1^s, \cdots, \phi_{21}(F_{11}) \times \mathbf{e}_{\nu_{11}-\nu_{21}}^s]\}^{-1} \times [\phi_{21}(F_{11})\mathbf{e}_1^s, \cdots, \phi_{21}(F_{11})\mathbf{e}_{\nu_{11}-\nu_{21}}^s]^T\gamma_{21}^{111}$, where $\gamma_{21}^1 = \mathbf{T}_1 C_1^{21T} = [\gamma_{21}^{111}, \mathbf{0}]^T$, $\gamma_{21}^{111} = [\mathbf{e}_1^s, F_{11}\mathbf{e}_1^s, \cdots, F_{11}^{\nu_{11}-1}\mathbf{e}_1^s]C_1^{21}$.
Once the $\mathbf{d}_{11}^1$ is found, $\mathbf{d}_{1j+1}^1$ are found recursively by the following equation:
$\mathbf{d}_{1j+1}^1 = F_{\beta I}\mathbf{d}_{1j}^1 - F_{\gamma I}(\mathbf{b}_{1j}^1 - \mathbf{e}_{\nu_{11}+\cdots+\nu_{1j}+1})$ for j $(1 \leq j \leq q_2)$.

③-2 We consider $\mathbf{T}_2 \in K^{n \times \nu_2}$ of T_s.
$\mathbf{t}_{2j}^1$ is given by the relation $\mathbf{t}_{2j}^1 = (\mathbf{td})_{1j}^1 + \mathbf{d}_{1j}^1$ for $2 \leq j \leq q_1$, where $(\mathbf{td})_{1j}^1 = F_{\beta I}(\mathbf{td})_{1j-1}^1 + F_{\gamma I}(\mathbf{d}_{1j-1}^1 - \mathbf{e}_{\nu_{11}+\cdots+\nu_{1j-1}+1})$ $(2 \leq j \leq p)$, with $(\mathbf{td})_{11}^1 = 0$.
Then $\mathbf{t}_{2j}^1$ can be expressed as $\mathbf{t}_{2j}^1 = [\mathbf{t}_{2j}^{11}, \mathbf{t}_{2j}^{12}, \mathbf{0}]^T \in K^n$, $\mathbf{t}_{2j}^{11} = [\mathbf{t}_{2j}^{111}, \cdots, \mathbf{t}_{2j}^{11j}, \mathbf{0}]^T \in K^{\nu_1}$, $\mathbf{t}_{2j}^{12} = [\mathbf{t}_{2j}^{121}, \cdots, \mathbf{t}_{2j}^{12j}, \mathbf{e}_1^s]^T \in K^{\nu_2}$, $\mathbf{t}_{2j}^{11\hat{j}} = [\mathbf{0}, \widetilde{\mathbf{t}_{2j}^{11\hat{j}}}, \mathbf{0}]^T \in K^{\nu_{1\hat{j}}}$, $\mathbf{t}_{2j}^{12\hat{j}} = [\mathbf{0}, \widetilde{\mathbf{t}_{2j}^{12\hat{j}}}, \mathbf{0}]^T \in K^{\nu_{2\hat{j}}}$ for $\hat{j}$ $(1 \leq \hat{j} \leq j)$, where $\widetilde{\mathbf{t}_{2j}^{11\hat{j}}} \in K^{\nu_{1j}-\nu_{2j}}$, $\widetilde{\mathbf{t}_{2j}^{12\hat{j}}} \in K^{\nu_{1j}-\nu_{2j}}$.

④

④-1 Now we consider $\mathbf{B}_2 \in K^{n \times \nu_2}$ of $F_{\beta I}$, where
$\mathbf{B}_2 = \left[\mathbf{B}_{21}, \mathbf{B}_{22}, \cdots, \mathbf{B}_{2q_2}\right]$.
$\mathbf{b}_{2j}^1$ can be expressed as $\mathbf{b}_{2j}^1 = [\mathbf{b}_{2j}^{11}, \mathbf{b}_{2j}^{12}, \mathbf{0}]^T \in K^n$, where
$\mathbf{b}_{2j}^{11} = [\mathbf{b}_{2j}^{111}, \cdots, \mathbf{b}_{2j}^{11j}, \mathbf{0}]^T \in K^{\nu_1}$, $\mathbf{b}_{2j}^{12} = [\mathbf{b}_{2j}^{121}, \cdots, \mathbf{b}_{2j}^{12j}, \mathbf{e}_1^s]^T \in K^{\nu_2}$,
$\mathbf{b}_{2j}^{11\hat{j}} := [\mathbf{0}, \widetilde{\mathbf{b}_{2j}^{11\hat{j}}}, \mathbf{0}]^T \in K^{\nu_{1\hat{j}}}$, $\widetilde{\mathbf{b}_{2j}^{11\hat{j}}} \in K^{\nu_{2j}-\nu_{2j+1}}$ $(2 \leq \hat{j} \leq j)$,
$\mathbf{b}_{2j}^{12\hat{j}} := [\mathbf{0}, \mathbf{b}_{2j}^{12\hat{j}}, \mathbf{0}]^T \in K^{\nu_{2\hat{j}}}$, $\widetilde{\mathbf{b}_{2j}^{12\hat{j}}} \in K^{\nu_{2j}-\nu_{2j+1}}$.
$\widetilde{\mathbf{b}_{2j}^{11\hat{j}}} \in K^{\nu_{2j}-\nu_{2j+1}}$ and $\widetilde{\mathbf{b}_{2j}^{12\hat{j}}} \in K^{\nu_{2j}-\nu_{2j+1}}$ $(1 \leq \hat{j} \leq j)$ are found as:

$$\begin{aligned}
&\widetilde{\mathbf{b}_{2j}^{11\hat{j}}} \\
&= \{[\phi_{2j+1}(F_{1\hat{j}})\mathbf{e}_{\nu_{1\hat{j}}-\nu_{2j}+1}^s, \cdots, \phi_{2j+1}(F_{1\hat{j}})\mathbf{e}_{\nu_{1\hat{j}}-\nu_{2j+1}}^s]^T \\
&\times [\phi_{2j+1}(F_{1\hat{j}})\mathbf{e}_{\nu_{1\hat{j}}-\nu_{2j}+1}^s, \cdots, \phi_{2j+1}(F_{1\hat{j}})\mathbf{e}_{\nu_{1\hat{j}}-\nu_{2j+1}}^s]\}^{-1} \\
&\times [\phi_{2j+1}(F_{1\hat{j}})\mathbf{e}_{\nu_{1\hat{j}}-\nu_{2j}+1}^s, \cdots, \phi_{2j+1}(F_{1\hat{j}})\mathbf{e}_{\nu_{1\hat{j}}-\nu_{2j+1}}^s]^T\{\beta_{1j}^{11\hat{j}} - \phi_{2j+1}(F_{1\hat{j}})
\end{aligned}$$

$\widetilde{(\mathbf{tb})_{2j}^{11\hat{j}}}\}$,

$\widetilde{\mathbf{b}_{2j}^{12\hat{j}}} = \{[\phi_{2j+1}(F_{2\hat{j}})\mathbf{e}^s_{\nu_{2\hat{j}}-\nu_{2j}+1}, \cdots, \phi_{2j+1}(F_{2\hat{j}})\mathbf{e}^s_{\nu_{2\hat{j}}-\nu_{2j+1}}]^T$
$\times [\phi_{2j+1}(F_{2\hat{j}})\mathbf{e}^s_{\nu_{2\hat{j}}-\nu_{2j}+1}, \cdots, \phi_{2j+1}(F_{2\hat{j}})\mathbf{e}^s_{\nu_{2\hat{j}}-\nu_{2j+1}}]\}^{-1}$
$\times [\phi_{2j+1}(F_{2\hat{j}})\mathbf{e}^s_{\nu_{2\hat{j}}-\nu_{2j}+1}, \cdots, \phi_{2j+1}(F_{2\hat{j}})\mathbf{e}^s_{\nu_{2\hat{j}}-\nu_{2j+1}}]^T\{\beta_{2j}^{12\hat{j}} - \phi_{2j+1}(F_{2\hat{j}})$
$(\mathbf{tb})_{2j}^{12\hat{j}}\}$. $\mathbf{b}^1_{2q_2}$ is calculated by $\mathbf{b}^1_{2q_2} = T_s\mathbf{c}^{2q_2+1} - (\mathbf{tb})^1_{2q_2}$, where
$(\mathbf{tb})^1_{2q_2} = F_{\beta I}^{q_2-1}(\mathbf{b}^1_{21} - \mathbf{e}_{\nu_1+\nu_{21}+1}) + F_{\beta I}^{q_2-2}(\mathbf{b}^1_{22} - \mathbf{e}_{\nu_1+\nu_{21}+\nu_{22}+1}) + \cdots$
$+ F_{\beta I}(\mathbf{b}^1_{2q_2-1} - \mathbf{e}_{\nu_1+\nu_{21}+\cdots+\nu_{2q_2-1}+1})$.

④-2 Let us consider $\mathbf{D}_2 \in K^{n\times\nu_2}$ of $F_{\gamma I}$, where
$\mathbf{D}_2 = \begin{bmatrix}\mathbf{D}_{21}, \mathbf{D}_{22}, \cdots, \mathbf{D}_{2q_2}\end{bmatrix}$.
$\mathbf{d}^1_{21}$ is expressed as $\mathbf{d}^1_{21} = [\mathbf{d}^{11}_{21}, \mathbf{d}^{12}_{21}, \mathbf{e}^s_1, \mathbf{0}]^T$, where $\mathbf{d}^{11}_{21} = [\mathbf{d}^{111}_{21}, \mathbf{0}]^T \in K^{\nu_1}$, $\mathbf{d}^{111}_{21} = [\mathbf{0}, \widetilde{\mathbf{d}^{111}_{21}}, \mathbf{0}]^T \in K^{\nu_{11}}$, $\widetilde{\mathbf{d}^{111}_{21}} \in K^{\nu_{21}-\nu_{31}}$, $\mathbf{d}^{12}_{21} = [\mathbf{d}^{121}_{21}, \mathbf{0}]^T \in K^{\nu_2}$, $\mathbf{d}^{121}_{21} = [\widetilde{\mathbf{d}^{121}_{21}}, \mathbf{0}]^T \in K^{\nu_{21}}$, $\widetilde{\mathbf{d}^{121}_{21}} \in K^{\nu_{21}-\nu_{31}}$, $\mathbf{e}^s_1 \in K^{\nu_3}$. $\widetilde{\mathbf{d}^{111}_{21}}$ and $\widetilde{\mathbf{d}^{121}_{21}}$ are represented as:

$\widetilde{\mathbf{d}^{111}_{21}} = \{[\phi_{31}(F_{11})\mathbf{e}^s_{\nu_{11}-\nu_{21}+1}, \cdots, \phi_{31}(F_{11})\mathbf{e}^s_{\nu_{11}-\nu_{31}}]^T$
$\times [\phi_{31}(F_{11})\mathbf{e}^s_{\nu_{11}-\nu_{21}+1}, \cdots, \phi_{31}(F_{11})\mathbf{e}^s_{\nu_{11}-\nu_{31}}]\}^{-1}$
$\times [\phi_{31}(F_{11})\mathbf{e}^s_{\nu_{11}-\nu_{21}+1}, \cdots, \phi_{31}(F_{11})\mathbf{e}^s_{\nu_{11}-\nu_{31}}]^T\gamma^{111}_{31}$,

$\widetilde{\mathbf{d}^{121}_{21}} = \{[\phi_{31}(F_{11})\mathbf{e}^s_1, \cdots, \phi_{31}(F_{11})\mathbf{e}^s_{\nu_{21}-\nu_{31}}]^T[\phi_{31}(F_{11})\mathbf{e}^s_1, \cdots, \phi_{31}(F_{11})$
$\times\mathbf{e}^s_{\nu_{21}-\nu_{31}}]\}^{-1} \times [\phi_{31}(F_{11})\mathbf{e}^s_1, \cdots, \phi_{31}(F_{11})\mathbf{e}^s_{\nu_{21}-\nu_{31}}]^T\gamma^{112}_{31}$, where $\gamma^1_{31} :=$
$[\mathbf{T}_1,\ \mathbf{T}_2][C^{31}_1, C^{31}_2]^T + [F_{\gamma I}(F_{\gamma I}\mathbf{e}_1 - \mathbf{e}_{\nu_1+1}), F_{\alpha I}F_{\gamma I}(F_{\gamma I}\mathbf{e}_1 - \mathbf{e}_{\nu_1+1})]C^{31}_{31}$,
$\gamma^1_{31} := [\gamma^{111}_{31}, \mathbf{0}, \gamma^{121}_{31}, \mathbf{0}, \gamma^{131}_{31}, \mathbf{0}]^T \in K^n$, $\gamma^{111}_{31} = [\mathbf{0}, \widetilde{\gamma^{111}_{31}}, \mathbf{0}]^T \in K^{\nu_{11}}$,
$\widetilde{\gamma^{111}_{31}} \in K^{\nu_{21}-\nu_{31}}$, $\gamma^{121}_{31} = [\mathbf{0}, \widetilde{\gamma^{121}_{31}}, \mathbf{0}]^T \in K^{\nu_{21}}$, $\widetilde{\gamma^{121}_{31}} \in K^{\nu_{21}-\nu_{31}}$.
$\mathbf{d}^1_{2j}$ can be calculated by:

$$\mathbf{d}^1_{2j+1} = F_{\beta I}\mathbf{d}^1_{2j} - F_{\gamma I}(\mathbf{b}^1_{2j} - \mathbf{e}_{\nu_1+\nu_{21}+\cdots+\nu_{2j}+1}) \text{ for } 2 \le j \le q_3 + 1.$$

$\mathbf{d}^1_{2j}$ for $q_3 + 2 \le j \le q_2$ can be determined by the following set of equations:

$$F_{\gamma I}(F_{\beta I}^{q_3-1}\mathbf{b}^1_{11} - \mathbf{e}_{\nu_1+\nu_{21}+\cdots+\nu_{2q_3}+1}) + \mathbf{d}^1_{2q_3+1} = T_s\mathbf{c}^{3q_3+1},$$
$$F_{\gamma I}(F_{\beta I}^{q_3}\mathbf{b}^1_{11} - \mathbf{e}_{\nu_1+\nu_{21}+\cdots+\nu_{2q_3+1}+1}) + \mathbf{d}^1_{2q_3+2} = F_{\beta I}T_s\mathbf{c}^{3q_3+1},$$
$$\vdots$$
$$F_{\gamma I}(F_{\beta I}^{q_2-2}\mathbf{b}^1_{11} - \mathbf{e}_{\nu_1+\nu_{21}+\cdots+\nu_{2q_2-1}+1}) + \mathbf{d}^1_{2q_2} = F_{\beta I}^{q_2-q_3-1}T_s\mathbf{c}^{3q_3+1}.$$

④-3 Now we consider $\mathbf{T}_3 \in K^{n\times\nu_3}$ of T_s. $\mathbf{t}^1_{3j}$ are found by:

$$\mathbf{t}^1_{3j} = (\mathbf{td})^1_{2j} + \mathbf{d}^1_{2j} \text{ for any } j\ (1 \le j \le q_3), \text{ where}$$

$$(\mathbf{td})^1_{2j} = F_{\beta I}(\mathbf{td})^1_{2j-1} + F_{\gamma I}(\mathbf{b}^1_{2j-1} - \mathbf{e}_{\nu_1++\nu_{21}+\cdots+\nu_{2j-1}+1}) \text{ with}$$

$$(\mathbf{td})^1_{21} = F_{\gamma I}(\mathbf{d}^1_{11} - \mathbf{e}_{\nu_1+1}).$$

⑤

⑤-1 Let us consider $\mathbf{B}_i \in K^{n\times\nu_i}$ of $F_{\beta I}$, where $\mathbf{B}_i = \left[\mathbf{B}_{i1}, \mathbf{B}_{i2}, \cdots, \mathbf{B}_{iq_i}\right]$ for i $(3 \le i \le p)$.
$\mathbf{b}^1_{ij} \in K^n$ is expressed as:
$\mathbf{b}^1_{ij} = [\mathbf{b}^{11}_{ij}, \mathbf{b}^{12}_{ij}, \mathbf{b}^{13}_{ij}, \cdots, \mathbf{b}^{1i-1}_{ij}, \mathbf{b}^{1i}_{ij}, \mathbf{0}]^T$ for i, j $(3 \le i \le p,\ 1 \le j \le q_i)$,
where $\mathbf{b}^{1\hat{i}}_{ij} := [\mathbf{b}^{1\hat{i}1}_{ij}, \mathbf{b}^{1\hat{i}2}_{ij}, \cdots, \mathbf{b}^{1\hat{i}j}_{ij}, \mathbf{0}]^T \in K^{\nu_{\hat{i}}}$ for $\hat{i}$ $(1 \le \hat{i} \le i-1)$,
$\mathbf{b}^{1\hat{i}\hat{j}}_{ij} := [\mathbf{0}, \widetilde{\mathbf{b}^{1\hat{i}\hat{j}}_{ij}}, \mathbf{0}]^T \in K^{\nu_{\hat{i}\hat{j}}}$, $\widetilde{\mathbf{b}^{1\hat{i}\hat{j}}_{ij}} \in K^{\nu_{ij}-\nu_{ij+1}}$ for $\hat{j}$ $(1 \le \hat{j} \le j)$.
In particular, $\mathbf{b}^{1i}_{ij} := [\mathbf{b}^{1i1}_{ij}, \mathbf{b}^{1i2}_{ij}, \cdots, \mathbf{b}^{1ij}_{ij}, \mathbf{b}^{1ij+1}_{ij}, \mathbf{0}]^T \in K^{\nu_i}$,
$\mathbf{b}^{1i\hat{j}}_{ij} := [\mathbf{0}, \widetilde{\mathbf{b}^{1i\hat{j}}_{ij}}, \mathbf{0}]^T \in K^{\nu_{i\hat{j}}}$, $\widetilde{\mathbf{b}^{1i\hat{j}}_{ij}} \in K^{\nu_{ij}-\nu_{ij+1}}$ for $\hat{j}$ $(1 \le \hat{j} \le j)$,
$\mathbf{b}^{1ij}_{ij} := [\mathbf{0}, \widetilde{\mathbf{b}^{1ij}_{ij}}, \mathbf{0}]^T \in K^{\nu_{ij}}$, $\mathbf{b}^{1ij+1}_{ij} = \mathbf{e}^s_1$.

$\widetilde{\mathbf{b}^{1\hat{i}\hat{j}}_{ij}}$ is represented as:

$$
\begin{aligned}
&\widetilde{\mathbf{b}^{1\hat{i}\hat{j}}_{ij}} \\
&= \{[\phi_{ij+1}(F_{\hat{i}\hat{j}})\mathbf{e}^s_{\nu_{\hat{i}\hat{j}}-\nu_{ij}+1}, \cdots, \phi_{ij+1}(F_{\hat{i}\hat{j}})\mathbf{e}^s_{\nu_{\hat{i}\hat{j}}-\nu_{ij+1}}]^T \\
&\times [\phi_{ij+1}(F_{\hat{i}\hat{j}})\mathbf{e}^s_{\nu_{\hat{i}\hat{j}}-\nu_{ij}+1}, \cdots, \phi_{ij+1}(F_{\hat{i}\hat{j}})\mathbf{e}^s_{\nu_{\hat{i}\hat{j}}-\nu_{ij+1}}]\}^{-1} \\
&\times [\phi_{ij+1}(F_{\hat{i}\hat{j}})\mathbf{e}^s_{\nu_{\hat{i}\hat{j}}-\nu_{ij}+1}, \cdots, \phi_{ij+1}(F_{\hat{i}\hat{j}})\mathbf{e}^s_{\nu_{\hat{i}\hat{j}}-\nu_{ij}}]^T \{\gamma^{1\hat{i}\hat{j}}_{ij} - \phi_{ij+1}(F_{\alpha I}) \\
&\times (\mathbf{tb})^{1\hat{i}\hat{j}}_{ij}\} \text{ for } 2 \le \hat{i} \le i,\ 2 \le \hat{j} \le j.
\end{aligned}
$$

$\mathbf{b}^1_{iq_i}$ for i $(1 \le i \le p)$ is calculated as:
$\mathbf{b}^1_{iq_i} = T_s\mathbf{c}^{iq_i+1} - (\mathbf{tb})^1_{iq_i}$, where

$$
\begin{aligned}
&(\mathbf{tb})^1_{iq_i} = \\
&F^{q_i-1}_{\beta I}(\mathbf{b}^1_{i1} - \mathbf{e}_{\nu_1+\cdots+\nu_{i-1}+\nu_{i1}+1}) + F^{q_i-2}_{\beta I}(\mathbf{b}^1_{i2} - \mathbf{e}_{\nu_1+\cdots+\nu_{i-1}+\nu_{i1}+\nu_{i2}+1}) \\
&+ \cdots + F_{\beta I}(\mathbf{b}^1_{iq_i-1} - \mathbf{e}_{\nu_1+\cdots+\nu_{i-1}+\nu_{i1}+\cdots+\nu_{iq_i-1}+1}).
\end{aligned}
$$

⑤-2 Consider $\mathbf{D}_i \in K^{n\times\nu_i}$ of $F_{\gamma I}$, where $\mathbf{D}_i = \left[\mathbf{D}_{i1}, \mathbf{D}_{i2}, \cdots, \mathbf{D}_{iq_i}\right]$ for i $(3 \le i \le p)$.
$\mathbf{d}^1_{i1} \in K^n$ $(3 \le i \le p-1)$ is expressed as
$\mathbf{d}^1_{i1} := [\mathbf{d}^{11}_{i1}, \mathbf{d}^{12}_{i1}, \cdots, \mathbf{d}^{1i}_{i1}, \mathbf{e}^s_1, \mathbf{0}]^T$, $\mathbf{d}^{1\hat{i}}_{i1} = [\mathbf{d}^{1\hat{i}1}_{i1}, \cdots, \mathbf{d}^{1\hat{i}j}_{i1}, \mathbf{0}]^T \in K^{\nu_{\hat{i}}}$ for $\hat{i}$, $\hat{j}$ $(1 \le \hat{i} \le i,\ 1 \le \hat{j} \le j)$, $\mathbf{e}^s_1 \in K^{\nu_{i+1}}$, $\mathbf{d}^{1\hat{i}\hat{j}}_{i1} = [\mathbf{0}, \widetilde{\mathbf{d}^{1\hat{i}\hat{j}}_{i1}}, \mathbf{0}]^T \in K^{\nu_{\hat{i}\hat{j}}}$, $\widetilde{\mathbf{d}^{1\hat{i}\hat{j}}_{i1}} \in K^{\nu_{i1}-\nu_{i+11}}$.

$\widetilde{\mathbf{d}^{1\hat{i}\hat{j}}_{i1}}$ for $2 \le \hat{i} \le i,\ 2 \le \hat{j} \le j$ is given as:

$$
\begin{aligned}
&\widetilde{\mathbf{d}^{1\hat{i}\hat{j}}_{i1}} \\
&= \{[\phi_{i+11}(F_{\hat{i}\hat{j}})\mathbf{e}^s_{\nu_{\hat{i}\hat{j}}-\nu_{i1}+1}, \cdots, \phi_{i+11}(F_{\hat{i}\hat{j}})\mathbf{e}^s_{\nu_{\hat{i}\hat{j}}-\nu_{i+11}}]^T \\
&\times [\phi_{i+11}(F_{\hat{i}\hat{j}})\mathbf{e}^s_{\nu_{\hat{i}\hat{j}}-\nu_{i1}+1}, \cdots, \phi_{i+11}(F_{\hat{i}\hat{j}})\mathbf{e}^s_{\nu_{\hat{i}\hat{j}}-\nu_{i+11}}]\}^{-1} \\
&\times [\phi_{i+11}(F_{\hat{i}\hat{j}})\mathbf{e}^s_{\nu_{\hat{i}\hat{j}}-\nu_{i1}+1}, \cdots, \phi_{i+11}(F_{\hat{i}\hat{j}})\mathbf{e}^s_{\nu_{\hat{i}\hat{j}}-\nu_{i+11}}]^T\ \gamma^{1\hat{i}\hat{j}}_{i+11},
\end{aligned}
$$

where $\gamma^1_{i+11} := [\mathbf{T}_1,\ \mathbf{T}_2,\ \cdots,\mathbf{T}_i][C_1^{i+11}, C_2^{i+11},\cdots,C_i^{i+11}]^T$
$+[F_{\gamma I}(F_{\gamma I}^{i-1}\mathbf{e}_1-\mathbf{e}_{\nu_1+\cdots+\nu_{i-1}+1}),\ F_{\alpha I}F_{\gamma I}(F_{\gamma I}^{i-1}\mathbf{e}_1-\mathbf{e}_{\nu_1+\cdots+\nu_{i-1}+1}),\cdots,$
$F_{\alpha I}^{\nu_{i+11}-1}F_{\gamma I}(F_{\gamma I}^{i-1}\mathbf{e}_1-\mathbf{e}_{\nu_1+\cdots+\nu_{i-1}+1})]C^{i+11}_{i+11}$.

Once the $\mathbf{d}^1_{i1}$ was found, the following equation determines $\mathbf{d}^1_{ij+1}$ as:

$\mathbf{d}^1_{ij+1} = F_{\beta I}\mathbf{d}^1_{ij} - F_{\gamma I}(\mathbf{b}^1_{ij} - \mathbf{e}_{\nu_1+\cdots+\nu_{i-1}+\nu_{i1}+\cdots+\nu_{ij}+1})$ for $1 \le j \le q_{i+1}$.

In particular, $\mathbf{d}^1_{pj+1}$ $(1 \le j \le q_p - 1)$ is determined by:

$\mathbf{d}^1_{pj+1} = F_{\beta I}\mathbf{d}^1_{pj} - F_{\gamma I}(\mathbf{b}^1_{pj} - \mathbf{e}_{\nu_1+\cdots+\nu_{p-1}+\nu_{p1}+\cdots+\nu_{pj}+1})$,

where $\mathbf{d}^1_{p1} = T_s\mathbf{c}^{p+1} - F_{\gamma I}(\mathbf{td})^1_{p-11} - F_{\gamma I}(\mathbf{d}^1_{p-11} - \mathbf{e}_{\nu_1+\cdots+\nu_{p-1}+1})$.
Once the $\mathbf{d}^1_{iq_{i+1}+1} \in K^n$ is found, $\mathbf{d}^1_{ij} \in K^n$ is found dependently by:
$\mathbf{d}^1_{ij} = [\mathbf{d}^{11}_{ij}, \mathbf{d}^{12}_{ij},\cdots,\mathbf{d}^{1j}_{ij},\mathbf{0}]^T$ for j, where $q_{i+1}+2 \le j \le q_i$.

⑤-3 Let us consider $\mathbf{T}_{i+1} \in K^{n\times\nu_2}$ of T_s for $3 \le i \le p-1$.

$\mathbf{t}^1_{i+1j}$ is calculated by:

$\mathbf{t}^1_{i+1j} = (\mathbf{td})^1_{ij} + \mathbf{d}^1_{ij}$, where

$(\mathbf{td})^1_{ij} = F_{\beta I}(\mathbf{td})^1_{ij-1} + F_{\gamma I}(\mathbf{b}^1_{ij-1} - \mathbf{e}_{\nu_1+\cdots+\nu_{i-1}+\nu_{i1}+\cdots+\nu_{ij-1}+1})$,
$(\mathbf{td})^1_{i1} = F_{\gamma I}^{i-1}(\mathbf{d}^1_{11} - \mathbf{e}_{\nu_1+1}) + F_{\gamma I}^{i-2}(\mathbf{d}^1_{21} - \mathbf{e}_{\nu_1+\nu_2+1}) + \cdots$
$+ F_{\gamma I}(\mathbf{d}^1_{i-11} - \mathbf{e}_{\nu_1+\cdots+\nu_{i-1}+1})$ for $2 \le i \le p$ with $(\mathbf{td})^1_{11} = 0$.

⑥ Let $T_s \in K^{n\times n}$ be $T_s := [\mathbf{T}_1, \mathbf{T}_2, \cdots, \mathbf{T}_p]$.
⑦ Let $F_{\beta I}$ be $F_{\beta I} := T_sF_{\beta s}T_s^{-1}$.
⑧ Let $F_{\gamma I}$ be $F_{\gamma I} := T_sF_{\gamma s}T_s^{-1}$.
⑨ Let h_I be $h_I := h_sT_s^{-1}$.

Proof. For this proof, see (7-A.21) in Appendix 7-A.

Remark: This effective encoding procedure is executed to obtain the Invariant Standard System from the Quasi-reachable Standard System. By virtue of the relation between an actual three-dimensional image and our mathematical model, the Quasi-reachable Standard System can be obtained from a given actual finite-sized three-dimensional image by using the partial realization algorithm discussed in Theorem (6.20). Therefore, we can obtain the Invariant Standard System from the finite-sized three-dimensional image.

The partial realization algorithm provides a method of encoding an information source in image processing. On the other hand, this procedure for obtaining the Invariant Standard System can be interpreted as an effective encoding of channel.

Example 7.4. Consider the image in Figure 7.2 treated in Example (6.22) of Chapter 6 as the first example of the encoding of a three-dimensional image. This image can be encoded as in the coding list shown in Figure 7.3. Let K be $N/3N$ which is the quotient field modulo the prime number 3, and let the set Y of output values be K. We have already found the

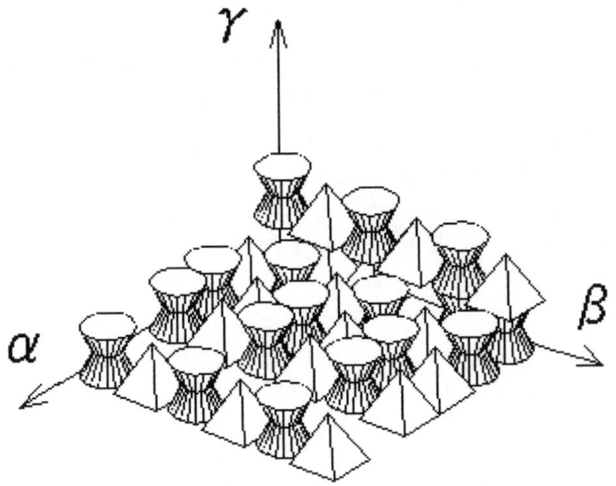

Fig. 7.2. The image treated in Example (6.22) of Chapter 6

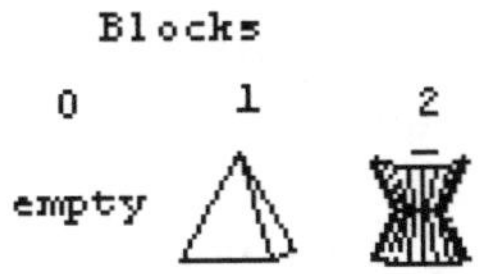

Fig. 7.3. The coding list for Figure 7.2

Quasi-reachable Standard System $\sigma_s = ((K^4, F_{\alpha s}, F_{\beta s}, F_{\gamma s}), \mathbf{e}_1, h_s)$, where

$$F_{\alpha s} = \begin{bmatrix} 0\,0\,0\,0 \\ 1\,0\,1\,2 \\ 0\,1\,2\,0 \\ 0\,0\,0\,0 \end{bmatrix}, \quad F_{\beta s} = \begin{bmatrix} 0\,0\,0\,2 \\ 0\,2\,0\,0 \\ 0\,0\,2\,0 \\ 1\,0\,0\,1 \end{bmatrix}, \quad F_{\gamma s} = \begin{bmatrix} 2\,0\,0\,1 \\ 0\,0\,0\,0 \\ 0\,0\,0\,0 \\ 2\,0\,0\,1 \end{bmatrix},$$

$h_s = [0, 2, 1, 2]$.

The vector index ν of this system is $\nu = (\nu_1)$, where $\nu_1 = (\nu_{11},\ \nu_{12})$, $\nu_{11} = 3,\ \nu_{12} = 1$. The integers q_1 and p are $q_1 = 2,\ p = 1$, respectively.

The coefficients of this system are:
$\mathbf{c}^{11} = [0,1,2]^T$, $\mathbf{c}^{12} = [0,2,0,0]^T$, $\mathbf{c}^{21} = [2,0,0,2]^T$,
$\mathbf{c}^{22} = [1,0,0,1]^T$, $\mathbf{c}^{2} = [2,0,0,2]^T$.

Invariant Standard System $\sigma_I = ((K^4, F_{\alpha I}, F_{\beta I}, F_{\gamma I}),\ \mathbf{e}_1, h_I)$ is derived by the effective encoding algorithm described in (7.3).

① $F_{\alpha s}$ leads to $\phi_{11}(\lambda) = \lambda(\lambda^2 - 2\lambda - 1)$ and $\phi_{12}(\lambda) = \lambda$. Then $\chi_{11}(\lambda) = \phi_{11}(\lambda)/\phi_{12}(\lambda) = \lambda^2 - 2\lambda - 1$ and $\chi_{12}(\lambda) = \phi_{12}(\lambda) = \lambda$. Hence

$$F_{11} = \begin{bmatrix} 0 & 1 \\ 1 & 2 \end{bmatrix}, \quad F_{12} = 0, \quad F_2 = 0.$$

Thus we arrive at $F_{\alpha I}$ in this way:

$$F_{\alpha I} = \begin{bmatrix} 0&1&0&0 \\ 1&2&0&0 \\ 0&1&0&0 \\ 0&0&0&0 \end{bmatrix},$$

where the integers ${}_1\ell_1$, ${}_1\ell_2$ and ${}_2\ell_1$ are found as:
${}_1\ell_1 = 2$, ${}_1\ell_2 = 1$, ${}_2\ell_1 = 1$.

②

②-1 The structure of $F_{\beta I}$ is represented as:

$$F_{\beta I} = \mathbf{B}_1 = [\mathbf{B}_{11}, \mathbf{B}_{12}],$$
$$\mathbf{B}_{11} = [\mathbf{b}_{11}^1\ F_{\alpha I}\mathbf{b}_{11}^1\ \mathbf{b}_{11}^2],\ \mathbf{b}_{11}^1 = [\widetilde{\mathbf{b}_{11}^{111}\ 0\ 1}]^T,$$
$$\mathbf{B}_{12} = \mathbf{b}_{12}^1,\ \mathbf{b}_{12}^1 = [0\ \widetilde{\mathbf{b}_{12}^{111}}\ 0\ \widetilde{\mathbf{b}_{12}^{121}}]^T.$$

Remark of Lemma (7-A.7) leads to $\mathbf{b}_{11}^1 = [2,0,0,1]^T$. $\mathbf{b}_{11}^2 = [0,0,2,2]^T$ follows from $\mathbf{b}_{11}^2 = \chi_{11}(F_{\alpha I})\mathbf{b}_{11}^1$ in Lemma (7-A.4).
Then $\mathbf{B}_{11}$ can be found as:

$$\mathbf{B}_{11} = \begin{bmatrix} 2&0&0 \\ 0&2&0 \\ 0&0&2 \\ 1&0&2 \end{bmatrix}.$$

②-2 The structure of the 3-Commutative Linear Representation System morphism $T_s\ : \sigma_s \to \sigma_I$ is expressed as:

$$T_s = \mathbf{T}_1 = [\mathbf{T}_{11}, \mathbf{T}_{12}],$$
$$\mathbf{T}_{11} = [\mathbf{t}_{11}^1\ F_{\alpha I}\mathbf{t}_{11}^1\ F_{\alpha I}^2\mathbf{t}_{11}^1],\ \mathbf{t}_{11}^1 = \mathbf{e}_1$$
$$\mathbf{T}_{12} = \mathbf{t}_{12}^1 = F_{\beta I}\mathbf{e}_1 = \mathbf{b}_{11}^1.$$

Then T_s can be obtained as:

$$T_s = \begin{bmatrix} 1\,0\,1\,2 \\ 0\,1\,2\,0 \\ 0\,0\,1\,0 \\ 0\,0\,0\,1 \end{bmatrix}.$$

Let us obtain $\mathbf{b}_{12}^1$. Since $\mathbf{b}_{12}^1 = T_s\mathbf{c}^{13} - F_{\beta I}(\mathbf{b}_{11}^1 - \mathbf{e}_4)$, substituting $\mathbf{c}^{13} = [2,3,0,1]^T$ yields $\mathbf{b}_{12}^1 = [0,0,0,2]^T$. Then

$$F_{\beta I} = \begin{bmatrix} 2\,0\,0\,0 \\ 0\,2\,0\,0 \\ 0\,0\,2\,0 \\ 1\,0\,2\,2 \end{bmatrix}.$$

③

③-1 The structure of $F_{\gamma I}$ is represented as:

$$F_{\gamma I} = \mathbf{D}_1 = [\mathbf{D}_{11}, \mathbf{D}_{12}],$$
$$\mathbf{D}_{11} = [\mathbf{d}_{11}^1\ F_{\alpha I}\mathbf{d}_{11}^1\ \mathbf{d}_{11}^2],\ \mathbf{d}_{11}^1 = \widetilde{\mathbf{d}_{11}^{111}}^T,$$
$$\mathbf{D}_{12} = \mathbf{d}_{12}^1,\ \mathbf{d}_{12}^1 = [\widetilde{\mathbf{d}_{12}^{111}}\ \widetilde{\mathbf{d}_{12}^{112}}]^T.$$

Since $\mathbf{d}_{11}^1 = T_s\mathbf{c}^2$, where $\mathbf{c}^2 = [2,0,0,2]^T$, we obtain $\mathbf{d}_{11}^1 = [0,0,0,2]^T$. $\mathbf{d}_{12}^1 = [0,0,0,0]^T$ follows from $\mathbf{d}_{12}^1 = F_{\beta I}T_s\mathbf{c}^2 - F_{\gamma I}(\mathbf{b}_{11}^1 - \mathbf{e}_4)$. Moreover $\mathbf{d}_{11}^2 = [0,0,0,1]^T$ follows from $\mathbf{d}_{11}^2 = \chi_{11}(F_{\alpha I})\mathbf{d}_{11}^1$ in Lemma (7-A.4).

Then $F_{\gamma I}$ is as:

$$F_{\gamma I} = \begin{bmatrix} 0\,0\,0\,0 \\ 0\,0\,0\,0 \\ 0\,0\,0\,0 \\ 2\,0\,1\,0 \end{bmatrix}.$$

④ and ⑤ are skipped because of the structure of $T_s, F_{\beta I}$ and $F_{\gamma I}$. Furthermore, ⑥ is also skipped since T_s has already given in ②.

⑦,⑧ $F_{\beta I}$ and $F_{\gamma I}$ given in this step are the same as those of ②-2 and ③-1.

⑨ $h_I = h_sT_s^{-1} = [0,2,0,2]$.
Consequently, $\sigma_I = ((K^4, F_{\alpha I}, F_{\beta I}, F_{\gamma I}), \mathbf{e}_1, h_I)$, and we have:

$$F_{\alpha I} = \begin{bmatrix} 0\,1\,0\,0 \\ 1\,2\,0\,0 \\ 0\,1\,0\,0 \\ 0\,0\,0\,0 \end{bmatrix}, \quad F_{\beta I} = \begin{bmatrix} 2\,0\,0\,0 \\ 0\,2\,0\,0 \\ 0\,0\,2\,0 \\ 1\,0\,2\,2 \end{bmatrix}, \quad F_{\gamma I} = \begin{bmatrix} 0\,0\,0\,0 \\ 0\,0\,0\,0 \\ 0\,0\,0\,0 \\ 2\,0\,1\,0 \end{bmatrix},$$

$$h_I = \begin{bmatrix} 0\,2\,0\,2 \end{bmatrix}.$$

Example 7.5. Consider the $6 \times 6 \times 6$ image depicted in Figure 7.4.

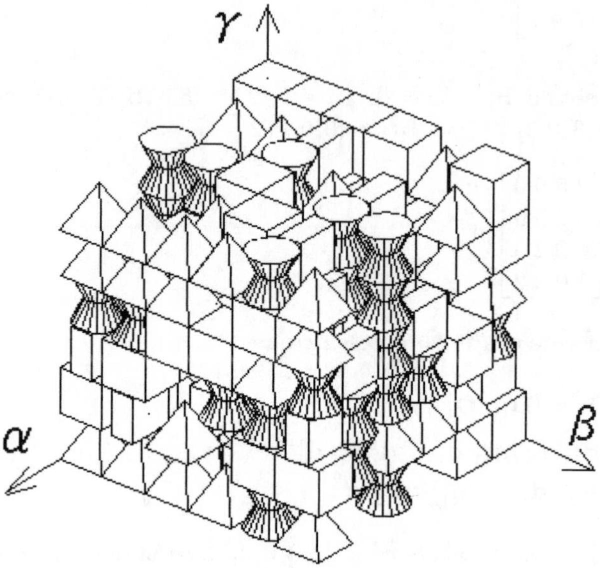

Fig. 7.4. The $6 \times 6 \times 6$ image for Example (7.5)

This image can be encoded as in the coding list shown in Figure 7.5.

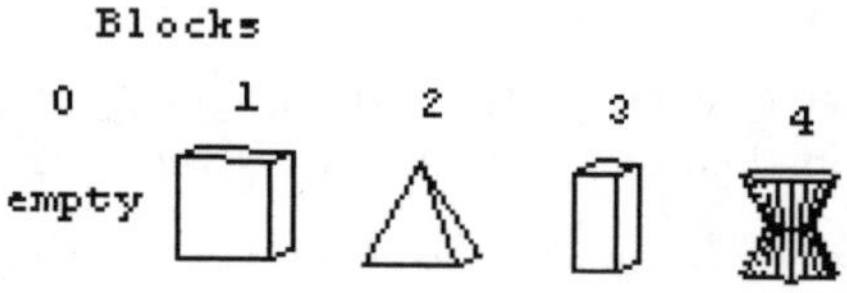

Fig. 7.5. The coding list for Figure 7.4

Let K be $N/5N$ which is the quotient field modulo the prime number 5, and let the set Y of output values be K. Applying the Theorem (6.20), the Quasi-reachable Standard System $\sigma_s = ((K^6, F_{\alpha s}, F_{\beta s}, F_{\gamma s}), \mathbf{e}_1, h_s)$ was found, where

$$F_{\alpha s} = \begin{bmatrix} 0 & 0 & 3 & 0 & 4 & 4 \\ 1 & 0 & 3 & 0 & 1 & 1 \\ 0 & 1 & 1 & 0 & 1 & 1 \\ 0 & 0 & 0 & 0 & 1 & 3 \\ 0 & 0 & 0 & 1 & 4 & 1 \\ 0 & 0 & 0 & 0 & 0 & 2 \end{bmatrix}, \quad F_{\beta s} = \begin{bmatrix} 0 & 0 & 4 & 2 & 0 & 3 \\ 0 & 0 & 1 & 1 & 2 & 1 \\ 0 & 0 & 1 & 0 & 1 & 1 \\ 1 & 0 & 1 & 2 & 0 & 4 \\ 0 & 1 & 4 & 0 & 2 & 1 \\ 0 & 0 & 0 & 0 & 0 & 1 \end{bmatrix},$$

$$F_{\gamma s} = \begin{bmatrix} 0 & 4 & 0 & 3 & 1 & 2 \\ 0 & 1 & 0 & 1 & 3 & 3 \\ 0 & 1 & 0 & 1 & 4 & 4 \\ 0 & 3 & 2 & 4 & 4 & 1 \\ 0 & 1 & 4 & 1 & 4 & 2 \\ 1 & 2 & 4 & 1 & 2 & 2 \end{bmatrix}, \quad h_s = [1, 2, 0, 1, 2, 3] \,.$$

The vector index ν of this system is $\nu = (\nu_1,\ \nu_2)$, where $\nu_1 = (\nu_{11},\ \nu_{12})$, $\nu_2 = (\nu_{21})$, $\nu_{11} = 3,\ \nu_{12} = 2,\ \nu_{21} = 1$. The integers q_1, q_2 and p are $q_1 = 2,\ q_2 = 1,\ p = 2$, respectively.

The coefficients in this system are:
$\mathbf{c}^{11} = [3, 3, 1]^T,\ \mathbf{c}^{12} = [4, 1, 1, 1, 4]^T,\ \mathbf{c}^{13} = [2, 1, 0, 2, 0]^T,$
$\mathbf{c}^{21} = [4, 1, 1, 3, 1, 2]^T,\ \mathbf{c}^{22} = [3, 1, 1, 4, 1, 1]^T,\ \mathbf{c}^{3} = [2, 3, 4, 1, 1, 2]^T.$

The effective encoding algorithm presented in (7.3) derives the Invariant Standard System $\sigma_I = ((K^6, F_{\alpha I}, F_{\beta I}, F_{\gamma I}), \mathbf{e}_1, h_I)$ in this way.

① $F_{\alpha s}$ leads to $\phi_{11}(\lambda) = (\lambda-2)^2(\lambda+3)$, $\phi_{12}(\lambda) = (\lambda-2)^2$ and $\phi_{21}(\lambda) = \lambda - 2$. Then $\chi_{11}(\lambda) = \phi_{11}(\lambda)/\phi_{12}(\lambda) = \lambda+3$, $\chi_{12}(\lambda) = \phi_{12}(\lambda) = (\lambda-2)^2$ and $\chi_{21}(\lambda) = \phi_{21}(\lambda) = \lambda - 2$. Hence

$$F_{11} = \begin{bmatrix} 2 & 0 & 0 \\ 1 & 0 & 1 \\ 0 & 1 & 4 \end{bmatrix}, \quad F_{12} = \begin{bmatrix} 0 & 1 \\ 1 & 4 \end{bmatrix}, \quad F_{21} = 2, \quad F_2 = 2.$$

Thus, $F_{\alpha I}$ can be found as:

$$F_{\alpha I} = \begin{bmatrix} 2 & 0 & 0 & 0 & 0 & 0 \\ 1 & 0 & 1 & 0 & 0 & 0 \\ 0 & 1 & 4 & 0 & 0 & 0 \\ 0 & 0 & 0 & 0 & 1 & 0 \\ 0 & 0 & 0 & 1 & 4 & 0 \\ 0 & 0 & 0 & 0 & 0 & 2 \end{bmatrix}.$$

The integers ${}_1\ell_1$, ${}_1\ell_2$ and ${}_2\ell_1$ are:
${}_1\ell_1 = 1,\ {}_1\ell_2 = 2,\ {}_2\ell_1 = 1.$

②

②-1 The structure of $F_{\beta I}$ is represented as:

$F_{\beta I} = [\mathbf{B}_1, \mathbf{B}_2] = [\mathbf{B}_{11}, \mathbf{B}_{12}, \mathbf{B}_{21}],$
$\mathbf{B}_{11} = [\mathbf{b}_{11}^1\ \mathbf{b}_{11}^2\ F_{\alpha I}\mathbf{b}_{11}^2],\ \mathbf{b}_{11}^1 = [\widetilde{\mathbf{b}_{11}^{111}}\ 0\ 0\ 1\ 0\ 0]^T,$

$\mathbf{b}_{11}^2 = \chi_{11}(F_{\alpha I})\mathbf{b}_{11}^1.$

The Remark of Lemma (7-A.7) results in $\mathbf{b}_{11}^1 = [1,0,0,1,0,0]^T$. Thus $\mathbf{b}_{11}^2 = [0,1,0,3,1,0]^T$. Then $\mathbf{B}_{11}$ is found as:

$$\mathbf{B}_{11} = \begin{bmatrix} 1\,0\,0 \\ 0\,1\,0 \\ 0\,0\,1 \\ 1\,3\,1 \\ 0\,1\,2 \\ 0\,0\,0 \end{bmatrix}.$$

②-2 The structure of the 3-Commutative Linear Representation System morphism $T_s\ : \sigma_s \to \sigma_I$ is expressed as:

$$\begin{aligned} &T_s = [\mathbf{T}_1, \mathbf{T}_2] = [\mathbf{T}_{11}, \mathbf{T}_{12}, \mathbf{T}_{21}], \\ &\mathbf{T}_{11} = [\mathbf{t}_{11}^1\ F_{\alpha I}\mathbf{t}_{11}^1\ F_{\alpha I}^2\mathbf{t}_{11}^1],\ \mathbf{t}_{11}^1 = \mathbf{e}_1, \\ &\mathbf{T}_{12} = [\mathbf{t}_{12}^1\ F_{\alpha I}\mathbf{t}_{12}^1], \qquad \mathbf{t}_{12}^1 = F_{\beta I}\mathbf{e}_1 = \mathbf{b}_{11}^1, \\ &\mathbf{T}_{21} = \mathbf{t}_{21}^1. \end{aligned}$$

$\mathbf{T}_{21}$ is derived from the constraints appeared in the proof of Proposition (7-A.11), that is, $\mathbf{T}_1 C_1^{21} = \phi_{21}(F_{\alpha I})\mathbf{t}_{21}^1$, and $C_1^{21} = [4,1,1,3,1]^T$. Then T_s is found as:

$$T_s = \begin{bmatrix} 1\,2\,4\,1\,2\,1 \\ 0\,1\,2\,0\,1\,1 \\ 0\,0\,1\,0\,0\,0 \\ 0\,0\,0\,1\,0\,1 \\ 0\,0\,0\,0\,1\,0 \\ 0\,0\,0\,0\,0\,1 \end{bmatrix}.$$

Let us find $\mathbf{B}_{12} = [\mathbf{b}_{12}^1\ F_{\alpha I}\mathbf{b}_{12}^1]$. Since $\mathbf{b}_{12}^1 = T_s\mathbf{c}^{13} - F_{\beta I}(\mathbf{b}_{11}^1 - \mathbf{e}_4)$, substituting $\mathbf{c}^{13} = [2,1,0,2,0,0]^T$ yields $\mathbf{b}_{12}^1 = [0,1,0,1,0,0]^T$. Then

$$\mathbf{B}_{12} = \begin{bmatrix} 0\,0 \\ 1\,0 \\ 0\,1 \\ 1\,0 \\ 0\,1 \\ 0\,0 \end{bmatrix}.$$

③

③-1 The structure of $F_{\gamma I}$ is represented as:

$$\begin{aligned} &F_{\gamma I} = [\mathbf{D}_1\ \mathbf{D}_2] = [\mathbf{D}_{11}\ \mathbf{D}_{12}\ \mathbf{D}_{21}], \\ &\mathbf{D}_{11} = [\mathbf{d}_{11}^1\ \mathbf{d}_{11}^2\ F_{\alpha I}\mathbf{d}_{11}^2],\ \mathbf{d}_{11}^1 = \mathbf{t}_{21}^1,\ \mathbf{d}_{11}^2 = \chi_{11}(F_{\alpha I})\mathbf{d}_{11}^1, \\ &\mathbf{D}_{12} = [\mathbf{d}_{12}^1\ F_{\alpha I}\mathbf{d}_{12}^1],\ \mathbf{d}_{12}^1 = T_s\mathbf{c}^{22} - F_{\gamma I}\mathbf{e}_1 = T_s\mathbf{c}^{22} - \mathbf{d}_{11}^1, \end{aligned}$$

$\mathbf{c}^{22} = [3, 1, 1, 4, 1, 1]^T$

Then $\mathbf{d}_{11}^1 = [1, 1, 0, 1, 0, 1]^T$, $\mathbf{d}_{11}^2 = [0, 4, 1, 3, 1, 0]^T$ and $\mathbf{d}_{12}^1 = [0, 4, 1, 4, 1, 0]^T$ are derived. Hence,

$$\mathbf{D}_1 = \begin{bmatrix} 1\,0\,0\,0\,0 \\ 1\,4\,1\,4\,1 \\ 0\,1\,3\,1\,3 \\ 1\,3\,1\,4\,1 \\ 0\,1\,2\,1\,3 \\ 1\,0\,0\,0\,0 \end{bmatrix}.$$

③-2 This step is skipped since $\mathbf{T}_2$ has already given in ②-2.

④

④-1 Let us find $\mathbf{B}_2$. According to the Proposition (7-A.12), $\mathbf{B}_2 = \mathbf{B}_{21} = \mathbf{b}_{21}^1 = F_{\gamma I}\mathbf{b}_{11}^1 - F_{\beta I}(\mathbf{d}_{11}^1 - \mathbf{e}_6)$. Then $\mathbf{B}_2 = [0, 3, 1, 0, 0, 1]$. Hence,

$$F_{\beta I} = \begin{bmatrix} 1\,0\,0\,0\,0\,0 \\ 0\,1\,0\,1\,0\,3 \\ 0\,0\,1\,0\,1\,1 \\ 1\,3\,1\,1\,0\,0 \\ 0\,1\,2\,0\,1\,0 \\ 0\,0\,0\,0\,0\,1 \end{bmatrix}.$$

④-2 Next, we will find $\mathbf{D}_2$. According to the Proposition (7-A.12), $\mathbf{D}_2 = \mathbf{D}_{21} = \mathbf{d}_{21}^1$. Multiplying $\mathbf{e}_6$ to the both side of the relation $T_s F_{\gamma s} = F_{\gamma I} T_s$ from the left yields
$\mathbf{d}_{21}^1 = T_s \mathbf{c}^3 - F_{\gamma I}(\mathbf{d}_{11}^1 - \mathbf{e}_6)$. Then $\mathbf{D}_2 = [0, 1, 2, 0, 0, 1]^T$. Hence,

$$F_{\gamma I} = \begin{bmatrix} 1\,0\,0\,0\,0\,0 \\ 1\,4\,1\,4\,1\,1 \\ 0\,1\,3\,1\,3\,2 \\ 1\,3\,1\,4\,1\,0 \\ 0\,1\,2\,1\,3\,0 \\ 1\,2\,0\,0\,0\,1 \end{bmatrix}.$$

Since $F_{\beta I}, F_{\gamma I}$ and T_s were already found, ⑤ and ⑥ are skipped.

⑦,⑧ $F_{\beta I}$ and $F_{\gamma I}$ given in this step are the same as those of ②-2 and ③-1.

⑨ $h_I = h_s T_s^{-1} = [1, 0, 1, 0, 0, 2]$.

Finally, $\sigma_I = ((K^6, F_{\alpha I}, F_{\beta I}, F_{\gamma I}), \mathbf{e}_1, h_I)$, and we have

$$F_{\alpha I} = \begin{bmatrix} 2&0&0&0&0&0 \\ 1&0&1&0&0&0 \\ 0&1&4&0&0&0 \\ 0&0&0&0&1&0 \\ 0&0&0&1&4&0 \\ 0&0&0&0&0&2 \end{bmatrix}, \quad F_{\beta I} = \begin{bmatrix} 1&0&0&0&0&0 \\ 0&1&0&1&0&3 \\ 0&0&1&0&1&1 \\ 1&3&1&1&0&0 \\ 0&1&2&0&1&0 \\ 0&0&0&0&0&1 \end{bmatrix},$$

$$F_{\gamma I} = \begin{bmatrix} 1&0&0&0&0&0 \\ 1&4&1&4&1&1 \\ 0&1&3&1&3&2 \\ 1&3&1&4&1&0 \\ 0&1&2&1&3&0 \\ 1&0&0&0&0&1 \end{bmatrix}, \quad h_I = \begin{bmatrix} 1&0&1&0&0&2 \end{bmatrix}.$$

7.3 Historical Notes and Concluding Remarks

We have derived a structure theorem for 3-Commutative Linear Representation Systems which is an extension of the theorem for discrete-time linear systems proposed in [Kalman, et al., 1969]. Employing the Invariant Standard System which has the simplest structure in the class of 3-Commutative Linear Representation Systems, we examined the structure problem and obtained new results. We derived an isomorphic Invariant Standard System from the Quasi-reachable Standard System which has a companion form. The Invariant Standard System has a companion form and it is a representative of the isomorphic class. Moreover, the system has the following properties:

① For the Invariant Standard System $\sigma_I = ((K^n, F_{\alpha I}, F_{\beta I}, F_{\gamma I}), \mathbf{e}_1, h_I)$, the state space can be represented as the direct sum of the invariant subspaces under the transition matrix $F_{\alpha I}$ in σ_I.
② The invariant subspaces can be characterized by their minimal polynomials.
③ Three transition matrices, $F_{\alpha I}$, $F_{\beta I}$ and $F_{\gamma I}$, are characterized by a minimum number of parameters, and the positions of these parameters in $F_{\alpha I}$, $F_{\beta I}$ and $F_{\gamma I}$ are also clear.

Properties ① and ② are the same as the structure theorem for discrete-time linear systems. However, property ③ is new. It was found by making use of the concept of quasi-reachability in our systems. Our results provide a new algebraic encoding method which makes the encoding of channel for three-dimensional image easy. Note that an algorithm to find the Quasi-reachable Standard System from a given three-dimensional image implies an encoding of information source.

Appendix to Chapter 7

This appendix will provide the proofs for the developments in the preceding sections. First, some facts which will be needed for the proofs are presented. Consider the pointed $\{\alpha, \beta, \gamma\}$-action $((X, F_\alpha, F_\beta, F_\gamma), x^0)$ which is expressed by the following equations:

$$\begin{cases} x(i+1,j,k) = F_\alpha x(i,j,k) \\ x(i,j+1,k) = F_\beta x(i,j,k) \\ x(i,j,k+1) = F_\gamma x(i,j,k) \\ x(0,0,0) \quad = x^0 , \end{cases}$$

for any $i, j \in N$, where $x(i,j) \in X$.

Now consider the quasi-reachable standard form $((K^n, F_{\alpha s}, F_{\beta s}, F_{\gamma s}), \mathbf{e}_1)$ which is the pointed $\{\alpha, \beta, \gamma\}$-action of Quasi-reachable Standard System $((K^n, F_{\alpha s}, F_{\beta s}, F_{\gamma s}), \mathbf{e}_1, h_s)$ with a vector index $\nu = (\nu_1, \nu_2, \cdots, \nu_p)$, which is discussed in Definition (6.10) and Definition (6-B.7).

Lemma 7-A.1. Let $((K^n, F_{\alpha s}, F_{\beta s}, F_{\gamma s}), \mathbf{e}_1)$ be the quasi-reachable standard form with a vector index $\nu = (\nu_1, \nu_2, \cdots, \nu_p)$. Then the polynomial $\phi_{11}(\lambda)$ is a minimal polynomial of the matrix $F_{\alpha s}$.

Proof. It is apparent that $\phi_{11}(F_{\alpha s})\mathbf{e}_1 = \mathbf{0}$ because of the selection of the indenpendent vectors in Quasi-reachable Standard System, where $\mathbf{e}_1 := [10\cdots0]^T \in K^n$. Then

$$\begin{aligned}
&\phi_{11}(F_{\alpha s})\mathbf{e}_2 = \phi_{11}(F_{\alpha s})F_{\alpha s}\mathbf{e}_1 = F_{\alpha s}\phi_{11}(F_{\alpha s})\mathbf{e}_1 = \mathbf{0}, \\
&\phi_{11}(F_{\alpha s})\mathbf{e}_3 = \phi_{11}(F_{\alpha s})F_{\alpha s}^2\mathbf{e}_1 = F_{\alpha s}^2\phi_{11}(F_{\alpha s})\mathbf{e}_1 = \mathbf{0}, \\
&\vdots \\
&\phi_{11}(F_{\alpha s})\mathbf{e}_{\nu_{11}+1} = \phi_{11}(F_{\alpha s})F_{\beta s}\mathbf{e}_1 = F_{\beta s}\phi_{11}(F_{\alpha s})\mathbf{e}_1 = \mathbf{0}, \\
&\vdots \\
&\phi_{11}(F_{\alpha s})\mathbf{e}_{\nu_{11}+\nu_{12}} = \phi_{11}(F_{\alpha s})F_{\alpha s}^{\nu_{12}-1}F_{\beta s}\mathbf{e}_1 = F_{\beta s}F_{\alpha s}^{\nu_{12}-1}\phi_{11}(F_{\alpha s})\mathbf{e}_1 = \mathbf{0}, \\
&\vdots \\
&\phi_{11}(F_{\alpha s})\mathbf{e}_{\nu_{11}+\cdots+\nu_{pq_p}} = \phi_{11}(F_{\alpha s})F_{\alpha s}^{\nu_p-1}F_{\beta s}^{q_p-1}F_{\gamma s}^{p-1}\mathbf{e}_1. \\
&= F_{\alpha s}^{\nu_p-1}F_{\beta s}^{q_p-1}F_{\gamma s}^{p-1}\phi_{11}(F_{\alpha s})\mathbf{e}_1 = \mathbf{0}
\end{aligned}$$

Therefore, we obtain $\phi_{11}(F_{\alpha s})\mathbf{e}_{\nu_1+\nu_2+\cdots+\nu_p} = \phi_{11}(F_{\alpha s})\mathbf{e}_n = \mathbf{0}$.

Hence, we can show that $\phi_{11}(F_{\alpha s})\mathbf{e}_i = \mathbf{0}$ for any i $(1 \le i \le n)$. We insist that $\phi_{11}(\lambda)$ is an annihilating polynomial of $F_{\alpha s}$.

Next, we will show that $\phi_{11}(\lambda)$ is a minimal polynomial of $F_{\alpha s}$. Let $\phi(\lambda)$ be a minimal polynomial of $F_{\alpha s}$ such that the degree of $\phi(\lambda)$ is less than the degree of $\phi_{11}(\lambda)$. Since $\phi(\lambda)$ is a minimal polynomial, $\phi(F_{\alpha s})\mathbf{e}_1 = \mathbf{0}$ holds. But $\phi_{11}(\lambda)$ is a minimal polynomial such that $\phi_{11}(F_{\alpha s})\mathbf{e}_1 = \mathbf{0}$ by the definition of the quasi-reachable standard form. This means that an assumption of $\phi(\lambda)$ is inconsistent with the definition of $\phi_{11}(\lambda)$.

Lemma 7-A.2. Let $((K^n, F_{\alpha s}, F_{\beta s}, F_{\gamma s}), \mathbf{e}_1)$ be the quasi-reachable standard form with a vector index $\nu = (\nu_1, \nu_2, \cdots, \nu_p)$. Let i and j be integers such that $1 \leq i \leq p$ and $1 \leq j \leq q_i$ hold.

Then $\phi_{ij}(\lambda)$ can be divided by $\phi_{ij+1}(\lambda)$. Therefore, there exist polynomials $\{\chi_{ij}(\lambda) : 1 \leq i \leq p, 1 \leq j \leq q_i\}$ such that $\phi_{ij}(\lambda) = \chi_{ij}(\lambda)\chi_{ij+1}(\lambda)\cdots\chi_{iq_i}(\lambda)$.

Moreover, $\phi_{ij}(\lambda)$ can be divided by $\phi_{i+1j}(\lambda)$ for $1 \leq i \leq p-1$, $1 \leq j \leq q_i$.

Furthermore, $\phi_{ij}(\lambda)$ can be expresed as:

$$\phi_{ij}(\lambda) =$$
$$\lambda^{\nu_{ij}} - c^{ij}_{\nu_1+\cdots+\nu_{i-1}+\nu_{i1}+\nu_{i2}+\cdots+\nu_{ij}}\lambda^{\nu_{ij}-1} - c^{ij}_{\nu_1+\cdots+\nu_{i-1}+\nu_{i1}+\nu_{i2}+\cdots+\nu_{ij}-1}\lambda^{\nu_{ij}-2} -$$
$$\cdots - c^{ij}_{\nu_1+\cdots+\nu_{i-1}+\nu_{i1}+\nu_{i2}+\cdots+\nu_{ij-1}+2}\lambda - c^{ij}_{\nu_1+\cdots+\nu_{i-1}+\nu_{i1}+\nu_{i2}+\cdots+\nu_{ij-1}+1}.$$

Proof. Each $\nu_i \times \nu_i$ $(1 \leq i \leq p)$ submatrix, which is a block-diagonal part in transition matrix $F_{\alpha s}$, has the companion form having the characteristic polynomial $\{\phi_{ij}(\lambda) : 1 \leq i \leq p,\ 1 \leq j \leq q_i\}$. Since $F_{\alpha s}$ is a block-upper triangular matrix, the characteristic polynomial of $F_{\alpha s}$ can be expressed as a product of the characteristic polynomials of each block-diagonal submatrices $\phi_{11}(\lambda)\phi_{12}(\lambda)\cdots\phi_{pq_p}(\lambda)$.

Next, consider any integers i and j. With respect to $F_{\alpha s}$ and $F_{\alpha I}$, K^n is represented as direct sum of K^{ν_1}, $K^{\nu_2}, \cdots K^{\nu_{p-1}}$ and K^{ν_p}, namely, $K^n = K^{\nu_1} \oplus K^{\nu_2} \oplus \cdots \oplus K^{\nu_p}$.

Then $\phi_{i1}(F_{\alpha s})\mathbf{e}_{\nu_1+\cdots+\nu_{i-1}+1} = \mathbf{0}$ holds because of the selection of independent vectors. Moreover, the following equations hold:

$$\phi_{i1}(F_{\alpha s})\mathbf{e}_{\nu_1+\cdots+\nu_{i-1}+2} = \phi_{i1}(F_{\alpha s})F_{\alpha s}\mathbf{e}_{\nu_1+\cdots+\nu_{i-1}+1}$$
$$= F_{\alpha s}\phi_{i1}(F_{\alpha s})\mathbf{e}_{\nu_1+\cdots+\nu_{i-1}+1} = \mathbf{0}.$$

$$\vdots$$

$$\phi_{i1}(F_{\alpha s})\mathbf{e}_{\nu_1+\cdots+\nu_{i-1}+\nu_{i1}+1} = \phi_{i1}(F_{\alpha s})F_{\beta s}\mathbf{e}_{\nu_1+\cdots+\nu_{i-1}+1}$$
$$= F_{\beta s}\phi_{i1}(F_{\alpha s})\mathbf{e}_{\nu_1+\cdots+\nu_{i-1}+1} = \mathbf{0}.$$

$$\vdots$$

$$\phi_{i1}(F_{\alpha s})\mathbf{e}_{\nu_{i1}+\cdots+\nu_{iq_i}} = \phi_{i1}(F_{\alpha s})F_{\beta s}^{q_i-1}F_{\alpha s}^{\nu_{iq_i}-1}\mathbf{e}_{\nu_1+\cdots+\nu_{i-1}+1} = \mathbf{0}.$$

Note that the determination of $\phi_{i1}(\lambda)$ implies that $\phi_{i1}(\lambda)$ is the minimal polynomial in K^{ν_i} of K^n. Hence, $\phi_{ij}(\lambda)$ can be divided by $\phi_{ij+1}(\lambda)$ in the K^{ν_i} of K^n for $1 \leq i \leq p, 1 \leq j \leq q_i - 1$. Otherwise, it cotradicts the minimality of $\phi_{i1}(\lambda)$.

Next, we prove the latter part. Suppose that $\phi_{ij}(F_{\alpha s})\mathbf{e}_r = \mathbf{0}$ hold for r, where $\nu_1 + \cdots + \nu_{i-1} + \nu_{i1} + \cdots + \nu_{ij-1} + 1 \leq r \leq \nu_{iq_i}$. Then $\phi_{ij}(F_{\alpha s})\mathbf{e}_{r+\nu_{iq_i}+\nu_{i+11}+\cdots+\nu_{i+1j-1}} = \phi_{ij}(F_{\alpha s})F_{\gamma s}\mathbf{e}_r = F_{\gamma s}\phi_{ij}(F_{\alpha s})\mathbf{e}_r = \mathbf{0}$ hold. Hence, $\phi_{ij}(\lambda)$ can be divided by $\phi_{i+1j}(\lambda)$ for i, j $(1 \leq i \leq p-1,\ 1 \leq j \leq q_i)$. Since each $\phi_{ij}(\lambda)$ is the characteristic polynomial of each block-diagonal matrix $F_{\alpha s}$, $\phi_{ij}(\lambda)$ can be expressed by using the coefficients $\{c^{ij}_{\bar{i}};\ \nu_1 + \cdots + \nu_{i-1} + \nu_{i1} + \nu_{i2} + \cdots + \nu_{ij-1} + 1 \leq \bar{i} \leq \nu_1 + \cdots + \nu_{i-1} + \nu_{i1} + \nu_{i2} + \cdots + \nu_{ij}\}$.

Remark : The characteristic polynomial of the transition matrix $F_{\alpha I}$ of the Invariant Standard System can be expressed by the product of the characteristic polynomial of the block-diagonal submatrices of $F_{\alpha I}$. Lemma (7-A.1) and Lemma (7-A.2) imply that the transition matrix $F_{\alpha s}$ of the Quasi-reachable Standard System is isomorphic to the transition matrix $F_{\alpha I}$ of Invariant Standard System.

Lemma 7-A.3. Let $\phi_{ij}(\lambda)$ be a characteristic polynomial of F_{ij}, which is the submatrix of $F_{\alpha I}$ presented in Figure 7.1, for $1 \le i \le p$, $1 \le j \le q_i$. Then rank $\phi_{i\hat{j}}(F_{ij}) = \nu_{ij} - \nu_{i\hat{j}}$ holds for $\hat{j} \ge j$.

Proof. Direct calculations supply this lemma.

Next, we will consider the other transition matrices $F_{\beta I}$, $F_{\gamma I} \in K^{n\times n}$ of the Invariant Standard System which satisfy the commutativity $F_{\alpha I}F_{\beta I} = F_{\beta I}F_{\alpha I}$, $F_{\beta I}F_{\gamma I} = F_{\gamma I}F_{\beta I}$ and $F_{\gamma I}F_{\alpha I} = F_{\alpha I}F_{\gamma I}$.

Lemma 7-A.4. Three matrices $F_{\alpha I}$, $F_{\beta I}$ and $F_{\gamma I}$ satisfy the commutativity $F_{\alpha I}F_{\beta I} = F_{\beta I}F_{\alpha I}$, $F_{\beta I}F_{\gamma I} = F_{\gamma I}F_{\beta I}$ and $F_{\gamma I}F_{\alpha I} = F_{\alpha I}F_{\gamma I}$ if and only if $F_{\beta I}$ and $F_{\gamma I}$ are represented as follows:

$F_{\beta I} = [\mathbf{B}_1, \mathbf{B}_2, \cdots, \mathbf{B}_p]$, $\quad \mathbf{B}_i = \left[\mathbf{B}_{i1}, \mathbf{B}_{i2}, \mathbf{B}_{i3}, \cdots, \mathbf{B}_{iq_i}\right] \in K^{n\times \nu_i}$,

$F_{\gamma I} = [\mathbf{D}_1, \mathbf{D}_2, \cdots, \mathbf{D}_p]$, $\quad \mathbf{D}_i = \left[\mathbf{D}_{i1}, \mathbf{D}_{i2}, \mathbf{D}_{i3}, \cdots, \mathbf{D}_{iq_i},\right] \in K^{n\times \nu_i}$,

where $\mathbf{B}_{ij}$ and $\mathbf{D}_{ij} \in K^{n\times \nu_{ij}}$ are expressed as:

$$\mathbf{B}_{ij} =$$
$$\left[\mathbf{b}_{ij}^1, \cdots, F_{\alpha I}^{\,i\ell_j - 1}\mathbf{b}_{ij}^1, \mathbf{b}_{ij}^2, \cdots, F_{\alpha I}^{\,i\ell_{j+1}-1}\mathbf{b}_{ij}^2, \cdots, \mathbf{b}_{ij}^{q_i-j+1}, \cdots, F_{\alpha I}^{\,i\ell_{q_i}-1}\mathbf{b}_{ij}^{q_i-j+1}\right],$$

$$\mathbf{D}_{ij} =$$
$$\left[\mathbf{d}_{ij}^1, \cdots, F_{\alpha I}^{\,i\ell_j - 1}\mathbf{d}_{ij}^1, \mathbf{d}_{ij}^2, \cdots, F_{\alpha I}^{\,i\ell_{j+1}-1}\mathbf{d}_{ij}^2, \cdots, \mathbf{d}_{ij}^{q_i-j+1}, \cdots, F_{\alpha I}^{\,i\ell_{q_i}-1}\mathbf{d}_{ij}^{q_i-j+1}\right].$$

$\mathbf{b}_{ij}^{\hat{j}}$ and $\mathbf{d}_{ij}^{\hat{j}}$ satisfy:

$$\mathbf{b}_{ij}^{\hat{j}} = \chi_{ij}(F_{\alpha I})\chi_{ij+1}(F_{\alpha I})\cdots\chi_{ij+\hat{j}-2}(F_{\alpha I})\mathbf{b}_{ij}^1 \text{ for } \hat{j}\ (2 \le \hat{j} \le q_i).$$
$$\mathbf{d}_{ij}^{\hat{j}} = \chi_{ij}(F_{\alpha I})\chi_{ij+1}(F_{\alpha I})\cdots\chi_{ij+\hat{j}-2}(F_{\alpha I})\mathbf{d}_{ij}^1 \text{ for } \hat{j}\ (2 \le \hat{j} \le q_i).$$

Furthermore,

$$\phi_{ij}(F_{\alpha I})\mathbf{b}_{ij}^1 = \chi_{ij}(F_{\alpha I})\chi_{ij+1}(F_{\alpha I})\cdots\chi_{iq_i}(F_{\alpha I})\mathbf{b}_{ij}^1 = \mathbf{0}\ (1 \le j \le q_i),$$
$$\phi_{ij}(F_{\alpha I})\mathbf{d}_{ij}^1 = \chi_{ij}(F_{\alpha I})\chi_{ij+1}(F_{\alpha I})\cdots\chi_{iq_i}(F_{\alpha I})\mathbf{d}_{ij}^1 = \mathbf{0}\ (1 \le j \le q_i).$$

Proof. Direct calculations lead to this lemma.

Let $\phi_{ij}(\lambda)$ be the minimal polynomial of the matrix F_{ij} in Figure 7.1. Then it can be expressed as:

$$\phi_{ij}(\lambda) =$$
$$\lambda^{\nu_{ij}} - c^{ij}_{\nu_1+\cdots+\nu_{i-1}+\nu_{i1}+\nu_{i2}+\cdots+\nu_{ij}}\lambda^{\nu_{ij}-1} - c^{ij}_{\nu_1+\cdots+\nu_{i-1}+\nu_{i1}+\nu_{i2}+\cdots+\nu_{ij}-1}\lambda^{\nu_{ij}-2}$$
$$- \cdots - c^{ij}_{\nu_1+\cdots+\nu_{i-1}+\nu_{i1}+\nu_{i2}+\cdots+\nu_{ij-1}+2}\lambda - c^{ij}_{\nu_1+\cdots+\nu_{i-1}+\nu_{i1}+\nu_{i2}+\cdots+\nu_{ij-1}+1}$$

For the $\mathbf{c}^{ij} := [c_1^{ij}, c_2^{ij}, \cdots, c_{\nu_1+\cdots+\nu_{i-1}+\nu_{i1}+\nu_{i2}+\cdots+\nu_{ij}}^{ij}, \mathbf{0}]^T$, the following notations are also used in the subsequent discussions:

$\mathbf{c}^{ij} := [C_1^{ij}, C_2^{ij}, \cdots, C_{i-1}^{ij}, C_{i1}^{ij}, C_{i2}^{ij}, \cdots, C_{ij}^{ij}, \mathbf{0}]^T,$
$C_{\bar{i}}^{ij} := [c_{\nu_1+\cdots+\nu_{\bar{i}-1}+1}^{ij}, \cdots, c_{\nu_1+\cdots+\nu_{\bar{i}-1}+\nu_{\bar{i}}}^{ij}]^T$ and
$C_{i\bar{j}}^{ij} := [c_{\nu_1+\cdots+\nu_{i-1}+\nu_{i1}+\cdots+\nu_{i\bar{j}-1}+1}^{ij}, \cdots, c_{\nu_1+\cdots+\nu_{i-1}+\nu_{i1}+\cdots+\nu_{i\bar{j}}}^{ij}]^T$

for $1 \leq \bar{i} \leq i-1)$ and $1 \leq \bar{j} \leq j$.

Lemma 7-A.5. Let $(K^n, F_{\alpha s}, F_{\beta s}, F_{\gamma s})$ be $\{\alpha, \beta, \gamma\}$-action of the quasi-reachable standard form $((K^n, F_{\alpha s}, F_{\beta s}, F_{\gamma s}), \mathbf{e}_1)$ with a vector index $\nu = (\nu_1, \nu_2, \cdots, \nu_p)$ and let $(K^n, F_{\alpha s}, F_{\beta s}, F_{\gamma s})$ be the $\{\alpha, \beta, \gamma\}$-action. Then T_s is the pointed $\{\alpha, \beta, \gamma\}$-morphism $T_s : ((K^n, F_{\alpha s}, F_{\beta s}, F_{\gamma s}), \mathbf{e}_1) \to ((K^n, F_{\alpha I}, F_{\beta I}, F_{\gamma I}), \mathbf{e}_1)$, that is which satisfies $F_{\alpha I}T_s = T_sF_{\alpha s}$, $F_{\beta I}T_s = T_sF_{\beta s}$, $F_{\gamma I}T_s = T_sF_{\gamma s}$ and $T_s\mathbf{e}_1 = \mathbf{e}_1$ if and only if the following three conditions hold:

① T_s is represented as:
$T_s = [\mathbf{T}_1, \mathbf{T}_2, \cdots, \mathbf{T}_p]$, $\mathbf{T}_i = [\mathbf{T}_{i1}, \mathbf{T}_{i2}, \cdots, \mathbf{T}_{iq_i}]$
$\mathbf{T}_{ij} = [\mathbf{t}_{ij}^1, \cdots, F_{\alpha I}^{\nu_{ij}-1}\mathbf{t}_{ij}^1]$, $\mathbf{t}_{ij}^1 = F_{\beta I}^{j-1}\mathbf{t}_{i1}^1$ and $\mathbf{t}_{11}^1 = \mathbf{e}_1$,
where $\mathbf{t}_{i1}^1 = F_{\gamma I}^{i-1}\mathbf{e}_1$ for $1 \leq i \leq p$, in particular, $\mathbf{t}_{12}^1 = \mathbf{b}_{11}^1$ and $\mathbf{t}_{21}^1 = \mathbf{d}_{11}^1$.

② For any $1 \leq i \leq p$ and $1 \leq j \leq q_i - 1$, the following two equations hold:
$[\mathbf{T}_1,\ \mathbf{T}_2,\ \cdots, \mathbf{T}_{i-1},\ \mathbf{T}_{i1}, \mathbf{T}_{i2}, \cdots, \mathbf{T}_{ij}, \mathbf{T}_{ij+1}]$
$\times [C_1^{ij+1}, C_2^{ij+1}, \cdots, C_{i-1}^{ij+1}, C_{i1}^{ij+1}, \cdots, C_{ij}^{ij+1} C_{ij+1}^{ij+1}]^T$
$= F_{\alpha I}^{\nu_{ij+1}} F_{\beta I}^{j} \mathbf{e}_{\nu_1+\cdots+\nu_{i-1}+1}$ and
$T_s\mathbf{c}^{iq_i+1} = F_{\beta I}^{q_i}\mathbf{e}_{\nu_1+\cdots+\nu_{i-1}+1}$ for $1 \leq i \leq p$.

③ For any $i,\ j\ (2 \leq i \leq p, 1 \leq j \leq q_i)$, the following equations hold:
$[\mathbf{T}_1,\ \mathbf{T}_2,\ \cdots, \mathbf{T}_{i-1},\ \mathbf{T}_{i1}, \mathbf{T}_{i2}, \cdots, \mathbf{T}_{ij-1}, \mathbf{T}_{ij}][C_1^{ij}, C_2^{ij}, \cdots, C_{ij}^{ij-1}, C_{ij}^{ij}]^T$
$= F_{\alpha I}^{\nu_{ij}} F_{\beta I}^{j-1} F_{\gamma I}\mathbf{e}_{\nu_1+\cdots+\nu_{i-2}+1}$ and $T_s\mathbf{c}^{iq_i+1} = F_{\beta I}^{q_i}F_{\gamma I}\mathbf{e}_{\nu_1+\cdots+\nu_{i-2}+1}$ for $2 \leq i \leq p$.

④ $T_s\mathbf{c}^{p+1} = F_{\gamma I}\mathbf{t}_{p1}^1$.

Proof. Direct calculations result in this lemma.

Remark: T_s is the regular matrix and the pointed $\{\alpha, \beta, \gamma\}$-morphism $T_s : ((K^n, F_{\alpha s}, F_{\beta s}, F_{\gamma s}), \mathbf{e}_1) \to ((K^n, F_{\alpha I}, F_{\beta I}, F_{\gamma I}), \mathbf{e}_1)$, that is which satisfy $F_{\alpha I}T_s = T_sF_{\alpha s}$, $F_{\beta I}T_s = T_sF_{\beta s}$ and $F_{\gamma I}T_s = T_sF_{\gamma s}$, if and only if T_s preserves the quasi-reachability of $((K^n, F_{\alpha s}, F_{\beta s}, F_{\gamma s}), \mathbf{e}_1)$.

Next, we only find T_s that is to be a regular matrix, since the Remarks of lemmas (7-A.11), (7-A.12) and Proposition (7-A.14) indicate that T_s is the right upper triangular matrix. Let T_s which satisfies Lemma (7-A.5) be $T_s = [\mathbf{T}_1, \mathbf{T}_2, \cdots, \mathbf{T}_p]$, $\mathbf{T}_i = [\mathbf{T}_{i1}, \mathbf{T}_{i2}, \cdots, \mathbf{T}_{iq_i}]$, $\mathbf{T}_{ij} = [\mathbf{t}_{ij}^1, \cdots, F_{\alpha I}^{\nu_{ij}-1}\mathbf{t}_{ij}^1]$, $\mathbf{t}_{ij}^1 = F_{\beta I}^{j-1}\mathbf{t}_{i1}^1$, where $\mathbf{t}_{i1}^1 = F_{\gamma I}^{i-1}\mathbf{e}_1$ for $1 \leq i \leq p$. In particular, $\mathbf{t}_{11}^1 = \mathbf{e}_1$.

Lemma 7-A.6. Let T_s be a regular matrix and a $\{\alpha, \beta, \gamma\}$-morphism $T_s : (K^n, F_{\alpha s}, F_{\beta s}, F_{\gamma s}) \to (K^n, F_{\alpha I}, F_{\beta I}, F_{\gamma I})$. Then the equations of ② and ③ in Lemma (7-A.5) are expressed as:

(1) $[\mathbf{T}_1,\ \mathbf{T}_2,\ \cdots, \mathbf{T}_{i-1},\ \mathbf{T}_{i1}, \mathbf{T}_{i2}, \cdots, \mathbf{T}_{ij}]$
$\times [C_1^{ij+1}, C_2^{ij+1}, \cdots, C_{i-1}^{ij+1}, C_{i1}^{ij+1}, C_{i2}^{ij+1} \cdots, C_{ij}^{ij+1}]^T$
$= \phi_{ij+1}(F_{\alpha I}) F_{\beta I}^{j-1} F_{\beta I} \mathbf{e}_{\nu_1 + \cdots + \nu_{i-1}+1} - [F_{\beta I}^j (F_{\gamma I}^{i-1} \mathbf{e}_1 - \mathbf{e}_{\nu_1 + \cdots + \nu_{i-1}+1}),$
$F_{\alpha I} F_{\beta I}^j (F_{\gamma I}^{i-1} \mathbf{e}_1 - \mathbf{e}_{\nu_1 + \cdots + \nu_{i-1}+1}), \cdots,$
$F_{\alpha I}^{\nu_{ij+1}-1} F_{\beta I}^j (F_{\gamma I}^{i-1} \mathbf{e}_1 - \mathbf{e}_{\nu_1 + \cdots + \nu_{i-1}+1})] C_{ij+1}^{ij+1}$
for any $1 \le i \le p,\ 1 \le j \le q_i - 1$.

(2) $[\mathbf{T}_1,\ \mathbf{T}_2,\ \cdots, \mathbf{T}_{i-1},\ \mathbf{T}_{i1}, \mathbf{T}_{i2}, \cdots, \mathbf{T}_{ij-1}][C_1^{ij}, C_2^{ij}, \cdots, C_{ij-1}^{ij}]^T$
$= \phi_{ij}(F_{\alpha I}) F_{\beta I}^{j-1} F_{\gamma I} \mathbf{e}_{\nu_1 + \cdots + \nu_{i-2}+1} - [F_{\gamma I} F_{\beta I}^{j-1} (F_{\gamma I}^{i-2} \mathbf{e}_1 - \mathbf{e}_{\nu_1 + \cdots + \nu_{i-2}+1}),$
$F_{\alpha I} F_{\gamma I} F_{\beta I}^{j-1} (F_{\gamma I}^{i-2} \mathbf{e}_1 - \mathbf{e}_{\nu_1 + \cdots + \nu_{i-2}+1}), \cdots,$
$F_{\alpha I}^{\nu_{ij}-1} F_{\gamma I} F_{\beta I}^{j-1} (F_{\gamma I}^{i-2} \mathbf{e}_1 - \mathbf{e}_{\nu_1 + \cdots + \nu_{i-2}+1})] C_{ij}^{ij}$
for any $2 \le i \le p,\ 1 \le j \le q_i$.

Proof. This lemma is obtained by using the relations:
$\mathbf{t}_{ij+1}^1 = F_{\beta I}^j F_{\gamma I}^{i-1} \mathbf{e}_1,\ \mathbf{t}_{ij}^1 = F_{\beta I}^{j-1} F_{\gamma I}^{i-1} \mathbf{e}_1$ and
$F_{\beta I}^{j-1} \mathbf{b}_{i1}^1 = F_{\beta I}^{j-1} F_{\beta I} \mathbf{e}_{\nu_1 + \cdots + \nu_{i-1}+1}$
$= F_{\beta I}(F_{\beta I}^{j-2} \mathbf{b}_{i1}^1 - \mathbf{e}_{\nu_1 + \cdots + \nu_{i-1} + \nu_{i1} + \cdots \nu_{ij-1}+1}) + \mathbf{b}_{ij}^1$.

Lemma 7-A.7. $F_{\beta I}^{j-1} F_{\beta I} \mathbf{e}_{\nu_1 + \cdots + \nu_{i-1}+1}$ and $F_{\gamma I} \mathbf{e}_{\nu_1 + \cdots + \nu_{i-2}+1}$ in Lemma (7-A.6) satisfy:

(1) $F_{\beta I}^{j-1} F_{\beta I} \mathbf{e}_{\nu_1 + \cdots + \nu_{i-1}+1} = (\mathbf{tb})_{ij}^1 + \mathbf{b}_{ij}^1$ for $1 \le i \le p, 1 \le j \le q_i - 1$, where $(\mathbf{tb})_{ij}^1$ is given by the following recursive equation:
$(\mathbf{tb})_{ij}^1 = F_{\beta I} (\mathbf{tb})_{ij-1}^1 + F_{\beta I}(\mathbf{b}_{ij-1}^1 - \mathbf{e}_{\nu_1 + \cdots + \nu_{i-1} + \nu_{i1} + \cdots + \nu_{ij-1}+1})$ with $(\mathbf{tb})_{i1}^1 = 0$,

(2) $F_{\gamma I} \mathbf{e}_{\nu_1 + \cdots + \nu_{i-2}+1} = \mathbf{d}_{i-11}^1$.

Proof. Direct calculations lead to this lemma.

Lemma 7-A.8. Under the conditions of Lemma (7-A.6) and (7-A.7), let β_{ij}^1 and γ_{i1}^1 be:

$$\begin{aligned}\beta_{ij}^1 := {} & [\mathbf{T}_1,\ \mathbf{T}_2,\ \cdots, \mathbf{T}_{i-1},\ \mathbf{T}_{i1}, \mathbf{T}_{i2}, \cdots, \mathbf{T}_{ij}] \\ & \times [C_1^{ij+1}, C_2^{ij+1}, \cdots, C_{i-1}^{ij+1}, C_{i1}^{ij+1}, C_{i2}^{ij+1} \cdots, C_{ij}^{ij+1}]^T \\ & + [F_{\beta I}^j (F_{\gamma I}^{i-1} \mathbf{e}_1 - \mathbf{e}_{\nu_1 + \cdots + \nu_{i-1}+1}), F_{\alpha I} F_{\beta I}^j (F_{\gamma I}^{i-1} \mathbf{e}_1 - \mathbf{e}_{\nu_1 + \cdots + \nu_{i-1}+1}), \cdots, \\ & \quad F_{\alpha I}^{\nu_{ij+1}-1} F_{\beta I}^j (F_{\gamma I}^{i-1} \mathbf{e}_1 - \mathbf{e}_{\nu_1 + \cdots + \nu_{i-1}+1})] C_{ij+1}^{ij+1},\end{aligned}$$

$$\begin{aligned}\gamma_{i1}^1 := {} & [\mathbf{T}_1,\ \mathbf{T}_2,\ \cdots, \mathbf{T}_{i-1}][C_1^{i1}, C_2^{i1}, \cdots, C_{i-1}^{i1}]^T \\ & + [F_{\gamma I}(F_{\gamma I}^{i-2} \mathbf{e}_1 - \mathbf{e}_{\nu_1 + \cdots + \nu_{i-2}+1}), F_{\alpha I} F_{\gamma I}(F_{\gamma I}^{i-2} \mathbf{e}_1 - \mathbf{e}_{\nu_1 + \cdots + \nu_{i-2}+1}), \cdots, \\ & \quad F_{\alpha I}^{\nu_{i1}-1} F_{\gamma I}(F_{\gamma I}^{i-2} \mathbf{e}_1 - \mathbf{e}_{\nu_1 + \cdots + \nu_{i-2}+1})] C_{i1}^{i1}.\end{aligned}$$

Then the following conditions hold:

(1) β_{ij}^1 $(= \phi_{ij+1}(F_{\alpha I})F_{\beta I}^{j-1}F_{\beta I}\mathbf{e}_{\nu_1+\cdots+\nu_{i-1}+1})$ and
γ_{i1}^1 $(= \phi_{i1}(F_{\alpha I})F_{\gamma I}\mathbf{e}_{\nu_1+\cdots+\nu_{i-2}+1}) \in K^n$ are expressed as:
$\beta_{ij}^1 := [\beta_{ij}^{11}, \beta_{ij}^{12}, \cdots, \beta_{ij}^{1j}, \mathbf{0}]^T$, $\gamma_{i1}^1 := [\gamma_{i1}^{11}, \gamma_{i1}^{12}, \cdots, \gamma_{i1}^{1j}, \mathbf{0}]^T$, where $\beta_{ij}^{11} \in K^{\nu_1}$, $\beta_{ij}^{11} = [\beta_{ij}^{111}, \cdots, \beta_{ij}^{11j}, \mathbf{0}]^T$, $\beta_{ij}^{11\hat{j}} \in K^{\nu_{1\hat{j}}}$ $(1 \le \hat{j} \le j)$, $\cdots$, $\beta_{ij}^{1j} \in K^{\nu_j}$, $\beta_{ij}^{1j} = [\beta_{ij}^{1j1}, \cdots, \beta_{ij}^{1jj}, \mathbf{0}]^T$, $\beta_{ij}^{1j\hat{j}} \in K^{\nu_{j\hat{j}}}(1 \le \hat{j} \le j)$, $\gamma_{i1}^{11} \in K^{\nu_1}$, $\gamma_{i1}^{11} = [\gamma_{i1}^{111}, \cdots, \gamma_{i1}^{11j}, \mathbf{0}]^T$, $\gamma_{i1}^{11\hat{j}} \in K^{\nu_{1\hat{j}}}$ $(1 \le \hat{j} \le j)$, $\cdots$, $\gamma_{i1}^{1j} \in K^{\nu_j}$, $\gamma_{i1}^{1j} = [\gamma_{i1}^{1j1}, \cdots, \gamma_{i1}^{1jj}, \mathbf{0}]^T$, $\gamma_{i1}^{1j\hat{j}} \in K^{\nu_{j\hat{j}}}$ $(1 \le \hat{j} \le j)$.

(2) $\mathbf{b}_{ij}^1$ are obtained by the solutions of the following equation:
$\phi_{ij+1}(F_{\alpha I})\mathbf{b}_{ij}^1 = \beta_{ij}^1 - \phi_{ij+1}(F_{\alpha I})(\mathbf{tb})_{ij}^1$, where
$\mathbf{b}_{ij}^1 := [\mathbf{b}_{ij}^{11}, \mathbf{b}_{ij}^{12}, \cdots, \mathbf{b}_{ij}^{1i}, \mathbf{0}]^T \in K^n$, $\mathbf{b}_{ij}^{1\hat{i}} = [\mathbf{b}_{ij}^{1\hat{i}1}, \cdots, \mathbf{b}_{ij}^{1\hat{i}j}, \mathbf{0}]^T \in K^{\nu_{\hat{i}}}$, $\mathbf{b}_{ij}^{1\hat{i}\hat{j}} = [\mathbf{0}, \widetilde{\mathbf{b}}_{ij}^{1\hat{i}\hat{j}}, \mathbf{0}] \in K^{\nu_{\hat{i}\hat{j}}}$, $\widetilde{\mathbf{b}}_{ij}^{1\hat{i}\hat{j}} \in K^{\nu_{ij}-\nu_{ij+1}}$ for $1 \le \hat{i} \le i-1$ and $1 \le \hat{j} \le j$.
In particular, $\mathbf{b}_{ij}^{1i} := [\mathbf{b}_{ij}^{1i1}, \mathbf{b}_{ij}^{1i2}, \cdots, \mathbf{b}_{ij}^{1ij}, \mathbf{e}_1^s, \mathbf{0}]^T \in K^{\nu_i}$, $\mathbf{e}_1^s \in K^{\nu_{ij+1}}$.

(3) $\mathbf{d}_{i-11}^1$ is expressed as:
$\phi_{i1}(F_{\alpha I})\mathbf{d}_{i-11}^1 = \gamma_{i1}^1$, where $\mathbf{d}_{i-11}^1 :=$
$[\mathbf{d}_{i-11}^{11}, \mathbf{d}_{i-11}^{12}, \cdots, \mathbf{d}_{i-11}^{1i-1}, \mathbf{e}_1^s, \mathbf{0}]^T$, $\mathbf{d}_{i-11}^{1\hat{i}} = [\mathbf{d}_{i-11}^{1\hat{i}1}, \cdots, \mathbf{d}_{i-11}^{1\hat{i}j}, \mathbf{0}]^T \in K^{\nu_{\hat{i}}}$ $(1 \le \hat{i} \le i-1)$, $\mathbf{e}_1^s \in K^{\nu_i}$, $\mathbf{d}_{i-11}^{1\hat{i}\hat{j}} = [\mathbf{0}, \widetilde{\mathbf{d}}_{i-11}^{1\hat{i}\hat{j}}, \mathbf{0}]^T \in K^{\nu_{\hat{i}\hat{j}}}$, $\widetilde{\mathbf{d}}_{i-11}^{1\hat{i}\hat{j}} \in K^{\nu_{i-11}-\nu_{i1}}$.

Proof. $\phi_{ij}(F_{\alpha I})\mathbf{b}_{ij}^1 = 0$ and $\phi_{i-11}(F_{\alpha I})\mathbf{d}_{i-11}^1 = 0$ hold from Lemmas (7-A.4) and (7-A.7). Then $\beta_{ij}^1 \in \text{im}\ \phi_{ij+1}(F_{\alpha I})$ and $\gamma_{i1}^1 \in \text{im}\ \phi_{i1}(F_{\alpha I})$ result in this lemma.

Lemma 7-A.9. Let T_s be a regular matrix and a $\{\alpha, \beta, \gamma\}$-morphism $T_s : (K^n, F_{\alpha s}, F_{\beta s}, F_{\gamma s}) \to (K^n, F_{\alpha I}, F_{\beta I}, F_{\gamma I})$. Then the following relations among column vectors $\mathbf{t}_{ij}^1$ of T_s, $\mathbf{b}_{1j-1}^1$ of $F_{\beta I}$ and column vectors $\mathbf{d}_{i-1j}^1$ of $F_{\gamma I}$ are obtained:

(1) $\mathbf{t}_{11}^1 = \mathbf{e}_1$, and $\mathbf{t}_{1j}^1 = (\mathbf{tb})_{1j-1}^1 + \mathbf{b}_{1j-1}^1$ for $2 \le j \le q_1$, where
$(\mathbf{tb})_{1j-1}^1 = F_{\beta I}(\mathbf{tb})_{1j-2}^1 + F_{\beta I}(\mathbf{b}_{1j-2}^1 - \mathbf{e}_{\nu_1+\cdots+\nu_{i-1}+\nu_{i1}+\cdots+\nu_{ij-2}+1})$ with $(\mathbf{tb})_{11}^1 = 0$ for $3 \le j \le q_1$,

(2) $\mathbf{t}_{i1}^1 = (\mathbf{td})_{i-11}^1 + \mathbf{d}_{i-11}^1$ for $2 \le i \le p$, where
$(\mathbf{td})_{i-11}^1 = F_{\gamma I}(\mathbf{td})_{i-21}^1 + F_{\gamma I}(\mathbf{d}_{i-21}^1 - \mathbf{e}_{\nu_1+\cdots+\nu_{i-2}+1})$ with $(\mathbf{td})_{11}^1 = 0$ for $3 \le i \le p$,

(3) $\mathbf{t}_{ij}^1 = (\mathbf{td})_{i-1j}^1 + \mathbf{d}_{i-1j}^1$ for $2 \le i \le p$ and $2 \le j \le q_i$, where
$(\mathbf{td})_{i-1j}^1 = F_{\beta I}(\mathbf{td})_{i-1j-1}^1 + F_{\gamma I}$
$\times (\mathbf{b}_{i-1j-1}^1 - \mathbf{e}_{\nu_1+\cdots+\nu_{i-2}+\nu_{i-11}+\cdots+\nu_{i-1j-1}+1})$, and
$(\mathbf{td})_{i-11}^1 = F_{\gamma I}^{i-2}(\mathbf{d}_{11}^1 - \mathbf{e}_{\nu_1+1}) + F_{\gamma I}^{i-3}(\mathbf{d}_{21}^1 - \mathbf{e}_{\nu_1+\nu_2+1}) + \cdots$
$+ F_{\gamma I}(\mathbf{d}_{i-21}^1 - \mathbf{e}_{\nu_1+\cdots+\nu_{i-2}+1})$ for $3 \le i \le p$. Note that this $(\mathbf{td})_{i-11}^1$ is another equivalent expression of one in (2).

Proof. Direct calculations using the relation $\mathbf{t}_{ij}^1 = F_{\gamma I}^{i-1} F_{\beta I}^{j-1} \mathbf{e}_1$ for $1 \le i \le p$ and $1 \le j \le q_i$ result in this lemma.

Next, the number of parameters of $\{\alpha, \beta, \gamma\}$-actions is investigated.

Lemma 7-A.10. Let T_s be a regular matrix and a $\{\alpha, \beta, \gamma\}$-morphism $T_s : (K^n, F_{\alpha s}, F_{\beta s}, F_{\gamma s}) \rightarrow (K^n, F_{\alpha I}, F_{\beta I}, F_{\gamma I})$. Then $F_{\alpha I}$ is uniquely determined from $F_{\alpha s}$ which has ν_{11} independent parameters.

Proof. Note that the characteristic polynomial of $F_{\alpha s}$ is $\phi_{11}(\lambda) \cdots \phi_{1q_1}(\lambda)\phi_{21}(\lambda) \cdots \phi_{2q_2}(\lambda) \cdots \phi_{p1}(\lambda) \cdots \phi_{pq_p}(\lambda)$. Each $\phi_{ij}(\lambda)$ is expressed as $\phi_{ij}(\lambda) = \chi_{ij}(\lambda)\chi_{ij+1}(\lambda) \cdots \chi_{iq_i}(\lambda)$ for $1 \le i \le p$ and $1 \le j \le q_i$ by Lemma (7-A.2).

Once the polynomials of $\phi_{iq_i}(\lambda)$, $\phi_{iq_i-1}(\lambda)$, $\cdots$, $\phi_{i2}(\lambda)$ and $\phi_{i1}(\lambda)$ are calculated for $1 \le i \le p$ in turn, we obtain $\chi_{iq_i}(\lambda)$, $\chi_{iq_i-1}(\lambda)$, $\cdots$, $\chi_{i2}(\lambda)$ and $\chi_{i1}(\lambda)$ in turn. This implies that $F_{\alpha I}$ can be uniquely determined.

Each polynomial $\chi_{ij}(\lambda)$ can be determined by the ${}_i\ell_j$ parameters, where $1 \le i \le p$ and $1 \le j \le q_i$. Thus the polynomials $\chi_{iq_i}(\lambda), \chi_{iq_i-1}(\lambda), \cdots, \chi_{i2}(\lambda)$ and $\chi_{i1}(\lambda)$ can be determined by ν_{i1} $(= {}_i\ell_{q_i} + {}_i\ell_{q_i-1} + \cdots + {}_i\ell_2 + {}_i\ell_1)$ parameters. Therefore, the matrix $F_{\alpha I}$ is determined by ν_{11} $(= \sum_{i=1}^{q_1} {}_1\ell_i)$ parameters, where ${}_i\ell_j := \nu_{ij} - \nu_{ij+1}$.

Remark : Suppose Lemma (7-A.10) hold. Since $\phi_{ij}(\lambda)$ is a characteristic polynomial of $F_{ij} \in K^{\nu_{ij} \times \nu_{ij}}$, $K^{\nu_{ij}} = K^{\,{}_i\ell_j} \oplus K^{\,{}_i\ell_{j+1}} \oplus + \cdots \oplus K^{\,{}_i\ell_{q_i}}$, thus $K^{\nu_i} = K^{\nu_{i1}} \oplus K^{\nu_{i2}} \oplus + \cdots \oplus K^{\nu_{iq_i}}$ and $K^n = K^{\nu_1} \oplus K^{\nu_2} \oplus + \cdots \oplus K^{\nu_p}$ hold. Then $\mathbf{b}_{ij}^1$ $(1 \le i \le p, 1 \le j \le q_i)$ in Lemma (7-A.4) has non-zero vectors of $K^{\nu_{ij}} = K^{\,{}_i\ell_j} \oplus K^{\,{}_i\ell_{j+1}} \oplus + \cdots \oplus K^{\,{}_i\ell_{q_i}}$ in every subspace K^{ν_i} $(1 \le i \le p)$, and the other vectors become zero vectors.

Lemma 7-A.11. Let T_s be a regular matrix and let a $\{\alpha, \beta, \gamma\}$-morphism $T_s : (K^n, F_{\alpha s}, F_{\beta s}, F_{\gamma s}) \rightarrow (K^n, F_{\alpha I}, F_{\beta I}, F_{\gamma I})$. Then $\mathbf{b}_{11}^1 \in K^n$ of $\mathbf{B}_1$ which is the submatrix of $F_{\beta I}$ is determined by ${}_1\ell_1$ minimal parameters.

Proof. Suppose the given conditions hold. According to Lemma (7-A.2) and the condition (1) with setting $i = 1$ and $j = 1$ in Lemma (7-A.6), $[\mathbf{e}_1, \cdots, F_{\alpha I}^{\nu_{11}-1}\mathbf{e}_1]C_{11}^{12} = \phi_{12}(F_{\alpha I})\mathbf{b}_{11}^1 = \chi_{12}(F_{\alpha I})\chi_{13}(F_{\alpha I}) \cdots \chi_{1q_1}(F_{\alpha I})\mathbf{b}_{11}^1$ holds. Because that $\phi_{1j}(\lambda)$ $(2 \le j \le q_1)$ is a characteristic polynomial of F_{1j} which is the submatrix of $F_{\alpha I}$ and from Remark in Lemma (7-A.10), a vector C_{11}^{12} depends only on the elements of $\mathbf{b}_{11}^1$ which belong to the subspace $K^{\,{}_1\ell_1}$ of $K^{\nu_{11}}$ and does not depend on the other elements of $\mathbf{b}_{11}^1$. Hence, the vector $\mathbf{b}_{11}^1$ is determined by ${}_1\ell_1$ parameters, where ${}_1\ell_1 = \nu_{11} - \nu_{12}$.

Remark : Under the constraints in Lemma (7-A.11), we will find a $\{\alpha, \beta, \gamma\}$-morphism $T_s : (K^n, F_{\alpha s}, F_{\beta s}, F_{\gamma s}) \rightarrow \widetilde{(K^n, F_{\alpha I}, F_{\beta I}, F_{\gamma I})}$. Set $\widetilde{\mathbf{b}_{11}^1} \in K^n$ as $\mathbf{b}_{11}^1 = [\mathbf{b}_{11}^{111}, \mathbf{e}_1^s, \mathbf{0}]^T$, where $\mathbf{b}_{11}^{111} = [\widetilde{\mathbf{b}_{11}^{111}}, \mathbf{0}]^T \in K^{\nu_{11}}$, $\widetilde{\mathbf{b}_{11}^{111}} \in K^{\,{}_1\ell_1}$ and $\mathbf{e}_1^s \in K^{\nu_{12}}$, ${}_1\ell_1 = \nu_{11} - \nu_{12}$.

The relation $[\mathbf{e}_1, \cdots, F_{\alpha I}^{\nu_{11}-1}\mathbf{e}_1]C_{11}^{12} = \phi_{12}(F_{\alpha I})\mathbf{b}_{11}^1$ presented (1) in Lemma (7-A.6) results in $[\mathbf{e}_1^s, \cdots, F_{11}^{\nu_{11}-1}\mathbf{e}_1^s]C_{11}^{12} = \phi_{12}(F_{11})\widetilde{\mathbf{b}_{11}^{111}}$.

Applying (1) in Lemma (7-A.8) to above equation yields $\beta_{11}^1 = \phi_{12}(F_{\alpha I})\mathbf{b}_{11}^1$ and $\beta_{11}^{111} = \phi_{12}(F_{11})\widetilde{\mathbf{b}_{11}^{111}}$, where $\beta_{11}^1 = [\beta_{11}^{111}, \mathbf{0}]^T \in K^n$ is given by $\beta_{11}^1 = [\mathbf{e}_1, \cdots, F_{\alpha I}^{\nu_{11}-1}\mathbf{e}_1]C_{11}^{12}$ and $\beta_{11}^{111} \in K^{\nu_{11}}$ is given by $\beta_{11}^{111} = [\mathbf{e}_1^s, \cdots, F_{11}^{\nu_{11}-1}\mathbf{e}_1^s]C_{11}^{12}$.

Therefore, $\widetilde{\mathbf{b}_{11}^{111}}$ is represented by

$\widetilde{\mathbf{b}_{11}^{111}} = \{[\phi_{12}(F_{11})\mathbf{e}_1^s, \cdots, \phi_{12}(F_{11})\mathbf{e}^s_{\,_1\ell_1}]^T[\phi_{12}(F_{11})\mathbf{e}_1^s, \cdots, \phi_{12}(F_{11})\mathbf{e}^s_{\,_1\ell_1}]\}^{-1}$
$\times [\phi_{12}(F_{11})\mathbf{e}_1^s, \cdots, \phi_{12}(F_{11})\mathbf{e}^s_{\,_1\ell_1}]^T\beta_{11}^{111}$, where $\beta_{11}^{111} = [\mathbf{e}_1^s, \cdots, F_{11}^{\nu_{11}-1}\mathbf{e}_1^s]C_{11}^{12}$.

According to ① of Lemma (7-A.5), $\mathbf{t}_{21}^1 = \mathbf{b}_{11}^1$ holds. Hence, all the column vectors from the first column to the $(\nu_{11}+\nu_{12})$-th column of T_s are mutually independent.

Lemma 7-A.12. Let T_s be a regular matrix and let a $\{\alpha, \beta, \gamma\}$-morphism $T_s : (K^n, F_{\alpha s}, F_{\beta s}, F_{\gamma s}) \to (K^n, F_{\alpha I}, F_{\beta I}, F_{\gamma I})$. Let $\mathbf{b}_{11}^1 \in K^n$ be $\mathbf{b}_{11}^1 = [\mathbf{b}_{11}^{111}, \mathbf{e}_1^s, \mathbf{0}]^T$, where $\mathbf{b}_{11}^{111} = [\widetilde{\mathbf{b}_{11}^{111}}, \mathbf{0}]^T \in K^{\nu_{11}}$, $\widetilde{\mathbf{b}_{11}^{111}} \in K^{\nu_{11}-\nu_{12}}$ and $\mathbf{e}_1^s \in K^{\nu_{12}}$. Then $\mathbf{b}_{12}^1 \in K^n$ of $\mathbf{B}_1$, which is the submatrix of $F_{\beta I}$, is determined by the $2 \times \,_1\ell_2$ minimal parameters, where $_1\ell_2 = \nu_{12} - \nu_{13}$. Moreover, $\mathbf{T}_1$ of $\{\alpha, \beta, \gamma\}$-morphism T_s can be obtained with full rank property.

Proof. Lemma (7-A.8) implies that $\phi_{1j+1}(F_{\alpha I})\mathbf{b}_{1j}^1 = \beta_{1j}^1 - \phi_{1j+1}(F_{\alpha I})(\mathbf{tb})_{1j}^1$ holds for $1 \le j \le q_1 - 1$, where
$\beta_{1j}^1 := [\mathbf{T}_{11}, \ \mathbf{T}_{12}, \ \cdots, \ \mathbf{T}_{1j}][C_{11}^{1j+1}, C_{12}^{1j+1}, \cdots, C_{1j}^{1j+1}]^T$. $(\mathbf{tb})_{1j}^1$ is given by (1) of Lemma (7-A.7).

Hence, $\mathbf{b}_{1j}^{11\hat{j}} = [\mathbf{0}, \widetilde{\mathbf{b}_{1j}^{11\hat{j}}}, \mathbf{e}_1^s, \mathbf{0}]^T \in K^{\nu_{1\hat{j}}}$ of $\mathbf{b}_{1j}^1 = [\mathbf{b}_{1j}^{111}, \mathbf{b}_{1j}^{112}, \cdots, \mathbf{b}_{1j}^{11j}, \mathbf{0}]^T$ is expressed by:

$$\begin{aligned}\widetilde{\mathbf{b}_{1j}^{11\hat{j}}} &= \{[\phi_{1j+1}(F_{1\hat{j}})\mathbf{e}^s_{\nu_{1\hat{j}}-\nu_{1j}+1}, \cdots, \phi_{1j+1}(F_{1\hat{j}})\mathbf{e}^s_{\nu_{1\hat{j}}-\nu_{1j+1}}]^T \\ &\quad \times [\phi_{1j+1}(F_{1\hat{j}})\mathbf{e}^s_{\nu_{1\hat{j}}-\nu_{1j}+1}, \cdots, \phi_{1j+1}(F_{1\hat{j}})\mathbf{e}^s_{\nu_{1\hat{j}}-\nu_{1j+1}}]\}^{-1} \\ &\quad \times [\phi_{1j+1}(F_{1\hat{j}})\mathbf{e}^s_{\nu_{1\hat{j}}-\nu_{1j}+1}, \cdots, \phi_{1j+1}(F_{1\hat{j}})\mathbf{e}^s_{\nu_{1\hat{j}}-\nu_{1j+1}}]^T \\ &\quad \times \{\beta_{1j}^{11\hat{j}} - \phi_{1j+1}(F_{1\hat{j}})(\mathbf{tb})_{11}^{11\hat{j}}\},\end{aligned}$$

where $\beta_{1j}^1 = [\beta_{1j}^{111}, \cdots, \beta_{1j}^{11j}, \mathbf{0}]^T$, $\beta_{1j}^{111} \in K^{\nu_{11}}, \cdots, \beta_{1j}^{11j} \in K^{\nu_{1j}}$.

$\mathbf{b}_{11}^1 = [\widetilde{\mathbf{b}_{11}^{111}}, \mathbf{0}, \mathbf{e}_1^s, \mathbf{0}]^T$ and Lemma (7-A.9) result in $\mathbf{t}_{1j}^1 = (\mathbf{tb})_{1j-1}^1 + \mathbf{b}_{1j-1}^1$ for $2 \le j \le q_1$. Thus, $\mathbf{t}_{1j}^1$ can be found in turn for $2 \le j \le q_1$ from $\mathbf{b}_{1j-1}^1$.

Note that $\mathbf{t}_{1j+1}^1$ is represented by $\mathbf{t}_{1j+1}^1 = [\mathbf{t}_{1j+1}^{11}, \mathbf{0}]^T$, where $\mathbf{t}_{1j+1}^{11} = [\mathbf{t}_{1j+1}^{111}, \mathbf{t}_{1j+1}^{112}, \cdots, \mathbf{t}_{1j+1}^{11j}, \mathbf{e}_1^s, \mathbf{0}]^T \in K^{\nu_1}$, $\mathbf{e}_1^s \in K^{\nu_{1j+1}}$.
Thus, the full rank property of $\mathbf{T}_1$ is preserved.

Furthermore, $\mathbf{t}_{1j+1}^{11\hat{j}}$ $(1 \le \hat{j} \le j) \in K^{\nu_{1\hat{j}}}$ is represented as:

$\mathbf{t}_{1j+1}^{11\hat{j}} = [\mathbf{0}, \widetilde{\mathbf{t}_{1j+1}^{11\hat{j}}}, \mathbf{0}]^T$, $\widetilde{\mathbf{t}_{1j+1}^{11\hat{j}}} \in K^{\nu_{1j}-\nu_{1j+1}}$,

$$\widetilde{\mathbf{t}_{1j+1}^{11\hat{j}}} = \{[\phi_{1j+1}(F_{1\hat{j}})\mathbf{e}^s_{\nu_{1\hat{j}}-\nu_{1j}+1}, \cdots, \phi_{1j+1}(F_{1\hat{j}})\mathbf{e}^s_{\nu_{1\hat{j}}-\nu_{1j+1}}]^T \times [\phi_{1j+1}(F_{1\hat{j}})\mathbf{e}^s_{\nu_{1\hat{j}}-\nu_{1j}+1}, \cdots, \phi_{1j+1}(F_{1\hat{j}})\mathbf{e}^s_{\nu_{1\hat{j}}-\nu_{1j+1}}]\}^{-1} \times [\phi_{1j+1}(F_{1\hat{j}})\mathbf{e}^s_{\nu_{1\hat{j}}-\nu_{1j}+1}, \cdots, \phi_{1j+1}(F_{1\hat{j}})\mathbf{e}^s_{\nu_{1\hat{j}}-\nu_{1j+1}}]^T \beta_{1j}^{11\hat{j}},$$

where $\beta_{1j}^1 = [\beta_{1j}^{111}, \cdots, \beta_{1j}^{11j}, \mathbf{0}]^T$, $\beta_{1j}^{111} \in K^{\nu_{11}}, \cdots, \beta_{1j}^{11j} \in K^{\nu_{1j}}$.

Remark: The vector $\mathbf{b}_{12}^1 \in K^n$ depends only on the elements of the subspace $K^{\ _1\ell_2}$ of $K^{\nu_{11}}$ and $K^{\nu_{12}}$ because of $\phi_{1j}(\lambda)$ $(3 \le j \le q_1)$ being a characteristic polynomial of F_{1j} in $F_{\alpha I}$, Remark in Lemma (7-A.5), the qeuation $\mathbf{t}_{13}^1 = (\mathbf{tb})_{12}^1 + \mathbf{b}_{12}^1$ and the equation $[T_{11}\ T_{12}][[C_{11}^{13}, C_{12}^{13}]^T = \phi_{13}(F_{\alpha I})\mathbf{t}_{13}^1$. Hence, the vector $\mathbf{b}_{12}^1$ is determined by $2 \times\ _1\ell_2$ parameters, where $_1\ell_2 = \nu_{12} - \nu_{13}$.

We have already determined $\mathbf{b}_{1j}^1$ for $2 \le j \le q_1$. but, for the purpose of reference, we will present some of them again as the need arises.

First, we will show $\mathbf{b}_{12}^1 = [\mathbf{b}_{12}^{111}, \mathbf{b}_{12}^{112}, \mathbf{e}_1^s, \mathbf{0}]^T$, where $\mathbf{b}_{12}^{111} := [\mathbf{0}, \widetilde{\mathbf{b}_{12}^{111}}, \mathbf{0}]^T \in K^{\nu_{11}}, \widetilde{\mathbf{b}_{12}^{111}} \in K^{\nu_{12}-\nu_{13}}$, $\mathbf{b}_{12}^{112} := [\mathbf{0}, \widetilde{\mathbf{b}_{12}^{112}}, \mathbf{0}]^T \in K^{\nu_{12}}$, $\widetilde{\mathbf{b}_{12}^{112}} \in K^{\nu_{12}-\nu_{13}}$ and $\mathbf{e}_1^s \in K^{\nu_{13}}$. $\widetilde{\mathbf{b}_{12}^{111}} \in K^{\nu_{12}-\nu_{13}}$ and $\widetilde{\mathbf{b}_{12}^{112}} \in K^{\nu_{12}-\nu_{13}}$ are expressed as:

$$\widetilde{\mathbf{b}_{12}^{111}} = \{[\phi_{13}(F_{11})\mathbf{e}^s_{\nu_{11}-\nu_{12}+1}, \cdots, \phi_{13}(F_{11})\mathbf{e}^s_{\nu_{11}-\nu_{13}}]^T \times [\phi_{13}(F_{11})\mathbf{e}^s_{\nu_{11}-\nu_{12}+1}, \cdots, \phi_{13}(F_{11})\mathbf{e}^s_{\nu_{11}-\nu_{13}}]\}^{-1} \times [\phi_{13}(F_{11})\mathbf{e}^s_{\nu_{11}-\nu_{12}+1}, \cdots, \phi_{13}(F_{11})\mathbf{e}^s_{\nu_{11}-\nu_{13}}]^T \{\beta_{12}^{111} - \phi_{13}(F_{11})(\mathbf{tb})_{12}^{111}\},$$

$$\widetilde{\mathbf{b}_{12}^{112}} = \{[\phi_{13}(F_{12})\mathbf{e}_1^s, \cdots, \phi_{13}(F_{12})\mathbf{e}^s_{\nu_{12}-\nu_{13}}]^T [\phi_{13}(F_{12})\mathbf{e}_1^s, \cdots, \phi_{13}(F_{12}) \times \mathbf{e}^s_{\nu_{12}-\nu_{13}}]\}^{-1} \times [\phi_{13}(F_{12})\mathbf{e}_1^s, \cdots, \phi_{13}(F_{12})\mathbf{e}^s_{\nu_{12}-\nu_{13}}]^T \{\beta_{12}^{112} - \phi_{13}(F_{12}) \times (\mathbf{tb})_{12}^{112}\},$$

where $\beta_{12}^{111} = [\mathbf{e}_1^s, \cdots, F_{11}^{\nu_{11}-1}\mathbf{e}_1^s]C_{11}^{13} + [\mathbf{t}_{12}^1, \cdots, F_{11}^{\nu_{12}-1}\mathbf{t}_{12}^1]C_{12}^{13}$,
$\beta_{12}^{112} = [\mathbf{t}_{12}^1, \cdots, F_{12}^{\nu_{12}-1}\mathbf{t}_{12}^1]C_{12}^{13}$,
$\beta_{12}^1 = [\beta_{12}^{111}, \beta_{12}^{112}, \mathbf{0}]^T \in K^n$, $(\mathbf{tb})_{12}^1 = [(\mathbf{tb})_{12}^{111}, (\mathbf{tb})_{12}^{112}, \mathbf{0}]^T \in K^n$,
$(\mathbf{tb})_{12}^{111} = [\mathbf{e}_1^s, \mathbf{e}_2^s, \cdots, \mathbf{e}^s_{\nu_{11}}, \mathbf{0}]\mathbf{B}_{11}\mathbf{b}_{12}^{111} \in K^{\nu_{11}}$, $(\mathbf{tb})_{12}^{112} = \mathbf{0} \in K^{\nu_{11}}$.

Next, we will show $\mathbf{b}_{13}^1 \in K^n$ concretely.
$\mathbf{b}_{13}^1 := [\mathbf{b}_{13}^{111}, \mathbf{b}_{13}^{112}, \mathbf{b}_{13}^{113}, \mathbf{e}_1^s, \mathbf{0}]^T$ is determined such that the all column vectors from the first to the $(\nu_{11} + \nu_{12} + \nu_{13})$-th column of T_s are mutually independent, where $\mathbf{b}_{13}^{111} := [\mathbf{0}, \widetilde{\mathbf{b}_{13}^{111}}, \mathbf{0}]^T \in K^{\nu_{11}}, \widetilde{\mathbf{b}_{13}^{111}} \in K^{\ _1\ell_3}$, $\mathbf{b}_{13}^{112} := [\mathbf{0}, \widetilde{\mathbf{b}_{13}^{112}}, \mathbf{0}]^T \in K^{\nu_{12}}$, $\widetilde{\mathbf{b}_{13}^{112}} \in K^{\ _1\ell_3}$ and $\mathbf{b}_{13}^{113} := [\mathbf{0}, \widetilde{\mathbf{b}_{13}^{113}}, \mathbf{0}]^T \in K^{\nu_{13}}, \widetilde{\mathbf{b}_{13}^{113}} \in K^{\ _1\ell_3}$, $_1\ell_3 = \nu_{13} - \nu_{14}$.

Furthermore, $\widetilde{\mathbf{b}_{13}^{111}}$, $\widetilde{\mathbf{b}_{13}^{112}}$ and $\widetilde{\mathbf{b}_{13}^{113}} \in K^{\ _1\ell_3}$ are expressed as:

$$\widetilde{\mathbf{b}_{13}^{111}} = \{[\phi_{14}(F_{11})\mathbf{e}^s_{\nu_{11}-\nu_{13}+1}, \cdots, \phi_{14}(F_{11})\mathbf{e}^s_{\nu_{11}-\nu_{14}}]^T \times [\phi_{14}(F_{11})\mathbf{e}^s_{\nu_{11}-\nu_{13}+1}, \cdots, \phi_{14}(F_{11})\mathbf{e}^s_{\nu_{11}-\nu_{14}}]\}^{-1} \times [\phi_{14}(F_{11})\mathbf{e}^s_{\nu_{11}-\nu_{13}+1}, \cdots, \phi_{14}(F_{11})\mathbf{e}^s_{\nu_{11}-\nu_{14}}]^T \{\beta_{13}^{111} - \phi_{14}(F_{11})(\mathbf{tb})_{13}^{111}\},$$

$$\widetilde{\mathbf{b}_{13}^{112}} = \{[\phi_{14}(F_{12})\mathbf{e}^s_{\nu_{12}-\nu_{13}+1}, \cdots, \phi_{14}(F_{12})\mathbf{e}^s_{\nu_{12}-\nu_{14}}]^T \times [\phi_{14}(F_{12})\mathbf{e}^s_{\nu_{12}-\nu_{13}+1}, \cdots, \phi_{14}(F_{12})\mathbf{e}^s_{\nu_{12}-\nu_{14}}]\}^{-1} \times [\phi_{14}(F_{12})\mathbf{e}^s_{\nu_{12}-\nu_{13}+1}, \cdots, \phi_{14}(F_{12})\mathbf{e}^s_{\nu_{12}-\nu_{14}}]^T\{\beta_{13}^{112} - \phi_{14}(F_{12})(\mathbf{tb})_{13}^{112}\},$$

$$\widetilde{\mathbf{b}_{13}^{113}} = \{[\phi_{14}(F_{13})\mathbf{e}_1^s, \cdots, \phi_{14}(F_{13})\mathbf{e}^s_{\nu_{13}-\nu_{14}}]^T[\phi_{14}(F_{13})\mathbf{e}_1^s, \cdots, \phi_{14}(F_{13}) \times \mathbf{e}^s_{\nu_{13}-\nu_{14}}]\}^{-1} \times [\phi_{14}(F_{13})\mathbf{e}_1^s, \cdots, \phi_{14}(F_{13})\mathbf{e}^s_{\nu_{13}-\nu_{14}}]^T\{\beta_{13}^{113} - \phi_{14}(F_{13}) (\mathbf{tb})_{13}^{113}\},$$

where

$$\beta_{13}^{111} = [\mathbf{t}_{11}^1, \cdots, F_{11}^{\nu_{11}-1}\mathbf{t}_{11}^1]C_{11}^{14} + [\mathbf{t}_{12}^1, \cdots, F_{11}^{\nu_{11}-1}\mathbf{t}_{12}^1]C_{12}^{14} + [\mathbf{t}_{13}^1, \cdots, F_{11}^{\nu_{11}-1}\mathbf{t}_{13}^1]C_{13}^{14},$$

$$\beta_{13}^{112} = [\mathbf{t}_{12}^1, \cdots, F_{12}^{\nu_{12}-1}\mathbf{t}_{12}^1]C_{12}^{14} + [\mathbf{t}_{13}^1, \cdots, F_{12}^{\nu_{12}-1}\mathbf{t}_{13}^1]C_{13}^{14},$$

$$\beta_{13}^{113} = [\mathbf{t}_{13}^1, \cdots, F_{13}^{\nu_{13}-1}\mathbf{t}_{13}^1]C_{13}^{14}, \quad \beta_{13}^1 = [\beta_{13}^{111}, \beta_{13}^{112}, \beta_{13}^{113}, \mathbf{0}]^T \in K^n,$$

$$(\mathbf{tb})_{13}^1 = [(\mathbf{tb})_{13}^{111}, (\mathbf{tb})_{13}^{112}, (\mathbf{tb})_{13}^{113}, \mathbf{0}]^T \in K^n,$$

$$(\mathbf{tb})_{13}^{111} = [\mathbf{e}_1^s, \mathbf{e}_2^s, \cdots, \mathbf{e}^s_{\nu_{11}}, \mathbf{0}]\mathbf{B}_{11}([\mathbf{e}_1^s, \mathbf{e}_2^s, \cdots, \mathbf{e}^s_{\nu_{11}}, \mathbf{0}]\mathbf{B}_{11}\mathbf{b}_{11}^{111} + \mathbf{b}_{12}^{111}) \in K^{\nu_{11}},$$

$$(\mathbf{tb})_{13}^{112} = [\mathbf{0}, \mathbf{e}^s_{\nu_{11}+1}, \cdots, \mathbf{e}^s_{\nu_{12}}, \mathbf{0}]\mathbf{B}_{12}\mathbf{b}_{12}^{112} \in K^{\nu_{12}}, \ (\mathbf{tb})_{13}^{113} = 0 \in K^{\nu_{13}}.$$

Since $\mathbf{b}_{13}^1$ was already presented as $\mathbf{b}_{13}^1 = [\mathbf{b}_{13}^{111}, \mathbf{b}_{13}^{112}, \mathbf{b}_{13}^{113}, \mathbf{e}_1^s, \mathbf{0}]^T \in K^n$, we could obtain $(\mathbf{tb})_{13}^1 = [(\mathbf{tb})_{13}^{111}, (\mathbf{tb})_{13}^{112}, (\mathbf{tb})_{13}^{113}, \mathbf{0}]^T \in K^n$.

By (1) of lemmas (7-A.6) and (7-A.7), we will determine $\mathbf{b}_{1j}^1$ and $\mathbf{t}_{1j+1}^1 = F_{\beta I}^{j-1}F_{\beta I}\mathbf{e}_1 = F_{\beta I}^{j-1}\mathbf{b}_{11}^1 = (\mathbf{tb})_{1j}^1 + \mathbf{b}_{1j}^1$ for $4 \leq j \leq q_1 - 1$ such that the all column vectors from the first to the $(\nu_{11} + \nu_{12} + \nu_{13} + \cdots + \nu_{1q_1})$-th column of T_s are mutually independent.

Let us present $(\mathbf{tb})_{1j}^1$ $(2 \leq j \leq q_1 - 1)$.

Since $(\mathbf{tb})_{1j}^1$ is given by the recursive equation in (1) of Lemma (7-A.7), $(\mathbf{tb})_{1j}^1$ becomes $(\mathbf{tb})_{1j}^1 = [(\mathbf{tb})_{1j}^{111}(\mathbf{tb})_{1j}^{112}, \cdots, (\mathbf{tb})_{1j}^{11j}, \mathbf{0}, \cdots, \mathbf{0}]^T \in K^n$, where $(\mathbf{tb})_{1j}^{111} \in K^{\nu_{11}}, (\mathbf{tb})_{1j}^{112} \in K^{\nu_{12}}, \cdots, (\mathbf{tb})_{1j}^{11j} \in K^{\nu_{1j}}$.

Next, the number of parameters of $\mathbf{b}_{1j}^1$ is investigated. By (1) of lemmas (7-A.7) and (7-A.8), set $\mathbf{b}_{1j}^1$ $(1 \leq j \leq q_i - 1)$ as:

$$\mathbf{b}_{1j}^1 = [\mathbf{b}_{1j}^{111}, \mathbf{b}_{1j}^{112}, \cdots, \mathbf{b}_{1j}^{11j}, \mathbf{e}_1^s, \mathbf{0}]^T \in K^n,$$

where $\mathbf{b}_{1j}^{111} := [\mathbf{0}, \widetilde{\mathbf{b}_{1j}^{111}}, \mathbf{0}]^T \in K^{\nu_{11}}, \ \widetilde{\mathbf{b}_{1j}^{111}} \in K^{\ {}_1\ell_j}$,

$\mathbf{b}_{1j}^{112} := [\mathbf{0}, \widetilde{\mathbf{b}_{1j}^{112}}, \ \mathbf{0}]^T \in K^{\nu_{12}}, \widetilde{\mathbf{b}_{1j}^{112}} \in K^{\ {}_1\ell_j}, \ \cdots, \ \mathbf{b}_{1j}^{11\hat{j}} := [\mathbf{0}, \widetilde{\mathbf{b}_{1j}^{11\hat{j}}}, \mathbf{0}]^T \in K^{\nu_{1\hat{j}}}$,

$\widetilde{\mathbf{b}_{1j}^{11\hat{j}}} \in K^{\ {}_1\ell_j}$ and $\mathbf{e}_1^s \in K^{\nu_{1j+1}}$.

Therefore, the number of parameters of $\mathbf{b}_{1j}^1$ becomes $j \times {}_1\ell_j$ which is apparently minimal for each $4 \leq j \leq q_1 - 1$, where ${}_1\ell_j := \nu_{1j} - \nu_{1j+1}$.

Let us determine the number of parameters of $\mathbf{b}_{1q_1}^1$ which appears in $\mathbf{B}_{1q_1}$ of $F_{\beta I} = [\mathbf{B}_1, \mathbf{B}_2, \cdots, \mathbf{B}_p]$, where $\mathbf{B}_1 = [\mathbf{B}_{11}, \mathbf{B}_{12}, \cdots, \mathbf{B}_{1q_1}]$ and $\mathbf{B}_{1q_1} = [\mathbf{b}_{1q_1}^1, F_{\alpha I}\mathbf{b}_{1q_1}^1, \cdots, F_{\alpha I}^{\ {}_1\ell_{q_1}-1}\mathbf{b}_{1q_1}^1]$.

Lemma 7-A.13. Let T_s be a regular matrix and be a $\{\alpha, \beta, \gamma\}$-morphism $T_s : (K^n, F_{\alpha s}, F_{\beta s}, F_{\gamma s}) \to (K^n, F_{\alpha I}, F_{\beta I}, F_{\gamma I})$.

Let $\mathbf{b}^1_{1j} \in K^n$ be $\mathbf{b}^1_{1j} = [\mathbf{b}^{111}_{1j}, \mathbf{b}^{112}_{1j}, \cdots, \mathbf{b}^{11j}_{1j}, \mathbf{e}^s_1, \mathbf{0}]^T \in K^n$ for $1 \leq j \leq q_1 - 1$, where the number of parameters of $\mathbf{b}^1_{1j}$ is $j \times {}_1\ell_j$ for $1 \leq j \leq q_1 - 1$, $\mathbf{b}^{111}_{1j} := [\mathbf{0}, \widetilde{\mathbf{b}^{111}_{1j}}, \mathbf{0}]^T \in K^{\nu_{11}}, \widetilde{\mathbf{b}^{111}_{1j}} \in K^{\ {}_1\ell_j}$, $\mathbf{b}^{112}_{1j} := [\mathbf{0}, \widetilde{\mathbf{b}^{112}_{1j}}, \mathbf{0}]^T \in K^{\nu_{12}}$, $\widetilde{\mathbf{b}^{112}_{1j}} \in K^{\ {}_1\ell_j}$, $\cdots$, $\mathbf{b}^{11j}_{1j} := [\mathbf{0}, \widetilde{\mathbf{b}^{11j}_{1j}}, \mathbf{0}]^T \in K^{\nu_{1j}}, \widetilde{\mathbf{b}^{11j}_{1j}} \in K^{\ {}_1\ell_j}, \mathbf{e}^s_1 \in K^{\nu_{1j+1}}$, ${}_1\ell_j = \nu_{1j} - \nu_{1j+1}$. In particular, $\mathbf{b}^{111}_{11} := [\widetilde{\mathbf{b}^{111}_{11}}, \mathbf{0}]^T \in K^{\nu_{11}}$.

Then $\mathbf{b}^1_{1q_1}$ $(= T_s\mathbf{c}^{1q_1+1} - (\mathbf{tb})^1_{1q_1}) \in K^n$ of $F_{\beta I}$ is determined by $q_1 \times {}_1\ell_{q_1}$ parameters, which is the minimal number of parameters, where ${}_1\ell_{q_1} = \nu_{1q_1}$.

Proof. The number of parameters of $\mathbf{b}^1_{1j}$ $(1 \leq j \leq q_1 - 1)$ is already presented in Remark of Lemma (7-A.12).

Let us investigate the number of parameters of $\mathbf{b}^1_{1q_1}$. By lemmas (7-A.5) and (7-A.7), the equation $T_sC^{1q_1+1} = F^{q_1-1}_{\beta I}F_{\beta I}\mathbf{e}_1 = (\mathbf{tb})^1_{1q_1} + \mathbf{b}^1_{1q_1}$ holds. Hence $\mathbf{b}^1_{1q_1} = T_sC^{1q_1+1} - (\mathbf{tb})^1_{1q_1}$ holds, where $(\mathbf{tb})^1_{1q_1}$ is expressed as:

$$(\mathbf{tb})^1_{1q_1} = F^{q_1-1}_{\beta I}(\mathbf{b}^1_{11} - \mathbf{e}_{\nu_{11}+1}) + F^{q_1-2}_{\beta I}(\mathbf{b}^1_{12} - \mathbf{e}_{\nu_{11}+\nu_{12}+1}) + \cdots$$
$$+ F_{\beta I}(\mathbf{b}^1_{1q_1-1} - \mathbf{e}_{\nu_{11}+\cdots+\nu_{1q_1-1}+1}).$$

By virtue of $\phi_{1q_1}(F_{\alpha I})\mathbf{b}^1_{1q_1} = 0$, Lemma (7-A.4) and Remark in Lemma (7-A.5), the vector $\mathbf{b}^1_{1q_1}$ is determined by $q_1 \times {}_1\ell_{q_1}$ parameters.

Lemma (7-A.4) through Lemma (7-A.13) and their remarks result in Proposition (7-A.14) which gives $\mathbf{B}_1 \in K^{n\times\nu_1}$ of $F_{\beta I}$ and $\mathbf{T}_1 \in K^{n\times\nu_1}$ of T_s.

Proposition 7-A.14. Let T_s be a regular matrix and a $\{\alpha, \beta, \gamma\}$-morphism $T_s : (K^n, F_{\alpha s}, F_{\beta s}, F_{\gamma s}) \to (K^n, F_{\alpha I}, F_{\beta I}, F_{\gamma I})$.

Then $\mathbf{b}^1_{1j}$ $(1 \leq j \leq q_1 - 1)$ and $\mathbf{b}^1_{1q_1}$ of $F_{\beta I}$ in $\{\alpha, \beta, \gamma\}$-action $(K^n, F_{\alpha I}, F_{\beta I}, F_{\gamma I})$ satisfy the following conditions (1) and (2). Furthermore, $\mathbf{T}_1$ of the $\{\alpha, \beta, \gamma\}$-morphism $T_s : (K^n, F_{\alpha s}, F_{\beta s}, F_{\gamma s}) \to (K^n, F_{\alpha I}, F_{\beta I}, F_{\gamma I})$ has the full rank property.

(1) $\mathbf{b}^1_{1j} \in K^n$ is $\mathbf{b}^1_{1j} = [\mathbf{b}^{111}_{1j}, \mathbf{b}^{112}_{1j}, \cdots, \mathbf{b}^{11j}_{1j}, \mathbf{e}^s_1, \mathbf{0}]^T \in K^n$ for $1 \leq j \leq q_1 - 1$,
where $\mathbf{b}^{111}_{1j} := [\mathbf{0}, \widetilde{\mathbf{b}^{111}_{1j}}, \mathbf{0}]^T \in K^{\nu_{11}}, \widetilde{\mathbf{b}^{111}_{1j}} \in K^{\ {}_1\ell_j}$,
$\mathbf{b}^{112}_{1j} := [\mathbf{0}, \widetilde{\mathbf{b}^{112}_{1j}}, \mathbf{0}]^T \in K^{\nu_{12}}, \widetilde{\mathbf{b}^{112}_{1j}} \in K^{\ {}_1\ell_j}, \cdots,$
$\mathbf{b}^{11j}_{1j} := [\mathbf{0}, \widetilde{\mathbf{b}^{11j}_{1j}}, \mathbf{0}]^T \in K^{\nu_{1j}}, \widetilde{\mathbf{b}^{11j}_{1j}} \in K^{\ {}_1\ell_j}, \mathbf{e}^s_1 \in K^{\nu_{1j+1}}$, ${}_1\ell_j = \nu_{1j} - \nu_{1j+1}$.

(2) $\mathbf{b}^1_{1q_1} \in K^n$ is $\mathbf{b}^1_{1q_1} = [\mathbf{b}^{111}_{1q_1}, \mathbf{b}^{112}_{1q_1}, \cdots, \mathbf{b}^{11q_1}_{1q_1}\mathbf{0}]^T \in K^n$,
where $\mathbf{b}^{11\hat{j}}_{1q_1} := [\mathbf{0}, \widetilde{\mathbf{b}^{11\hat{j}}_{1q_1}}]^T \in K^{\nu_{1\hat{j}}}, \widetilde{\mathbf{b}^{11\hat{j}}_{1q_1}} \in K^{\ {}_1\ell_{q_1}}$ for $1 \leq \hat{j} \leq q_1$,
$\mathbf{b}^{11q_1}_{1q_1} := \widetilde{\mathbf{b}^{11q_1}_{1q_1}} \in K^{\ {}_1\ell_{q_1}}$, ${}_1\ell_{q_1} = \nu_{1q_1}$.
$\mathbf{b}^{11q_1}_{1q_1}$ is calculated by the equation $\mathbf{b}^{11q_1}_{1q_1} = T_sC^{1q_1+1} - (\mathbf{tb})^1_{1q_1}$.

Remark: $\mathbf{B}_1 \in K^{n\times\nu_1}$ of $F_{\beta I}$ in $\{\alpha, \beta, \gamma\}$-action $(K^n, F_{\alpha I}, F_{\beta I}, F_{\gamma I})$ obtained from Proposition (7-A.14) is determined by ν_1 $(= {}_1\ell_1 + 2 \times {}_1\ell_2 + 3 \times {}_1\ell_3 + \cdots + q_1 \times {}_1\ell_{q_1})$ parameters, where ${}_1\ell_j := \nu_{1j} - \nu_{1j+1}$ for

$1 \leq j \leq q_1 - 1$, ${}_1\ell_{q_1} := \nu_{1q_1}$. In addition, $\mathbf{T}_1 \in K^{n \times \nu_1}$ in $\{\alpha, \beta, \gamma\}$-morphism T_s has non-zero values only in $K^{\nu_1 \times \nu_1}$.

Next, let us find $\mathbf{D}_1 \in K^{n \times \nu_1}$ of $F_{\gamma I}$ and $\mathbf{T}_2 \in K^{n \times \nu_2}$ of T_s.

Proposition 7-A.15. Let T_s be a regular matrix and a $\{\alpha, \beta, \gamma\}$-morphism $T_s : (K^n, F_{\alpha s}, F_{\beta s}, F_{\gamma s}) \to (K^n, F_{\alpha I}, F_{\beta I}, F_{\gamma I})$.

Then $\mathbf{d}^1_{1j} \in K^n$ $(1 \leq j \leq q_1)$ of $F_{\gamma I}$ in $\{\alpha, \beta, \gamma\}$-action $(K^n, F_{\alpha I}, F_{\beta I}, F_{\gamma I})$ satisfy the following conditions (1) and (2). Moreover, $\mathbf{T}_2$ of the $\{\alpha, \beta, \gamma\}$-morphism T_s given by (2) has a full rank property.

(1)

(1-1) $\mathbf{d}^1_{11} \in K^n$ is expressed as $\mathbf{d}^1_{11} = [\mathbf{d}^{11}_{11}, \mathbf{e}^s_1, \mathbf{0}]^T \in K^n$, where $\mathbf{d}^{11}_{11} = [\mathbf{d}^{111}_{11}, \mathbf{0}]^T \in K^{\nu_1}$, $\mathbf{d}^{111}_{11} := [\widetilde{\mathbf{d}^{111}_{11}}, \mathbf{0}]^T \in K^{\nu_{11}}$, $\widetilde{\mathbf{d}^{111}_{11}} \in K^{\nu_{11}-\nu_{21}}$, $\mathbf{e}^s_1 \in K^{\nu_2}$. $\widetilde{\mathbf{d}^{111}_{11}}$ is given as:

$$\widetilde{\mathbf{d}^{111}_{11}} = \{[\phi_{21}(F_{11})\mathbf{e}^s_1, \cdots, \phi_{21}(F_{11})\mathbf{e}^s_{\nu_{11}-\nu_{21}}]^T \times [\phi_{21}(F_{11})\mathbf{e}^s_1, \cdots, \phi_{21}(F_{11})\mathbf{e}^s_{\nu_{11}-\nu_{21}}]\}^{-1} \times [\phi_{21}(F_{11})\mathbf{e}^s_1, \cdots, \phi_{21}(F_{11})\mathbf{e}^s_{\nu_{11}-\nu_{21}}]^T \gamma^{111}_{21}, \text{ where } \gamma^1_{21} = [\gamma^{111}_{21}, \mathbf{0}]^T.$$

$\mathbf{t}^1_{11} = \mathbf{e}_1$ leads to $\gamma^{111}_{21} = [\mathbf{e}^s_1, F_{11}\mathbf{e}^s_1, \cdots, F^{\nu_{11}-1}_{11}\mathbf{e}^s_1]C^{21}_1$.

Once $\mathbf{d}^1_{11}$ is found, $\mathbf{d}^1_{1j+1}$ can be recursively obtained in turn by: $\mathbf{d}^1_{1j+1} = F_{\beta I}\mathbf{d}^1_{1j} - F_{\gamma I}(\mathbf{b}^1_{1j} - \mathbf{e}_{\nu_{11}+\cdots+\nu_{1j}+1})$ for $1 \leq j \leq q_2$.

(1-2) From the determined $\mathbf{d}^1_{1q_2+1}$, $\mathbf{d}^1_{1j} \in K^n$ can be dependently obtained by $\mathbf{d}^1_{1j} = [\mathbf{d}^{11}_{1j}, \mathbf{d}^{12}_{1j}, \cdots, \mathbf{d}^{1j}_{1j}, \mathbf{0}]^T \in K^n$ for $q_2 + 2 \leq j \leq q_1$.

(2) $\mathbf{t}^1_{2j}$ is given by the relation $\mathbf{t}^1_{2j} = (\mathbf{td})^1_{1j} + \mathbf{d}^1_{1j}$ for $2 \leq j \leq q_1$, where $(\mathbf{td})^1_{1j} = F_{\beta I}(\mathbf{td})^1_{1j-1} + F_{\gamma I}(\mathbf{d}^1_{1j-1} - \mathbf{e}_{\nu_{11}+\cdots+\nu_{1j-1}+1})$ for $2 \leq j \leq p$, with $(\mathbf{td})^1_{11} = 0$. See (3) of Lemma (7-A.9).

$\mathbf{t}^1_{2j}$ is expressed as:

$\mathbf{t}^1_{2j} = [\mathbf{t}^{11}_{2j}, \mathbf{t}^{12}_{2j}, \mathbf{0}]^T \in K^n$, where

$\mathbf{t}^{11}_{2j} = [\mathbf{t}^{111}_{2j}, \cdots, \mathbf{t}^{11j}_{2j}, \mathbf{0}]^T \in K^{\nu_1}$, $\mathbf{t}^{12}_{2j} = [\mathbf{t}^{121}_{2j}, \cdots, \mathbf{t}^{12j}_{2j}, \mathbf{e}^s_1]^T \in K^{\nu_2}$,

$\mathbf{t}^{11\hat{j}}_{2j} = [\mathbf{0}, \widetilde{\mathbf{t}^{11\hat{j}}_{2j}}, \mathbf{0}]^T \in K^{\nu_{1\hat{j}}}$, $\mathbf{t}^{12\hat{j}}_{2j} = [\mathbf{0}, \widetilde{\mathbf{t}^{12\hat{j}}_{2j}}, \mathbf{0}]^T \in K^{\nu_{2\hat{j}}}$ for $1 \leq \hat{j} \leq j$,

$\widetilde{\mathbf{t}^{11\hat{j}}_{2j}} \in K^{\nu_{1j}-\nu_{2j}}$, $\widetilde{\mathbf{t}^{12\hat{j}}_{2j}} \in K^{\nu_{1j}-\nu_{2j}}$.

Proof. Set $i = 2$. The Lemmas (7-A.4), (7-A.7) and (7-A.8) result in $\mathbf{T}_1 C^{21T}_1 = \phi_{21}(F_{\alpha I})\mathbf{d}^1_{11} = \gamma^1_{21}$. Hence, $\mathbf{d}^1_{11}$ $(= \mathbf{t}^1_{21})$ is expressed as:

$\mathbf{d}^1_{11} = [\mathbf{d}^{11}_{11}, \mathbf{e}^s_1, \mathbf{0}]^T \in K^n$, $\mathbf{d}^{11}_{11} = [\mathbf{d}^{111}_{11}, \mathbf{0}]^T \in K^{\nu_1}$,

$\mathbf{d}^{111}_{11} = [\widetilde{\mathbf{d}^{111}_{11}}, \mathbf{0}]^T \in K^{\nu_{11}}, \widetilde{\mathbf{d}^{111}_{11}} \in K^{\nu_{11}-\nu_{21}}, \mathbf{e}^s_1 \in K^{\nu_2}$,

$$\widetilde{\mathbf{d}^{111}_{11}} = \{[\phi_{21}(F_{11})\mathbf{e}^s_1, \cdots, \phi_{21}(F_{11})\mathbf{e}^s_{\nu_{11}-\nu_{21}}]^T \times [\phi_{21}(F_{11})\mathbf{e}^s_1, \cdots, \phi_{21}(F_{11})\mathbf{e}^s_{\nu_{11}-\nu_{21}}]\}^{-1} \times [\phi_{21}(F_{11})\mathbf{e}^s_1, \cdots, \phi_{21}(F_{11})\mathbf{e}^s_{\nu_{11}-\nu_{21}}]^T \gamma^{111}_{21}, \text{ where } \gamma^1_{21} = [\gamma^{111}_{21}, \mathbf{0}]^T.$$

$\mathbf{t}^1_{11} = \mathbf{e}_1$ leads to $\gamma^{111}_{21} = [\mathbf{e}^s_1, F_{11}\mathbf{e}^s_1, \cdots, F^{\nu_{11}-1}_{11}\mathbf{e}^s_1]C^{21}_1$.

Next, we will find $\mathbf{d}^1_{1j}$ and $\mathbf{t}^1_{2j}$ for $2 \le j \le q_1$.

The commutativity of matrices $F_{\beta I}$ and $F_{\gamma I}$, that is $F_{\beta I}F_{\gamma I} = F_{\gamma I}F_{\beta I}$, leads to $F_{\beta I}\mathbf{d}^1_{1j} = F_{\gamma I}(\mathbf{b}^1_{1j} - \mathbf{e}_{\nu_{11}+\cdots+\nu_{1j}+1}) + \mathbf{d}^1_{1j+1}$ for $1 \le j \le q_2$.

Thus, once $\mathbf{d}^1_{11}$ is found, then we can dependently obtain $\mathbf{d}^1_{1j}$ in turn for $2 \le j \le q_2+1$. Let us find $\mathbf{d}^1_{1j}$ for $q_2+2 \le j \le q_1$.

We refer the following relations presented in Lemma (7-A.5).

$$\begin{cases} F_{\gamma I}T_s\mathbf{e}_{\nu_{11}+\cdots+\nu_{1q_2}+1} = F_{\gamma I}F_{\beta I}^{q_2}\mathbf{e}_1 \\ \qquad\qquad = F_{\gamma I}(F_{\beta I}^{q_2-1}\mathbf{b}^1_{11} - \mathbf{e}_{\nu_{11}+\cdots+\nu_{1q_2}+1}) + \mathbf{d}^1_{1q_2+1} \\ T_sF_{\gamma s}\mathbf{e}_{\nu_{11}+\cdots+\nu_{1q_2}+1} = T_s\mathbf{c}^{2q_2+1} \end{cases}$$

$$\begin{cases} F_{\gamma I}T_s\mathbf{e}_{\nu_{11}+\cdots+\nu_{1q_2+1}+1} = F_{\gamma I}F_{\beta I}^{q_2+1}\mathbf{e}_1 \\ \qquad\qquad = F_{\gamma I}(F_{\beta I}^{q_2}\mathbf{b}^1_{11} - \mathbf{e}_{\nu_{11}+\cdots+\nu_{1q_2+1}+1}) + \mathbf{d}^1_{1q_2+2} \\ T_sF_{\gamma s}\mathbf{e}_{\nu_{11}+\cdots+\nu_{1q_2+1}+1} = T_sF_{\beta s}\mathbf{c}^{2q_2+1} = F_{\beta I}T_s\mathbf{c}^{2q_2+1} \end{cases}$$

$$\vdots$$

$$\begin{cases} F_{\gamma I}T_s\mathbf{e}_{\nu_{11}+\cdots+\nu_{1q_1-1}+1} = F_{\gamma I}F_{\beta I}^{q_1-1}\mathbf{e}_1 \\ \qquad\qquad = F_{\gamma I}(F_{\beta I}^{q_1-2}\mathbf{b}^1_{11} - \mathbf{e}_{\nu_{11}+\cdots+\nu_{1q_1-1}+1}) + \mathbf{d}^1_{1q_1} \\ T_sF_{\gamma s}\mathbf{e}_{\nu_{11}+\cdots+\nu_{1q_1-1}+1} = T_sF_{\beta s}^{q_1-q_2-1}\mathbf{c}^{2q_2+1} = F_{\beta I}^{q_1-q_2-1}T_s\mathbf{c}^{2q_2+1} \end{cases}$$

Then once $\mathbf{d}^1_{1q_2+1}$ is found, $\mathbf{d}^1_{1q_2+2}, \mathbf{d}^1_{1q_2+3}, \cdots \mathbf{d}^1_{1q_1}$ are dependently found as:

$$F_{\gamma I}(F_{\beta I}^{q_2-1}\mathbf{b}^1_{11} - \mathbf{e}_{\nu_{11}+\cdots+\nu_{1q_2}+1}) + \mathbf{d}^1_{1q_2+1} = T_s\mathbf{c}^{2q_2+1},$$
$$F_{\gamma I}(F_{\beta I}^{q_2}\mathbf{b}^1_{11} - \mathbf{e}_{\nu_{11}+\cdots+\nu_{1q_2+1}+1}) + \mathbf{d}^1_{1q_2+2} = F_{\beta I}T_s\mathbf{c}^{2q_2+1},$$
$$\vdots$$
$$F_{\gamma I}(F_{\beta I}^{q_1-2}\mathbf{b}^1_{11} - \mathbf{e}_{\nu_{11}+\cdots+\nu_{1q_1-1}+1}) + \mathbf{d}^1_{1q_1} = F_{\beta I}^{q_1-q_2-1}T_s\mathbf{c}^{2q_2+1}.$$

Note that the relation $\mathbf{t}^1_{2j} = (\mathbf{td})^1_{1j} + \mathbf{d}^1_{1j}$ holds for $2 \le j \le q_1$ as we see in (3) of Lemma (7-A.9), where $(\mathbf{td})^1_{1j} = F_{\beta I}(\mathbf{td})^1_{1j-1} + F_{\gamma I}(\mathbf{d}^1_{1j-1} - \mathbf{e}_{\nu_{11}+\cdots+\nu_{1j-1}+1})$ for $2 \le j \le p$ with $(\mathbf{td})^1_{11} = 0$. Thus, $\mathbf{d}^1_{1j}$ and $\mathbf{t}^1_{2j}$ are found recursively in turn for $2 \le j \le q_2$, where $\mathbf{t}^1_{2j} = [\mathbf{t}^{111}_{2j}, \cdots, \mathbf{t}^{11j}_{2j}, \mathbf{0}, \mathbf{t}^{121}_{2j}, \cdots, \mathbf{t}^{12j-1}_{2j}, \mathbf{e}^s_1, \mathbf{0}]^T \in K^n$, $\mathbf{e}^s_1 \in K^{\nu_{2j}}$ and $\mathbf{t}^{11\hat{j}}_{2j} \in K^{\nu_{1\hat{j}}}$ $(1 \le \hat{j} \le j)$.

Remark: $\mathbf{D}_1 \in K^{n\times\nu_1}$ of $F_{\gamma I}$ in $\{\alpha,\beta,\gamma\}$-action $(K^n, F_{\alpha I}, F_{\beta I}, F_{\gamma I})$ presented in Proposition (7-A.15) is determined by $\nu_{11} - \nu_{21}$ parameters.

As a consequence, $\mathbf{B}_1$ of $F_{\beta I}$,$\mathbf{T}_1, \mathbf{T}_2$ of T_s and $\mathbf{D}_1$ of $F_{\gamma I}$ are determined by Propositions (7-A.14) and (7-A.15).

Next, we will determine $\mathbf{B}_2 \in K^{n\times\nu_2}$ of $F_{\beta I}$, $\mathbf{D}_2 \in K^{n\times\nu_2}$ of $F_{\gamma I}$ and $\mathbf{T}_3 \in K^{n\times\nu_3}$ of T_s.

Proposition 7-A.16. Let T_s be a regular matrix and a $\{\alpha,\beta,\gamma\}$-morphism $T_s : (K^n, F_{\alpha s}, F_{\beta s}, F_{\gamma s}) \to (K^n, F_{\alpha I}, F_{\beta I}, F_{\gamma I})$. Then $\mathbf{b}^1_{2j}$ of $F_{\beta I}$ and $\mathbf{d}^1_{2j}$ of $F_{\gamma I}$ in $\{\alpha,\beta,\gamma\}$-action $(K^n, F_{\alpha I}, F_{\beta,I}, F_{\gamma I})$ satisfy the following conditions (1) and (2). Moreover, $\mathbf{T}_3$ of the $\{\alpha,\beta,\gamma\}$-morphism T_s given by (3) has a full rank property.

(1)

$\mathbf{b}^1_{2j}$ is expressed as $\mathbf{b}^1_{2j} = [\mathbf{b}^{11}_{2j}, \mathbf{b}^{12}_{2j}, \mathbf{0}]^T \in K^n$, where
$\mathbf{b}^{11}_{2j} = [\mathbf{b}^{111}_{2j}, \cdots, \mathbf{b}^{11j}_{2j}, \mathbf{0}]^T \in K^{\nu_1}$, $\mathbf{b}^{12}_{2j} = [\mathbf{b}^{121}_{2j}, \cdots, \mathbf{b}^{12j}_{2j}, \mathbf{e}^s_1]^T \in K^{\nu_2}$,
$\mathbf{b}^{11\hat{j}}_{2j} := [\mathbf{0}, \widetilde{\mathbf{b}^{11\hat{j}}_{2j}}, \mathbf{0}]^T \in K^{\nu_{1\hat{j}}}$, $\widetilde{\mathbf{b}^{11\hat{j}}_{2j}} \in K^{\nu_{2j}-\nu_{2j+1}}$ $(2 \le \hat{j} \le j)$,
$\mathbf{b}^{12\hat{j}}_{2j} := [\mathbf{0}, \widetilde{\mathbf{b}^{12\hat{j}}_{2j}}, \mathbf{0}]^T \in K^{\nu_{2\hat{j}}}$, $\widetilde{\mathbf{b}^{12\hat{j}}_{2j}} \in K^{\nu_{2j}-\nu_{2j+1}}$.

$\widetilde{\mathbf{b}^{11\hat{j}}_{2j}} \in K^{\nu_{2j}-\nu_{2j+1}}$ and $\widetilde{\mathbf{b}^{12\hat{j}}_{2j}} \in K^{\nu_{2j}-\nu_{2j+1}}$ $(1 \le \hat{j} \le j)$ are represented as:

$$\widetilde{\mathbf{b}^{11\hat{j}}_{2j}} = \{[\phi_{2j+1}(F_{1\hat{j}})\mathbf{e}^s_{\nu_{1\hat{j}}-\nu_{2j}+1}, \cdots, \phi_{2j+1}(F_{1\hat{j}})\mathbf{e}^s_{\nu_{1\hat{j}}-\nu_{2j+1}}]^T \times [\phi_{2j+1}(F_{1\hat{j}})\mathbf{e}^s_{\nu_{1\hat{j}}-\nu_{2j}+1}, \cdots, \phi_{2j+1}(F_{1\hat{j}})\mathbf{e}^s_{\nu_{1\hat{j}}-\nu_{2j+1}}]\}^{-1} \times [\phi_{2j+1}(F_{1\hat{j}})\mathbf{e}^s_{\nu_{1\hat{j}}-\nu_{2j}+1}, \cdots, \phi_{2j+1}(F_{1\hat{j}})\mathbf{e}^s_{\nu_{1\hat{j}}-\nu_{2j+1}}]^T \{\beta^{11\hat{j}}_{1j} - \phi_{2j+1}(F_{1\hat{j}}) \times (\mathbf{tb})^{11\hat{j}}_{2j}\},$$

$$\widetilde{\mathbf{b}^{12\hat{j}}_{2j}} = \{[\phi_{2j+1}(F_{2\hat{j}})\mathbf{e}^s_{\nu_{2\hat{j}}-\nu_{2j}+1}, \cdots, \phi_{2j+1}(F_{2\hat{j}})\mathbf{e}^s_{\nu_{2\hat{j}}-\nu_{2j+1}}]^T \times [\phi_{2j+1}(F_{2\hat{j}})\mathbf{e}^s_{\nu_{2\hat{j}}-\nu_{2j}+1}, \cdots, \phi_{2j+1}(F_{2\hat{j}})\mathbf{e}^s_{\nu_{2\hat{j}}-\nu_{2j+1}}]\}^{-1} \times [\phi_{2j+1}(F_{2\hat{j}})\mathbf{e}^s_{\nu_{2\hat{j}}-\nu_{2j}+1}, \cdots, \phi_{2j+1}(F_{2\hat{j}})\mathbf{e}^s_{\nu_{2\hat{j}}-\nu_{2j+1}}]^T \{\beta^{12\hat{j}}_{2j} - \phi_{2j+1}(F_{2\hat{j}}) \times (\mathbf{tb})^{12\hat{j}}_{2j}\}.$$

$\mathbf{b}^1_{2q_2}$ is given by the equation $\mathbf{b}^1_{2q_2} = T_s C^{2q_2+1} - (\mathbf{tb})^1_{2q_2}$, where
$(\mathbf{tb})^1_{2q_2} = F^{q_2-1}_{\beta I}(\mathbf{b}^1_{21} - \mathbf{e}_{\nu_1+\nu_{21}+1}) + F^{q_2-2}_{\beta I}(\mathbf{b}^1_{22} - \mathbf{e}_{\nu_1+\nu_{21}+\nu_{22}+1}) + \cdots + F_{\beta I}(\mathbf{b}^1_{2q_2-1} - \mathbf{e}_{\nu_1+\nu_{21}+\cdots+\nu_{2q_2-1}+1})$.

(2)

(2-1) $\mathbf{d}^1_{21}$ is expressed as $\mathbf{d}^1_{21} = [\mathbf{d}^{11}_{21}, \mathbf{d}^{12}_{21}, \mathbf{e}^s_1, \mathbf{0}]^T$, where
$\mathbf{d}^{11}_{21} = [\mathbf{d}^{111}_{21}, \mathbf{0}]^T \in K^{\nu_1}$, $\mathbf{d}^{111}_{21} = [\mathbf{0}, \widetilde{\mathbf{d}^{111}_{21}}, \mathbf{0}]^T \in K^{\nu_{11}}$, $\widetilde{\mathbf{d}^{111}_{21}} \in K^{\nu_{21}-\nu_{31}}$,
$\mathbf{d}^{12}_{21} = [\mathbf{d}^{121}_{21}, \mathbf{0}]^T \in K^{\nu_2}$, $\mathbf{d}^{121}_{21} = [\widetilde{\mathbf{d}^{121}_{21}}, \mathbf{0}]^T \in K^{\nu_{21}}$, $\widetilde{\mathbf{d}^{121}_{21}} \in K^{\nu_{21}-\nu_{31}}$,
$\mathbf{e}^s_1 \in K^{\nu_3}$.

$\widetilde{\mathbf{d}^{111}_{21}}$ and $\widetilde{\mathbf{d}^{121}_{21}}$ are represented as:

$$\widetilde{\mathbf{d}^{111}_{21}} = \{[\phi_{31}(F_{11})\mathbf{e}^s_{\nu_{11}-\nu_{21}+1}, \cdots, \phi_{31}(F_{11})\mathbf{e}^s_{\nu_{11}-\nu_{31}}]^T \times [\phi_{31}(F_{11})\mathbf{e}^s_{\nu_{11}-\nu_{21}+1}, \cdots, \phi_{31}(F_{11})\mathbf{e}^s_{\nu_{11}-\nu_{31}}]\}^{-1}$$

$\times [\phi_{31}(F_{11})\mathbf{e}^s_{\nu_{11}-\nu_{21}+1}, \cdots, \phi_{31}(F_{11})\mathbf{e}^s_{\nu_{11}-\nu_{31}}]^T \gamma^{111}_{31}$,

$\widetilde{\mathbf{d}^{121}_{21}} = \{[\phi_{31}(F_{11})\mathbf{e}^s_1, \cdots, \phi_{31}(F_{11})\mathbf{e}^s_{\nu_{21}-\nu_{31}}]^T$
$\times [\phi_{31}(F_{11})\mathbf{e}^s_1, \cdots, \phi_{31}(F_{11})\mathbf{e}^s_{\nu_{21}-\nu_{31}}]\}^{-1}$
$\times [\phi_{31}(F_{11})\mathbf{e}^s_1, \cdots, \phi_{31}(F_{11})\mathbf{e}^s_{\nu_{21}-\nu_{31}}]^T \gamma^{112}_{31}$,
where
$\gamma^1_{31} := [\mathbf{T}_1\ \mathbf{T}_2][C^{31}_1, C^{31}_2]^T + [F_{\gamma I}(F_{\gamma I}\mathbf{e}_1 - \mathbf{e}_{\nu_1+1}), F_{\alpha I}F_{\gamma I}$
$\times (F_{\gamma I}\mathbf{e}_1 - \mathbf{e}_{\nu_1+1})]C^{31}_{31}$.
Furthermore, γ^1_{31} is expressed as $\gamma^1_{31} := [\gamma^{111}_{31}, \mathbf{0}, \gamma^{121}_{31}, \mathbf{0}, \gamma^{131}_{31}, \mathbf{0}]^T \in K^n$,

$\gamma^{111}_{31} = [\mathbf{0}, \widetilde{\gamma^{111}_{31}}, \mathbf{0}]^T \in K^{\nu_{11}}$, $\widetilde{\gamma^{111}_{31}} \in K^{\nu_{21}-\nu_{31}}$,
$\gamma^{121}_{31} = [\mathbf{0}, \widetilde{\gamma^{121}_{31}}, \mathbf{0}]^T \in K^{\nu_{21}}$, $\widetilde{\gamma^{121}_{31}} \in K^{\nu_{21}-\nu_{31}}$.

Once $\mathbf{d}^1_{21}$ is found, $\mathbf{d}^1_{2j+1}$ is dependently found recursively by:

$\mathbf{d}^1_{2j+1} = F_{\beta I}\mathbf{d}^1_{2j} - F_{\gamma I}(\mathbf{b}^1_{2j} - \mathbf{e}_{\nu_1+\nu_{21}+\cdots+\nu_{2j}+1})$ for $1 \le j \le q_3$.

(2-2) From the determined $\mathbf{d}^1_{2q_3+1}$, $\mathbf{d}^1_{2j} \in K^n$ is dependently found as:
$\mathbf{d}^1_{2j} = [\mathbf{d}^{11}_{2j}, \mathbf{d}^{12}_{2j}, \cdots, \mathbf{d}^{1j}_{2j}, \mathbf{0}]^T \in K^n$ for $q_3 + 2 \le j \le q_2$:

(3)

$\mathbf{t}^1_{3j}$ is calculated in turn by using the equation $\mathbf{t}^1_{3j} = (\mathbf{td})^1_{2j} + \mathbf{d}^1_{2j}$ for any $1 \le j \le q_3$, where
$(\mathbf{td})^1_{2j} = F_{\beta I}(\mathbf{td})^1_{2j-1} + F_{\gamma I}(\mathbf{b}^1_{2j-1} - \mathbf{e}_{\nu_1++\nu_{21}+\cdots+\nu_{2j-1}+1})$,
$(\mathbf{td})^1_{21} = F_{\gamma I}(\mathbf{d}^1_{11} - \mathbf{e}_{\nu_1+1})$.

Proof. Lemma (7-A.4) and (2) of Lemma (7-A.8) lead to $\phi_{2j}(F_{\alpha I})\mathbf{b}^1_{2j} = 0$ and $\phi_{2j+1}(F_{\alpha I})\mathbf{b}^1_{2j} = \beta^1_{2j} - \phi_{2j+1}(F_{\alpha I})(\mathbf{tb})^1_{2j}$, where $(\mathbf{tb})^1_{2j} = F_{\beta I}(\mathbf{tb})^1_{2j-1} + F_{\beta I}(\mathbf{b}^1_{2j-1} - \mathbf{e}_{\nu_1+\nu_{21}+\cdots+\nu_{2j-1}+1})$, $(\mathbf{tb})^1_{21} = 0$, $\beta^1_{21} := [\mathbf{T}_1\mathbf{T}_{21}][C^{22}_1, C^{22}_{21}]^T + [F_{\beta I}(F_{\gamma I}\mathbf{e}_1 - \mathbf{e}_{\nu_1+1}), F_{\alpha I}F_{\beta I}(F_{\gamma I}\mathbf{e}_1 - \mathbf{e}_{\nu_1+1}), \cdots, F^{\nu_{22}-1}_{\alpha I}F_{\beta I}(F_{\gamma I}\mathbf{e}_1 - \mathbf{e}_{\nu_1+1})]$ C^{22}_{22}, $\beta^1_{21} = [\beta^{111}_{21}, \beta^{112}_{21}, \mathbf{0}, \beta^{121}_{21}, \beta^{122}_{21}, \mathbf{0}]^T$, $\beta^{111}_{21} \in K^{\nu_{11}}$, $\beta^{112}_{21} \in K^{\nu_{12}}$, $\beta^{121}_{21} \in K^{\nu_{21}}$, $\beta^{122}_{21} \in K^{\nu_{22}}$.

Therefore, $\mathbf{b}^1_{2j}$ is found by solving the equation:
$\phi_{2j+1}(F_{\alpha I})\mathbf{b}^1_{2j} = \beta^1_{2j} - \phi_{2j+1}(F_{\alpha I})(\mathbf{tb})^1_{2j}$ for $1 \le j \le q_2$.
$\mathbf{b}^1_{21} \in K^n$ is expressed as $\mathbf{b}^1_{21} = [\mathbf{b}^{111}_{21}, \mathbf{b}^{112}_{21}, \mathbf{0}, \mathbf{b}^{121}_{21}, \mathbf{e}^s_1, \mathbf{0}]^T \in K^n$, where

$\mathbf{b}^{111}_{21} := [\mathbf{0}, \widetilde{\mathbf{b}^{111}_{21}}, \mathbf{0}]^T \in K^{\nu_{11}}$, $\widetilde{\mathbf{b}^{111}_{21}} \in K^{\nu_{21}-\nu_{22}}$,

$\mathbf{b}^{112}_{21} := [\mathbf{0}, \widetilde{\mathbf{b}^{112}_{21}}, \mathbf{0}]^T \in K^{\nu_{12}}, \widetilde{\mathbf{b}^{112}_{21}} \in K^{\nu_{21}-\nu_{22}}$,

$\mathbf{b}^{121}_{21} := [\mathbf{0}, \widetilde{\mathbf{b}^{121}_{21}}, \mathbf{0}]^T \in K^{\nu_{21}}$, $\widetilde{\mathbf{b}^{121}_{21}} \in K^{\nu_{21}-\nu_{22}}$,

$\widetilde{\mathbf{b}^{111}_{21}} = \{[\phi_{22}(F_{11})\mathbf{e}^s_{\nu_{11}-\nu_{21}+1}, \cdots, \phi_{22}(F_{11})\mathbf{e}^s_{\nu_{11}-\nu_{22}}]^T$
$\times [\phi_{22}(F_{11})\mathbf{e}^s_{\nu_{11}-\nu_{21}+1}, \cdots, \phi_{22}(F_{11})\mathbf{e}^s_{\nu_{11}-\nu_{22}}]\}^{-1}$
$\times [\phi_{22}(F_{11})\mathbf{e}^s_{\nu_{11}-\nu_{21}+1}, \cdots, \phi_{22}(F_{11})\mathbf{e}^s_{\nu_{11}-\nu_{22}}]^T \{\beta^{111}_{21} - \phi_{22}(F_{\alpha I})(\mathbf{tb})^{111}_{21}\}$.

$\widetilde{\mathbf{b}^{112}_{21}} = \{[\phi_{22}(F_{12})\mathbf{e}^s_{\nu_{12}-\nu_{21}+1}, \cdots, \phi_{22}(F_{12})\mathbf{e}^s_{\nu_{12}-\nu_{22}}]^T$
$\times [\phi_{22}(F_{12})\mathbf{e}^s_{\nu_{12}-\nu_{21}+1}, \cdots, \phi_{22}(F_{12})\mathbf{e}^s_{\nu_{12}-\nu_{22}}]\}^{-1}$
$\times [\phi_{22}(F_{12})\mathbf{e}^s_{\nu_{12}-\nu_{21}+1}, \cdots, \phi_{22}(F_{12})\mathbf{e}^s_{\nu_{12}-\nu_{22}}]^T \{\beta^{112}_{21} - \phi_{22}(F_{\alpha I})(\mathbf{tb})^{112}_{21}\}$.

$\widetilde{\mathbf{b}_{21}^{121}} = \{[\phi_{22}(F_{21})\mathbf{e}_1^s, \cdots, \phi_{22}(F_{21})\mathbf{e}_{\nu_{21}-\nu_{22}}^s]^T[\phi_{22}(F_{21})\mathbf{e}_1^s, \cdots, \phi_{22}(F_{21}) \times\mathbf{e}_{\nu_{21}-\nu_{22}}^s]\}^{-1}\times[\phi_{22}(F_{21})\mathbf{e}_1^s, \cdots, \phi_{22}(F_{21})\mathbf{e}_{\nu_{21}-\nu_{22}}^s]^T\{\beta_{21}^{121}-\phi_{22}(F_{\alpha I})(\mathbf{tb})_{21}^{121}\}$.

Next, let us find $\mathbf{b}_{2j}^1 \in K^n$ for $1 \le j \le q_2$.
By virtue of (2) of Lemma (7-A.6), $\mathbf{b}_{2j}^1$ satisfies the equation:
$\phi_{2j+1}(F_{\alpha I})\mathbf{b}_{2j}^1 = \beta_{2j}^1 - \phi_{2j+1}(F_{\alpha I})(\mathbf{tb})_{2j}^1$, where
$(\mathbf{tb})_{2j}^1 = F_{\beta I}(\mathbf{tb})_{2j-1}^1 + F_{\beta I}(\mathbf{b}_{2j-1}^1 - \mathbf{e}_{\nu_1+\nu_{21}+\cdots+\nu_{2j-1}+1})$ with $(\mathbf{tb})_{21}^1 = 0$.
Note that $\mathbf{b}_{2j}^1$ is expressed as

$$\mathbf{b}_{2j}^1 = [\mathbf{b}_{2j}^{111}, \cdots, \mathbf{b}_{2j}^{11j}, \mathbf{0}, \mathbf{b}_{2j}^{121}, \cdots, \mathbf{b}_{2j}^{12j}, \mathbf{e}_1^s, \mathbf{0}]^T \in K^n ,$$

where $\mathbf{b}_{2j}^{11\hat{j}} := [\mathbf{0}, \widetilde{\mathbf{b}_{2j}^{11\hat{j}}}, \mathbf{0}]^T \in K^{\nu_{1\hat{j}}}$, $\widetilde{\mathbf{b}_{2j}^{11\hat{j}}} \in K^{\nu_{2j}-\nu_{2j+1}}$,
$\mathbf{b}_{2j}^{12\hat{j}} := [\mathbf{0}, \widetilde{\mathbf{b}_{2j}^{12\hat{j}}}, \mathbf{0}]^T \in K^{\nu_{2\hat{j}}}$, $\widetilde{\mathbf{b}_{2j}^{12\hat{j}}} \in K^{\nu_{2j}-\nu_{2j+1}}$ for $2 \le \hat{j} \le j$.
$\widetilde{\mathbf{b}_{2j}^{11\hat{j}}} \in K^{\nu_{2j}-\nu_{2j+1}}$ and $\widetilde{\mathbf{b}_{2j}^{12\hat{j}}} \in K^{\nu_{2j}-\nu_{2j+1}}$ $(1 \le \hat{j} \le j)$ are expressed as:

$$\widetilde{\mathbf{b}_{2j}^{11\hat{j}}} = \{[\phi_{2j+1}(F_{1\hat{j}})\mathbf{e}_{\nu_{1\hat{j}}-\nu_{2j}+1}^s, \cdots, \phi_{2j+1}(F_{1\hat{j}})\mathbf{e}_{\nu_{1\hat{j}}-\nu_{2j+1}}^s]^T \times [\phi_{2j+1}(F_{1\hat{j}})\mathbf{e}_{\nu_{1\hat{j}}-\nu_{2j}+1}^s, \cdots, \phi_{2j+1}(F_{1\hat{j}})\mathbf{e}_{\nu_{1\hat{j}}-\nu_{2j+1}}^s]\}^{-1} \times [\phi_{2j+1}(F_{1\hat{j}})\mathbf{e}_{\nu_{1\hat{j}}-\nu_{2j}+1}^s, \cdots, \phi_{2j+1}(F_{1\hat{j}})\mathbf{e}_{\nu_{1\hat{j}}-\nu_{2j+1}}^s]^T\{\beta_{1j}^{11\hat{j}} - \phi_{2j+1}(F_{1\hat{j}}) \times (\mathbf{tb})_{2j}^{11\hat{j}}\}.$$

$$\widetilde{\mathbf{b}_{2j}^{12\hat{j}}} = \{[\phi_{2j+1}(F_{2\hat{j}})\mathbf{e}_{\nu_{2\hat{j}}-\nu_{2j}+1}^s, \cdots, \phi_{2j+1}(F_{2\hat{j}})\mathbf{e}_{\nu_{2\hat{j}}-\nu_{2j+1}}^s]^T \times [\phi_{2j+1}(F_{2\hat{j}})\mathbf{e}_{\nu_{2\hat{j}}-\nu_{2j}+1}^s, \cdots, \phi_{2j+1}(F_{2\hat{j}})\mathbf{e}_{\nu_{2\hat{j}}-\nu_{2j+1}}^s]\}^{-1} \times [\phi_{2j+1}(F_{2\hat{j}})\mathbf{e}_{\nu_{2\hat{j}}-\nu_{2j}+1}^s, \cdots, \phi_{2j+1}(F_{2\hat{j}})\mathbf{e}_{\nu_{2\hat{j}}-\nu_{2j+1}}^s]^T\{\beta_{2j}^{12\hat{j}} - \phi_{2j+1}(F_{2\hat{j}}) \times (\mathbf{tb})_{2j}^{12\hat{j}}\}.$$

Let us find $\mathbf{b}_{2q_2}^1$. Lemmas (7-A.5) and (7-A.7) leads to $T_s\mathbf{c}^{2q_2+1} = F_{\beta I}^{q_2-1}F_{\beta I}\mathbf{e}_{\nu_1+1} = (\mathbf{tb})_{2q_2}^1 + \mathbf{b}_{2q_2}^1$, where
$(\mathbf{tb})_{2q_2}^1 = F_{\beta I}^{q_2-1}(\mathbf{b}_{21}^1 - \mathbf{e}_{\nu_1+\nu_{21}+1}) + F_{\beta I}^{q_2-2}(\mathbf{b}_{22}^1 - \mathbf{e}_{\nu_1+\nu_{21}+\nu_{22}+1}) + \cdots + F_{\beta I}(\mathbf{b}_{2q_2-1}^1 - \mathbf{e}_{\nu_1+\nu_{21}+\cdots+\nu_{2q_2-1}+1})$.

Next, let us find $\mathbf{d}_{2j}^1$ $(1 \le j \le q_2)$ and $\mathbf{t}_{3j}^1$ $(1 \le j \le q_3)$ by using Lemmas (7-A.4), (7-A.7), (7-A.8) and (7-A.9). The relation $\phi_{31}(F_{\alpha I})\mathbf{d}_{21}^1 = \gamma_{31}^1$ leads to $\mathbf{d}_{21}^1 = [\mathbf{d}_{21}^{11}, \mathbf{d}_{21}^{12}, \mathbf{e}_1^s, \mathbf{0}]^T$, where

$\mathbf{d}_{21}^{11} = [\mathbf{d}_{21}^{111}, \mathbf{0}]^T \in K^{\nu_1}$, $\mathbf{d}_{21}^{111} = [\mathbf{0}, \widetilde{\mathbf{d}_{21}^{111}}, \mathbf{0}]^T \in K^{\nu_{11}}$, $\widetilde{\mathbf{d}_{21}^{111}} \in K^{\nu_{21}-\nu_{31}}$,
$\mathbf{d}_{21}^{12} = [\mathbf{d}_{21}^{121}, \mathbf{0}]^T \in K^{\nu_2}$, $\mathbf{d}_{21}^{121} = [\widetilde{\mathbf{d}_{21}^{121}}, \mathbf{0}]^T \in K^{\nu_{21}}$, $\widetilde{\mathbf{d}_{21}^{121}} \in K^{\nu_{21}-\nu_{31}}$,
$\mathbf{e}_1^s \in K^{\nu_3}$,

$$\widetilde{\mathbf{d}_{21}^{111}} = \{[\phi_{31}(F_{11})\mathbf{e}_{\nu_{11}-\nu_{21}+1}^s, \cdots, \phi_{31}(F_{11})\mathbf{e}_{\nu_{11}-\nu_{31}}^s]^T \times [\phi_{31}(F_{11})\mathbf{e}_{\nu_{11}-\nu_{21}+1}^s, \cdots, \phi_{31}(F_{11})\mathbf{e}_{\nu_{11}-\nu_{31}}^s]\}^{-1}$$

$\times [\phi_{31}(F_{11})\mathbf{e}^s_{\nu_{11}-\nu_{21}+1}, \cdots, \phi_{31}(F_{11})\mathbf{e}^s_{\nu_{11}-\nu_{31}}]^T \gamma^{111}_{31}$,

$\widetilde{\mathbf{d}^{121}_{21}} = \{[\phi_{31}(F_{11})\mathbf{e}^s_1, \cdots, \phi_{31}(F_{11})\mathbf{e}^s_{\nu_{21}-\nu_{31}}]^T[\phi_{31}(F_{11})\mathbf{e}^s_1, \cdots, \phi_{31}(F_{11})$
$\times \mathbf{e}^s_{\nu_{21}-\nu_{31}}]\}^{-1} \times [\phi_{31}(F_{11})\mathbf{e}^s_1, \cdots, \phi_{31}(F_{11})\mathbf{e}^s_{\nu_{21}-\nu_{31}}]^T \gamma^{112}_{31}$,

where
$\gamma^1_{31} := [\mathbf{T}_1\ \mathbf{T}_2][C^{31}_1, C^{31}_2]^T + [F_{\gamma I}(F_{\gamma I}\mathbf{e}_1 - \mathbf{e}_{\nu_1+1}), F_{\alpha I}F_{\gamma I}(F_{\gamma I}\mathbf{e}_1 - \mathbf{e}_{\nu_1+1})]C^{31}_{31}$,
$\gamma^1_{31} := [\gamma^{111}_{31}, \mathbf{0}, \gamma^{121}_{31}, \mathbf{0}, \gamma^{131}_{31}, \mathbf{0}]^T \in K^n$, $\gamma^{111}_{31} = [\mathbf{0}, \widetilde{\gamma^{111}_{31}}, \mathbf{0}]^T \in K^{\nu_{11}}$,
$\widetilde{\gamma^{111}_{31}} \in K^{\nu_{21}-\nu_{31}}$, $\gamma^{121}_{31} = [\mathbf{0}, \widetilde{\gamma^{121}_{31}}, \mathbf{0}]^T \in K^{\nu_{21}}$, $\widetilde{\gamma^{121}_{31}} \in K^{\nu_{21}-\nu_{31}}$.
$\mathbf{d}^1_{2j}$ $(2 \le j \le q_2)$ is given by $\mathbf{d}^1_{2j+1} = F_{\beta I}\mathbf{d}^1_{2j} - F_{\gamma I}(\mathbf{b}^1_{2j} - \mathbf{e}_{\nu_1+\nu_{21}+\cdots+\nu_{2j}+1})$
which is derived from the commutatibity of $F_{\beta I}$ and $F_{\gamma I}$, namely,
$F_{\gamma I}F_{\beta I}\mathbf{e}_{\nu_1+\nu_{21}+\cdots+\nu_{2j-1}+1} = F_{\gamma I}F_{\beta I}\mathbf{e}_{\nu_1+\nu_{21}+\cdots+\nu_{2j-1}+1}$ for $2 \le j \le q_3$.

Then, we can dependently obtain $\mathbf{d}^1_{2j}$ in turn for $2 \le j \le q_3 + 1$.
Let us find $\mathbf{d}^1_{2j}$ for $q_3+2 \le j \le q_2$. We refer the following equations presented in Lemma (7-A.5).

$$\begin{cases} F_{\gamma I}T_s\mathbf{e}_{\nu_1+\nu_{21}+\cdots+\nu_{2q_3}+1} = F_{\gamma I}F^{q_3}_{\beta I}\mathbf{e}_1 \\ \qquad\qquad = F_{\gamma I}(F^{q_3-1}_{\beta I}\mathbf{b}^1_{11} - \mathbf{e}_{\nu_1+\nu_{21}+\cdots+\nu_{2q_3}+1}) + \mathbf{d}^1_{2q_3+1} \\ T_sF_{\gamma s}\mathbf{e}_{\nu_1+\nu_{21}+\cdots+\nu_{2q_3}+1} = T_s\mathbf{c}^{3q_3+1} \end{cases}$$

$$\begin{cases} F_{\gamma I}T_s\mathbf{e}_{\nu_1+\nu_{21}+\cdots+\nu_{2q_3+1}+1} = F_{\gamma I}F^{q_3+1}_{\beta I}\mathbf{e}_1 \\ \qquad\qquad = F_{\gamma I}(F^{q_3}_{\beta I}\mathbf{b}^1_{11} - \mathbf{e}_{\nu_1+\nu_{21}+\cdots+\nu_{2q_3+1}+1}) + \mathbf{d}^1_{2q_3+2} \\ T_sF_{\gamma s}\mathbf{e}_{\nu_1+\nu_{21}+\cdots+\nu_{2q_3+1}+1} = T_sF_{\beta s}\mathbf{c}^{3q_3+1} = F_{\beta I}T_s\mathbf{c}^{3q_3+1} \end{cases}$$

$$\vdots$$

$$\begin{cases} F_{\gamma I}T_s\mathbf{e}_{\nu_1+\nu_{21}+\cdots+\nu_{2q_2-1}+1} = F_{\gamma I}F^{q_2-1}_{\beta I}\mathbf{e}_1 \\ \qquad\qquad = F_{\gamma I}(F^{q_2-2}_{\beta I}\mathbf{b}^1_{11} - \mathbf{e}_{\nu_1+\nu_{21}+\cdots+\nu_{2q_2-1}+1}) + \mathbf{d}^1_{2q_2} \\ T_sF_{\gamma s}\mathbf{e}_{\nu_1+\nu_{21}+\cdots+\nu_{2q_2-1}+1} = T_sF^{q_2-q_3-1}_{\beta s}\mathbf{c}^{3q_3+1} = F^{q_2-q_3-1}_{\beta I}T_s\mathbf{c}^{3q_3+1} \end{cases}$$

Then, the following equations are derived:

$$F_{\gamma I}(F^{q_3-1}_{\beta I}\mathbf{b}^1_{11} - \mathbf{e}_{\nu_1+\nu_{21}+\cdots+\nu_{2q_3}+1}) + \mathbf{d}^1_{2q_3+1} = T_s\mathbf{c}^{3q_3+1},$$
$$F_{\gamma I}(F^{q_3}_{\beta I}\mathbf{b}^1_{11} - \mathbf{e}_{\nu_1+\nu_{21}+\cdots+\nu_{2q_3+1}+1}) + \mathbf{d}^1_{2q_3+2} = F_{\beta I}T_s\mathbf{c}^{3q_3+1},$$
$$\vdots$$
$$F_{\gamma I}(F^{q_2-2}_{\beta I}\mathbf{b}^1_{11} - \mathbf{e}_{\nu_1+\nu_{21}+\cdots+\nu_{2q_2-1}+1}) + \mathbf{d}^1_{2q_2} = F^{q_2-q_3-1}_{\beta I}T_s\mathbf{c}^{3q_3+1}.$$

Therefore, once $\mathbf{d}^1_{2q_3+1}$ is determined, $\mathbf{d}^1_{2j}$ $(q_3 + 2 \le j \le q_2)$ is found in turn.

Next, we will obtain $\mathbf{t}^1_{3j}$ $(1 \le j \le q_3)$. Lemma (7-A.9) leads to $\mathbf{t}^1_{3j} = (\mathbf{td})^1_{2j} + \mathbf{d}^1_{2j}$ for $1 \le j \le q_3$. Hence, $\mathbf{t}^1_{3j}$ is calculated in turn by using the pre-determined $\mathbf{d}^1_{2j}$.

Remark 1: $\mathbf{B}_2 \in K^{n\times\nu_2}$ of $F_{\beta I}$ is determined by $2\times\nu_2$ $(= 2\times(\nu_{21}-\nu_{22})+2\times 2\times(\nu_{22}-\nu_{23})+2\times3\times(\nu_{23}-\nu_{24})+\cdots+2\times(q_2-1)\times(\nu_{2q_2-1}-\nu_{2q_2})+2\times q_2\times\nu_{2q_2})$ parameters.

Remark 2: $\mathbf{D}_2 \in K^{n\times\nu_2}$ of $F_{\gamma I}$ is determined by $2\times(\nu_{21}-\nu_{31})$ parameters.

Consequently, $\mathbf{B}_2 \in K^{n\times\nu_2}$ of $F_{\beta I}$, $\mathbf{T}_3 \in K^{n\times\nu_3}$ of T_s and $\mathbf{D}_2 \in K^{n\times\nu_2}$ of $F_{\gamma I}$ were determined by Proposition (7-A.16).

Let us determine $\mathbf{B}_i \in K^{n\times\nu_i}$ of $F_{\beta I}$, $\mathbf{D}_i \in K^{n\times\nu_i}$ of $F_{\gamma I}$ and $\mathbf{T}_{i+1} \in K^{n\times\nu_{i+1}}$ of T_s for $3\le i\le p$. Note that $\mathbf{T}_{p+1}$ does not exist.

Proposition 7-A.17. Let T_s be a regular matrix and a $\{\alpha,\beta,\gamma\}$-morphism $T_s : (K^n, F_{\alpha s}, F_{\beta s}, F_{\gamma s}) \to (K^n, F_{\alpha I}, F_{\beta I}, F_{\gamma I})$.

Then $\mathbf{b}^1_{ij}$ of $F_{\beta I}$ and $\mathbf{d}^1_{ij}$ of $F_{\gamma I}$ in $\{\alpha,\beta,\gamma\}$-action $(K^n, F_{\alpha I}, F_{\beta I}, F_{\gamma I})$ satisfy the following conditions (1) and (2). Moreover, the $\{\alpha,\beta,\gamma\}$-morphism T_s presented in (3) has the full rank property.

(1) $\mathbf{b}^1_{ij} \in K^n$ is expressed as:

$\mathbf{b}^1_{ij} = [\mathbf{b}^{11}_{ij}, \mathbf{b}^{12}_{ij}, \mathbf{b}^{13}_{ij}, \cdots, \mathbf{b}^{1i-1}_{ij}, \mathbf{b}^{1i}_{ij}, \mathbf{0}]^T$ for $3\le i\le p$, $1\le j\le q_i$,

where $\mathbf{b}^{1\hat{i}}_{ij} := [\mathbf{b}^{1\hat{i}1}_{ij}, \mathbf{b}^{1\hat{i}2}_{ij}, \cdots, \mathbf{b}^{1\hat{i}j}_{ij}, \mathbf{0}]^T \in K^{\nu_{\hat{i}}}$ for $1\le\hat{i}\le i-1$,

$\mathbf{b}^{1\hat{i}\hat{j}}_{ij} := [\mathbf{0}, \widetilde{\mathbf{b}^{1\hat{i}\hat{j}}_{ij}}, \mathbf{0}]^T \in K^{\nu_{\hat{i}\hat{j}}}$, $\widetilde{\mathbf{b}^{1\hat{i}\hat{j}}_{ij}} \in K^{\nu_{ij}-\nu_{ij+1}}$ for $1\le\hat{j}\le j$,

In particular, $\mathbf{b}^{1i}_{ij} := [\mathbf{b}^{1i1}_{ij}, \mathbf{b}^{1i2}_{ij}, \cdots, \mathbf{b}^{1ij}_{ij}, \mathbf{b}^{1ij+1}_{ij}, \mathbf{0}]^T \in K^{\nu_i}$,

$\mathbf{b}^{1i\hat{j}}_{ij} := [\mathbf{0}, \widetilde{\mathbf{b}^{1i\hat{j}}_{ij}}, \mathbf{0}]^T \in K^{\nu_{i\hat{j}}}$, $\widetilde{\mathbf{b}^{1i\hat{j}}_{ij}} \in K^{\nu_{ij}-\nu_{ij+1}}$ for $1\le\hat{j}\le j$,

$\mathbf{b}^{1ij}_{ij} := [\mathbf{0}, \widetilde{\mathbf{b}^{1ij}_{ij}}, \mathbf{0}]^T \in K^{\nu_{ij}}$, $\mathbf{b}^{1ij+1}_{ij} = \mathbf{e}^s_1$,

$$\widetilde{\mathbf{b}^{1\hat{i}\hat{j}}_{ij}} = \{[\phi_{ij+1}(F_{\hat{i}\hat{j}})\mathbf{e}^s_{\nu_{\hat{i}\hat{j}}-\nu_{ij}+1}, \cdots, \phi_{ij+1}(F_{\hat{i}\hat{j}})\mathbf{e}^s_{\nu_{\hat{i}\hat{j}}-\nu_{ij+1}}]^T \times [\phi_{ij+1}(F_{\hat{i}\hat{j}})\mathbf{e}^s_{\nu_{\hat{i}\hat{j}}-\nu_{ij}+1}, \cdots, \phi_{ij+1}(F_{\hat{i}\hat{j}})\mathbf{e}^s_{\nu_{\hat{i}\hat{j}}-\nu_{ij+1}}]\}^{-1} \times [\phi_{ij+1}(F_{\hat{i}\hat{j}})\mathbf{e}^s_{\nu_{\hat{i}\hat{j}}-\nu_{ij}+1}, \cdots, \phi_{ij+1}(F_{\hat{i}\hat{j}})\mathbf{e}^s_{\nu_{\hat{i}\hat{j}}-\nu_{ij}}]^T\{\gamma^{1\hat{i}\hat{j}}_{ij} - \phi_{ij+1}(F_{\alpha I}) \times (\mathbf{tb})^{1\hat{i}\hat{j}}_{ij}\},$$ where $2\le\hat{i}\le i$, $2\le\hat{j}\le j$.

$\mathbf{b}^1_{iq_i}$ is given by $\mathbf{b}^1_{iq_i} = T_sC^{iq_i+1} - (\mathbf{tb})^1_{iq_i}$ for $1\le i\le p$, where

$$(\mathbf{tb})^1_{iq_i} = F^{q_i-1}_{\beta I}(\mathbf{b}^1_{i1} - \mathbf{e}_{\nu_1+\cdots+\nu_{i-1}+\nu_{i1}+1}) + F^{q_i-2}_{\beta I}(\mathbf{b}^1_{i2} - \mathbf{e}_{\nu_1+\cdots+\nu_{i-1}+\nu_{i1}+\nu_{i2}+1}) + \cdots + F_{\beta I}(\mathbf{b}^1_{iq_i-1} - \mathbf{e}_{\nu_1+\cdots+\nu_{i-1}+\nu_{i1}+\cdots+\nu_{iq_i-1}+1}).$$

(2)

(2-1) $\mathbf{d}^1_{i1} \in K^n$ $(3\le i\le p-1)$ is expressed as

$\mathbf{d}^1_{i1} := [\mathbf{d}^{11}_{i1}, \mathbf{d}^{12}_{i1}, \cdots, \mathbf{d}^{1i}_{i1}, \mathbf{d}^{1i+1}_{i1}, \mathbf{0}]^T$, where $\mathbf{d}_{i1}^{1\hat{i}} = [\mathbf{d}_{i1}^{1\hat{i}1}, \cdots, \mathbf{d}_{i1}^{1\hat{i}j}, \mathbf{0}]^T \in K^{\nu_{\hat{i}}}$ for $1 \le \hat{i} \le i$,

$\mathbf{d}_{i1}^{1i+1} = \mathbf{e}_1^s \in K^{\nu_{i+1}}$, $\mathbf{d}_{i1}^{1\hat{i}\hat{j}} = [\mathbf{0}, \widetilde{\mathbf{d}_{i1}^{1\hat{i}\hat{j}}}, \mathbf{0}]^T \in K^{\nu_{\hat{i}\hat{j}}}$, $\widetilde{\mathbf{d}_{i1}^{1\hat{i}\hat{j}}} \in K^{\nu_{i1}-\nu_{i+11}}$.

$\widetilde{\mathbf{d}_{i1}^{1\hat{i}\hat{j}}}$ $(2 \le \hat{i} \le i,\ 2 \le \hat{j} \le j)$ is expressed as:

$$\begin{aligned}\widetilde{\mathbf{d}_{i1}^{1\hat{i}\hat{j}}} &= \{[\phi_{i+11}(F_{\hat{i}\hat{j}})\mathbf{e}^s_{\nu_{\hat{i}\hat{j}}-\nu_{i1}+1}, \cdots, \phi_{i+11}(F_{\hat{i}\hat{j}})\mathbf{e}^s_{\nu_{\hat{i}\hat{j}}-\nu_{i+11}}]^T \\ &\times [\phi_{i+11}(F_{\hat{i}\hat{j}})\mathbf{e}^s_{\nu_{\hat{i}\hat{j}}-\nu_{i1}+1}, \cdots, \phi_{i+11}(F_{\hat{i}\hat{j}})\mathbf{e}^s_{\nu_{\hat{i}\hat{j}}-\nu_{i+11}}]\}^{-1} \\ &\times [\phi_{i+11}(F_{\hat{i}\hat{j}})\mathbf{e}^s_{\nu_{\hat{i}\hat{j}}-\nu_{i1}+1}, \cdots, \phi_{i+11}(F_{\hat{i}\hat{j}})\mathbf{e}^s_{\nu_{\hat{i}\hat{j}}-\nu_{i+11}}]^T \gamma_{i+11}^{1\hat{i}\hat{j}},\end{aligned}$$

where $2 \le \hat{i} \le i,\ \ 2 \le \hat{j} \le j$,

$$\begin{aligned}\gamma_{i+11}^1 &:= [\mathbf{T}_1,\ \mathbf{T}_2,\ \cdots, \mathbf{T}_i][C_1^{i+11}, C_2^{i+11}, \cdots, C_i^{i+11}]^T \\ &+ [F_{\gamma I}(F_{\gamma I}^{i-1}\mathbf{e}_1 - \mathbf{e}_{\nu_1+\cdots+\nu_{i-1}+1}),\ F_{\alpha I}F_{\gamma I}(F_{\gamma I}^{i-1}\mathbf{e}_1 - \mathbf{e}_{\nu_1+\cdots+\nu_{i-1}+1}), \\ &\cdots, F_{\alpha I}^{\nu_{i+11}-1}F_{\gamma I}(F_{\gamma I}^{i-1}\mathbf{e}_1 - \mathbf{e}_{\nu_1+\cdots+\nu_{i-1}+1})]C_{i+11}^{i+11}.\end{aligned}$$

Once $\mathbf{d}_{i1}^1$ is found, then $\mathbf{d}_{ij+1}^1$ is dependently obtained in turn by

$\mathbf{d}_{ij+1}^1 = F_{\beta I}\mathbf{d}_{ij}^1 - F_{\gamma I}(\mathbf{b}_{ij}^1 - \mathbf{e}_{\nu_1+\cdots+\nu_{i-1}+\nu_{i1}+\cdots+\nu_{ij}+1})$ for $1 \le j \le q_{i+1}$.

In particular, $\mathbf{d}_{pj+1}^1$ $(1 \le j \le q_p - 1)$ is given by

$\mathbf{d}_{pj+1}^1 = F_{\beta I}\mathbf{d}_{pj}^1 - F_{\gamma I}(\mathbf{b}_{pj}^1 - \mathbf{e}_{\nu_1+\cdots+\nu_{p-1}+\nu_{p1}+\cdots+\nu_{pj}+1})$,

where $\mathbf{d}_{p1}^1 = T_s\mathbf{c}^{p+1} - F_{\gamma I}(\mathbf{td})_{p-11}^1 - F_{\gamma I}(\mathbf{d}_{p-11}^1 - \mathbf{e}_{\nu_1+\cdots+\nu_{p-1}+1})$.

(2-2) From the determined $\mathbf{d}_{iq_{i+1}+1}^1 \in K^n$, $\mathbf{d}_{ij}^1 \in K^n$ is dependently found and expressed as $\mathbf{d}_{ij}^1 = [\mathbf{d}_{ij}^{11}, \mathbf{d}_{ij}^{12}, \cdots, \mathbf{d}_{ij}^{1j}, \mathbf{0}]^T$ for $q_{i+1} + 2 \le j \le q_i$.

(3) $\mathbf{t}_{ij}^1 = (\mathbf{td})_{i-1j}^1 + \mathbf{d}_{i-1j}^1$ $(2 \le i \le p,\ 2 \le j \le q_i)$, where $(\mathbf{td})_{i-1j}^1 = F_{\beta I}(\mathbf{td})_{i-1j-1}^1 + F_{\gamma I}(\mathbf{b}_{i-1j-1}^1 - \mathbf{e}_{\nu_1+\cdots+\nu_{i-2}+\nu_{i-11}+\cdots+\nu_{i-1j-1}+1})$,

$(\mathbf{td})_{i-11}^1 = F_{\gamma I}^{i-2}(\mathbf{d}_{11}^1 - \mathbf{e}_{\nu_1+1}) + F_{\gamma I}^{i-3}(\mathbf{d}_{21}^1 - \mathbf{e}_{\nu_1+\nu_2+1}) + \cdots$
$+ F_{\gamma I}(\mathbf{d}_{i-21}^1 - \mathbf{e}_{\nu_1+\cdots+\nu_{i-2}+1})$ $(3 \le i \le p)$.

Proof. By virtue of Lemma (7-A.8), $\mathbf{b}_{ij}^1$ is obtained as the solutions of the following equation: $\phi_{ij+1}(F_{\alpha I})\mathbf{b}_{ij}^1 = \beta_{ij}^1 - \phi_{ij+1}(F_{\alpha I})(\mathbf{tb})_{ij}^1$, where

$\mathbf{b}_{ij}^1 := [\mathbf{b}_{ij}^{1i1}, \mathbf{b}_{ij}^{1i2}, \cdots, \mathbf{b}_{ij}^{1ii}, \mathbf{0}]^T$, $\quad \mathbf{b}_{ij}^{1i1} = [\mathbf{0}, \widetilde{\mathbf{b}_{ij}^{1i1}}, \mathbf{0}]^T \in K^{\nu_{i1}}$,

$\widetilde{\mathbf{b}_{ij}^{1i1}} \in K^{\nu_{i1}-\nu_{1i}}, \cdots, \mathbf{b}_{ij}^{1\hat{i}} = [\mathbf{b}_{ij}^{1\hat{i}1}, \cdots, \mathbf{b}_{ij}^{1\hat{i}j}, \mathbf{0}]^T \in K^{\nu_{\hat{i}}}$,

$\mathbf{b}_{ij}^{1i} := [\mathbf{b}_{ij}^{1i1}, \mathbf{b}_{ij}^{1i2}, \cdots, \mathbf{b}_{ij}^{1ij}, \mathbf{e}_1^s, \mathbf{0}]^T \in K^{\nu_i}$, $\mathbf{e}_1^s \in K^{\nu_{ij+1}}$,

$\mathbf{b}_{ij}^{1\hat{i}\hat{j}} = [\mathbf{0}, \widetilde{\mathbf{b}_{ij}^{1\hat{i}\hat{j}}}, \mathbf{0}]^T \in K^{\nu_{\hat{i}}}$, $\widetilde{\mathbf{b}_{ij}^{1\hat{i}\hat{j}}} \in K^{\nu_{ij}-\nu_{ij+1}}$ for $1 \le \hat{i} \le i-1,\ 1 \le \hat{j} \le j$.

β_{ij}^1 is expressed as:

$$\beta_{ij}^1 := [\mathbf{T}_1,\ \mathbf{T}_2,\ \cdots,\mathbf{T}_{i-1},\ \mathbf{T}_{i1},\mathbf{T}_{i2},\cdots,\mathbf{T}_{ij}]$$
$$\times[C_1^{ij+1},C_2^{ij+1},\cdots,C_{i-1}^{ij+1},C_{i1}^{ij+1},C_{i2}^{ij+1}\cdots,C_{ij}^{ij+1}]^T$$
$$+[F_{\beta I}^j(F_{\gamma I}^{i-1}\mathbf{e}_1-\mathbf{e}_{\nu_1+\cdots+\nu_{i-1}+1}),F_{\alpha I}F_{\beta I}^j(F_{\gamma I}^{i-1}\mathbf{e}_1-\mathbf{e}_{\nu_1+\cdots+\nu_{i-1}+1}),\cdots,$$
$$F_{\alpha I}^{\nu_{ij+1}-1}F_{\beta I}^j(F_{\gamma I}^{i-1}\mathbf{e}_1-\mathbf{e}_{\nu_1+\cdots+\nu_{i-1}+1})]C_{ij+1}^{ij+1}.$$

(1) of Lemma (7-A.7) leads to $(\mathbf{tb})_{ij}^1$ as:

$$(\mathbf{tb})_{ij}^1=F_{\beta I}(\mathbf{tb})_{ij-1}^1+F_{\beta I}(\mathbf{b}_{ij-1}^1-\mathbf{e}_{\nu_1+\cdots+\nu_{i-1}+\nu_{i1}+\cdots+\nu_{ij-1}+1}),\ (\mathbf{tb})_{i1}^1=0.$$

Thus $\widetilde{\mathbf{b}_{ij}^{1i1}}$ is found as the form as we see in this proposition.

Let us find $\mathbf{b}_{iq_i}^1$. Lemmas (7-A.5) and (7-A.7) result in:

$T_s\mathbf{c}^{iq_i+1}=F_{\beta I}^{q_i-1}F_{\beta I}\mathbf{e}_{\nu_1+\cdots+\nu_{i-1}+1}=(\mathbf{tb})_{iq_i}^1+\mathbf{b}_{iq_i}^1$ for $1\le i\le p$, where

$$(\mathbf{tb})_{iq_i}^1=F_{\beta I}^{q_i-1}(\mathbf{b}_{i1}^1-\mathbf{e}_{\nu_1+\cdots+\nu_{i-1}+\nu_{i1}+1})+F_{\beta I}^{q_i-2}(\mathbf{b}_{i2}^1-\mathbf{e}_{\nu_1+\cdots+\nu_{i-1}+\nu_{i1}+\nu_{i2}+1})$$
$$+\cdots+F_{\beta I}(\mathbf{b}_{iq_i-1}^1-\mathbf{e}_{\nu_1+\cdots+\nu_{i-1}+\nu_{i1}+\cdots+\nu_{iq_i-1}+1}).$$

This equation leads to $\mathbf{b}_{iq_i}^1$.

Next, let us find $\mathbf{d}_{ij}^1$ $(3\le i\le p-1,\ 1\le j\le q_i)$ and $\mathbf{t}_{i+1j}^1$ $(3\le i\le p-1,$ $1\le j\le q_{i+1})$.
By virtue of lemmas (7-A.4) and (7-A.8), $\mathbf{d}_{i1}^1$ is obtained as the solutions of the following equation: $\phi_{i+11}(F_{\alpha I})\mathbf{d}_{i1}^1=\gamma_{i+11}^1$, where
$\mathbf{d}_{i1}^1:=[\mathbf{d}_{i1}^{11},\mathbf{d}_{i1}^{12},\cdots,\mathbf{d}_{i1}^{1i},\mathbf{d}_{i1}^{1i+1},\mathbf{0}]^T$,
$\mathbf{d}_{i1}^{1\hat{i}}=[\mathbf{d}_{i1}^{1\hat{i}1},\cdots,\mathbf{d}_{i1}^{1\hat{i}j},\mathbf{0}]^T\in K^{\nu_{\hat{i}}}$ for $1\le\hat{i}\le i$,

$$\mathbf{d}_{i1}^{1i+1}=\mathbf{e}_1^s\in K^{\nu_{i+1}},\ \ \mathbf{d}_{i1}^{1\hat{i}\hat{j}}=[\mathbf{0},\ \widetilde{\mathbf{d}_{i1}^{1\hat{i}\hat{j}}},\mathbf{0}]^T\in K^{\nu_{\hat{i}\hat{j}}},\widetilde{\mathbf{d}_{i1}^{1\hat{i}\hat{j}}}\in K^{\nu_{i1}-\nu_{i+11}},$$
$$\gamma_{i+11}^1:=[\mathbf{T}_1,\ \mathbf{T}_2,\ \cdots,\mathbf{T}_i][C_1^{i+11},C_2^{i+11},\cdots,C_i^{i+11}]^T$$
$$+[F_{\gamma I}(F_{\gamma I}^{i-1}\mathbf{e}_1-\mathbf{e}_{\nu_1+\cdots+\nu_{i-1}+1}),F_{\alpha I}F_{\gamma I}(F_{\gamma I}^{i-1}\mathbf{e}_1-\mathbf{e}_{\nu_1+\cdots+\nu_{i-1}+1}),$$
$$\cdots,F_{\alpha I}^{\nu_{i+11}-1}F_{\gamma I}(F_{\gamma I}^{i-1}\mathbf{e}_1-\mathbf{e}_{\nu_1+\cdots+\nu_{i-1}+1})]C_{i+11}^{i+11}.$$

Thus $\widetilde{\mathbf{d}_{i1}^{1\hat{i}\hat{j}}}$ is given as the form presented in this proposition.
The commutativity of matrices $F_{\beta I}$ and $F_{\gamma I}$, that is, $F_{\beta I}F_{\gamma I}=F_{\gamma I}F_{\beta I}$ leads to: $F_{\beta I}\mathbf{d}_{ij}^1=F_{\gamma I}(\mathbf{b}_{ij}^1-\mathbf{e}_{\nu_1+\cdots+\nu_{i-1}+\nu_{i1}+\cdots+\nu_{ij}+1})+\mathbf{d}_{ij+1}^1$ for $1\le j\le q_i$.
Hence, $\mathbf{d}_{ij}^1$ is found in turn for $2\le j\le q_i+1$. Thus, $\mathbf{d}_{ij}^1$ for $1\le j\le q_{i+1}+1$ is found.

Let us find $\mathbf{d}_{ij}^1$ for $q_{i+1}+2\le j\le q_i$.
We refer the following equations obtained in Lemma (7-A.5).

$$\begin{cases} F_{\gamma I}T_s\mathbf{e}_{\nu_1+\cdots+\nu_{i-1}+\nu_{i1}+\cdots+\nu_{iq_{i+1}}+1}=F_{\gamma I}^{i-1}F_{\beta I}^{q_{i+1}}\mathbf{e}_1 \\ \qquad =F_{\gamma I}(F_{\gamma I}^{i-2}F_{\beta I}^{q_{i+1}-1}\mathbf{b}_{11}^1-\mathbf{e}_{\nu_1+\cdots+\nu_{i-1}+\nu_{i1}+\cdots+\nu_{iq_{i+1}}+1})+\mathbf{d}_{iq_{i+1}+1}^1 \\ T_sF_{\gamma s}\mathbf{e}_{\nu_1+\cdots+\nu_{i-1}+\nu_{i1}+\cdots+\nu_{iq_{i+1}}+1}=T_s\mathbf{c}^{iq_{i+1}+1} \end{cases}$$

$$\begin{cases} F_{\gamma I} T_s \mathbf{e}_{\nu_1+\cdots+\nu_{i-1}+\nu_{i1}+\cdots+\nu_{iq_{i+1}+1}+1} = F_{\gamma I}^{i-1} F_{\beta I}^{q_{i+1}+1} \mathbf{e}_1 \\ \qquad = F_{\gamma I}(F_{\gamma I}^{i-2} F_{\beta I}^{q_{i+1}+1} \mathbf{b}_{11}^1 - \mathbf{e}_{\nu_1+\cdots+\nu_{i-1}+\nu_{i1}+\cdots+\nu_{iq_{i+1}+1}+1}) + \mathbf{d}_{iq_{i+1}+2}^1 \\ T_s F_{\gamma s} \mathbf{e}_{\nu_1+\cdots+\nu_{i-1}+\nu_{i1}+\cdots+\nu_{iq_{i+1}+1}+1} = T_s F_{\beta s} \mathbf{c}^{iq_{i+1}+1} = F_{\beta I} T_s \mathbf{c}^{iq_{i+1}+1} \end{cases}$$

$\vdots$

$$\begin{cases} F_{\gamma I} T_s \mathbf{e}_{\nu_1+\cdots+\nu_{i-1}+\nu_{i1}+\cdots+\nu_{iq_i-1}+1} = F_{\gamma I}^{i-1} F_{\beta I}^{q_i-1} \mathbf{e}_1 \\ \qquad = F_{\gamma I}(F_{\gamma I}^{i-2} F_{\beta I}^{q_i-2} \mathbf{b}_{11}^1 - \mathbf{e}_{\nu_1+\cdots+\nu_{i-1}+\nu_{i1}+\cdots+\nu_{iq_i-1}+1}) + \mathbf{d}_{iq_i}^1 \\ T_s F_{\gamma s} \mathbf{e}_{\nu_1+\cdots+\nu_{i-1}+\nu_{i1}+\cdots+\nu_{iq_i-1}+1} = T_s F_{\beta s}^{q_{i-1}-q_i-1} \mathbf{c}^{iq_i+1} \\ \qquad = F_{\beta I}^{q_{i-1}-q_i-1} T_s \mathbf{c}^{iq_i+1} \end{cases}$$

Then the following equations are derived:

$$F_{\gamma I}(F_{\gamma I}^{i-2} F_{\beta I}^{q_{i+1}-1} \mathbf{b}_{11}^1 - \mathbf{e}_{\nu_1+\cdots+\nu_{i-1}+\nu_{i1}+\cdots+\nu_{iq_{i+1}}+1}) + \mathbf{d}_{iq_{i+1}+1}^1 = T_s \mathbf{c}^{iq_{i+1}+1},$$

$$F_{\gamma I}(F_{\gamma I}^{i-2} F_{\beta I}^{q_{i+1}+1} \mathbf{b}_{11}^1 - \mathbf{e}_{\nu_1+\cdots+\nu_{i-1}+\nu_{i1}+\cdots+\nu_{iq_{i+1}+1}+1}) + \mathbf{d}_{iq_{i+1}+2}^1 = F_{\beta I} T_s \mathbf{c}^{iq_{i+1}+1},$$

$\vdots$

$$F_{\gamma I}(F_{\gamma I}^{i-2} F_{\beta I}^{q_i-2} \mathbf{b}_{11}^1 - \mathbf{e}_{\nu_1+\cdots+\nu_{i-1}+\nu_{i1}+\cdots+\nu_{iq_i-1}+1}) + \mathbf{d}_{iq_i}^1 = F_{\beta I}^{q_{i-1}-q_i-1} T_s \mathbf{c}^{iq_i+1}.$$

Therefore, once $\mathbf{d}_{iq_{i+1}+1}^1$ is determined, $\mathbf{d}_{ij}^1$ $(q_{i+1}+2 \le j \le q_i)$ is found in turn. According to the relation;

$$F_{\beta I} F_{\gamma I} \mathbf{e}_{\nu_1+\cdots+\nu_{p-1}+\nu_{p1}+\cdots+\nu_{pj-1}+1} = F_{\gamma I} F_{\beta I} \mathbf{e}_{\nu_1+\cdots+\nu_{p-1}+\nu_{p1}+\cdots+\nu_{pj-1}+1}$$

for $1 \le j \le q_p - 1$, the vectors $\mathbf{d}_{pj}^1$ $(1 \le j \le q_p)$ of $\mathbf{D}_p$ are obtained in turn by using the following equation:

$$\mathbf{d}_{pj+1}^1 = F_{\beta I} \mathbf{d}_{pj}^1 - F_{\gamma I}(\mathbf{b}_{pj}^1 - \mathbf{e}_{\nu_1+\cdots+\nu_{p-1}+\nu_{p1}+\cdots+\nu_{pj}+1}), \text{ for } 1 \le j \le q_p - 1,$$

where $\mathbf{d}_{p1}^1 = T_s \mathbf{c}^{p+1} - F_{\gamma I}(\mathbf{td})_{p-11}^1 - F_{\gamma I}(\mathbf{d}_{p-11}^1 - \mathbf{e}_{\nu_1+\cdots+\nu_{p-1}+1})$.

Next, let us find $\mathbf{t}_{ij}^1$ for $3 \le i \le p,\ 1 \le j \le q_i$. Lemma (7-A.9) leads to $\mathbf{t}_{ij}^1 = (\mathbf{td})_{i-1j}^1 + \mathbf{d}_{i-1j}^1$ $(2 \le i \le p,\ 2 \le j \le q_i)$, where
$(\mathbf{td})_{i-1j}^1 = F_{\beta I}(\mathbf{td})_{i-1j-1}^1 + F_{\gamma I}(\mathbf{b}_{i-1j-1}^1 - \mathbf{e}_{\nu_1+\cdots+\nu_{i-2}+\nu_{i-11}+\cdots+\nu_{i-1j-1}+1})$,

$(\mathbf{td})_{i-11}^1 = F_{\gamma I}^{i-2}(\mathbf{d}_{11}^1 - \mathbf{e}_{\nu_1+1}) + F_{\gamma I}^{i-3}(\mathbf{d}_{21}^1 - \mathbf{e}_{\nu_1+\nu_2+1}) + \cdots$
$+ F_{\gamma I}(\mathbf{d}_{i-21}^1 - \mathbf{e}_{\nu_1+\cdots+\nu_{i-2}+1})$ $(3 \le i \le p)$.

Therefore, $\mathbf{t}_{ij}^1$ $(1 \le j \le q_i)$ is found in turn.

Remark 1: $\mathbf{B}_i \in K^{n \times \nu_i}$ is determined by $i \times \nu_i$ $(= \sum_{j=1}^{q_i} i \times j \times (\nu_{ij} - \nu_{ij+1}))$ parameters.

Remark 2: $\mathbf{D}_i \in K^{n \times \nu_i}$ is determined by $i \times (\nu_{i1} - \nu_{i+11})$ parameters.

Consequently, $\mathbf{B}_i \in K^{n \times \nu_i}$ of $F_{\beta I}$, $\mathbf{T}_{i+1} \in K^{n \times \nu_{i+1}}$ of T_s and $\mathbf{D}_i \in K^{n \times \nu_i}$ of $F_{\gamma I}$ for i $(3 \le i \le p)$ are determined by Proposition (7-A.17).

Proposition 7-A.18. Let $\sigma_I = ((K^n, F_{\alpha I}, F_{\beta I}, F_{\gamma I}), \mathbf{e}_1, h_I)$ be any Invariant Standard System. Then σ_I is determined by the minimal parameters, whose number is $\nu_{11} + \nu_1 + 2 \times \nu_2 + \cdots + i \times \nu_i + \cdots + p \times \nu_p + \nu_{11} + \nu_{21} + \cdots + \nu_{p1}$.

Proof. Lemma (7-A.10) presents that $F_{\alpha I}$ is determined by ν_{11} parameters. Lemmas (7-A.11), (7-A.12) and (7-A.13) indicate that $\mathbf{B}_1$ of $F_{\beta I}$ is determined by ν_1 $(= \sum_{j=1}^{q_1} j \times (\nu_{1j} - \nu_{1j+1}))$ parameters.

Propositions (7-A.16) and (7-A.17) indicate that $\mathbf{B}_i$ of $F_{\beta I}$ is determined by $i \times \nu_i$ $(= \sum_{j=1}^{q_i} i \times j \times (\nu_{ij} - \nu_{ij+1})$ parameters for $2 \le i \le p$. Hence, $F_{\beta I}$ is determined by $\nu_1 + 2 \times \nu_2 + 3 \times \nu_3 + \cdots + (p-1) \times \nu_{p-1} + p \times \nu_p$ $(= \sum_{i=1}^{p} i \times \nu_i)$ parameters.

Propositions (7-A.15), (7-A.16) and (7-A.17) indicate that $\mathbf{d}_{i1}^1 \in K^n$ is determined by $i \times (\nu_{i1} - \nu_{i+11})$ parameters for $1 \le i \le p$. Hence, vectors $\{\mathbf{d}_{i1}^1 \in K^n; 1 \le i \le p\}$ are determined by $\nu_{11} + \nu_{21} + \cdots + \nu_{p1}$ $(= \sum_{i=1}^{p} i \times (\nu_{i1} - \nu_{i+11}))$ parameters.

Propositions (7-A.15), (7-A.16) and (7-A.17) indicate that the vectors $\mathbf{d}_{ij}^1 \in K^n$ $(2 \le i \le p,\ 2 \le j \le q_i)$ are dependently found from the predetermined submatrices $\mathbf{B}_i$ $(1 \le i \le p)$ of $F_{\beta I}$ and the vectors $\mathbf{d}_{i1}^1 \in K^n$ $(1 \le i \le p)$ of $F_{\gamma I}$.

The following theorem makes the relation between Quasi-reachable Standard Systems and Invariant Standard Systems clear.

Theorem 7-A.19. There exists a unique Invariant Standard System $((K^n, F_{\alpha I}, F_{\beta I}, F_{\gamma I}), \mathbf{e}_1, h_I)$ which is isomorphic to any Quasi-reachable Standard System $((K^n, F_{\alpha s}, F_{\beta s}, F_{\gamma s}), \mathbf{e}_1, h_s)$.

Proof. Let $\sigma_s = ((K^n, F_{\alpha s}, F_{\beta s}, F_{\gamma s}), \mathbf{e}_1, h_s)$ be Quasi-reachable Standard System with the index $\nu = (\nu_1, \nu_2, \cdots, \nu_p)$. Let $(K^n, F_{\alpha I}, F_{\beta I}, F_{\gamma I})$ be the $\{\alpha, \beta, \gamma\}$-action which satisfies Proposition (7-A.14).

Set $h_I := h_s T_s^{-1}$ for the regular matrix T_s which is a $\{\alpha, \beta, \gamma\}$-morphism $T_s : (K^n, F_{\alpha s}, F_{\beta s}, F_{\gamma s}) \to (K^n, F_{\alpha I}, F_{\beta I}, F_{\gamma I})$ presented in propositions (7-A.14), (7-A.15) and (7-A.17). Since $T_s \mathbf{e}_1 = \mathbf{e}_1$ holds, T_s is a 3-Commutative Linear Representation System morphism $T_s : ((K^n, F_{\alpha s}, F_{\beta s}, F_{\gamma s}), \mathbf{e}_1, h_s) \to ((K^n, F_{\alpha I}, F_{\beta I}, F_{\gamma I}), \mathbf{e}_1, h_I)$. Since T_s is bijective, $((K^n, F_{\alpha I}, F_{\beta I}, F_{\gamma I}), \mathbf{e}_1, h_I)$ is canonical, and it is also the Invariant Standard System with the index $\nu = (\nu_1, \nu_2, \cdots, \nu_p)$.

Since the process of obtaining $F_{\beta I}$ of Lemma (7-A.13) picks up the minimal parts of $((K^n, F_{\alpha s}, F_{\beta s}, F_{\gamma s}), \mathbf{e}_1, h_s)$, the Invariant Standard System $((K^n, F_{\alpha I}, F_{\beta I}, F_{\gamma I}), \mathbf{e}_1, h_I)$ is clearly unique.

7-A.20. Proof of Theorem (7.2)
Let $((K^n, F_\alpha, F_\beta, F_\gamma), \mathbf{e}_1, h)$ be any n-dimensional canonical 3-Commutative Linear Representation System. Then Theorem (6.11) implies that there exists the unique Quasi-reachable Standard System $((K^n, F_{\alpha s}, F_{\beta s}, F_{\gamma s}), \mathbf{e}_1, h_s)$ and the unique 3-Commutative Linear Representation System morphism

$T : ((K^n, F_\alpha, F_\beta, F_\gamma), \mathbf{e}_1, h) \to ((K^n, F_{\alpha s}, F_{\beta s}, F_{\gamma s}), \mathbf{e}_1, h_s)$ which is isomorphic to $((K^n, F_\alpha, F_\beta, F_\gamma), \mathbf{e}_1, h)$.

Let T_s be a 3-Commutative Linear Representation System morphism $T_s : ((K^n, F_{\alpha s}, F_{\beta s}, F_{\gamma s}), \mathbf{e}_1, h_s) \to ((K^n, F_{\alpha I}, F_{\beta I}, F_{\gamma I}), \mathbf{e}_1, h_I)$ presented in the proof of Theorem (7-A.18). Then $T \cdot T_s$ is a 3-Commutative Linear Representation System morphism $T_s \cdot T : ((K^n, F_\alpha, F_\beta, F_\gamma), \mathbf{e}_1, h) \to ((K^n, F_{\alpha I}, F_{\beta I}, F_{\gamma I}), \mathbf{e}_1, h_I)$.

Clearly, $((K^n, F_{\alpha I}, F_{\beta I}, F_{\gamma I}), \mathbf{e}_1, h_I)$ is unique for $((K^n, F_\alpha, F_\beta, F_\gamma), \mathbf{e}_1, h)$. Moreover, the behavior of $((K^n, F_{\alpha I}, F_{\beta I}, F_{\gamma I}), \mathbf{e}_1, h_I)$ is the same as that of $((K^n, F_\alpha, F_\beta, F_\gamma), \mathbf{e}_1, h)$.

7-A.21. Proof of Procedure for an effective encoding (7.3)
Lemma (7-A.4) through (7-A.13), Proposition (7-A.14) through (7-A.18) and their remarks assert that the Invarinat Standard System $((K^n, F_{\alpha I}, F_{\beta I}, F_{\gamma I}), \mathbf{e}_1, h_I)$ which has minimal parameters is found.

8 Design for Three-Dimensional Images

It was newly proposed in Chapter 6 that any three-dimensional image can be realized by a mathematical model called the 3-Commutative Linear Representation System. Many methods of image generation are intended primarily for the reality of images. Our proposed method is a fundamentally different graphic generation method. We can remember that the usual treatment of three-dimensional images is intended primarily to display a picture on a two-dimensional screen in such a way that it looks as much like the three-dimensional picture as possible.

However, this geometrical pattern generation is a new attempt to design patterns or artifacts which come only from our imaginations. All the calculations consisting of linear operations, this generation method is very suitable for computer algorithms. It is also of major importance that our model can describe any geometrical pattern and reproduce it exactly. The method can be applied to patterns for tablewares or fabrics. A number of examples will be given to illustrate the effectiveness of this method.

Computer graphics have been used to present exact images of natural objects and phenomena [Schahter and Ahuja, 1975; Haruyama and Barsky, 1984; and Aono and Kunii, 1984]. They have grown to the level of computer graphic art [Machover, 1995] and various works are exhibited in many museums. We also remember that such a process of development has happened in pictorial art. Usually generation algorithms for graphic art are irregular and complex. On the contrary, for ordinary artifacts and fancy articles, regular and simple patterns may be preferable. [Meiszner, et al., 1998] treats an art of knitted fabrics in three-dimensional images. The method in that paper is executed by non-linear operations.

We know that any three-dimensional geometrical pattern can be modeled by the 3-Commutative Linear Representation System. For the simple design for a three-dimensional images, a direct idea for three-dimensional image is presented by using 3-Commutative Linear Representation Systems. Several examples of geometrical pattern generation are presented.

8.1 3-Commutative Linear Representation Systems for Design

Any finite-sized three-dimensional pattern can usually be decomposed into two categories.

(1) Patterns with no periodicity;
(2) Patterns with periodicity.

We will first discuss 3-Commutative Linear Representation Systems σ which realize patterns with no periodicity and then take up patterns with periodicity.

The patterns treated will be $(\underline{L}+1) \times (\underline{M}+1) \times (\underline{N}+1)$-sized three-dimensional images for positive integers $\underline{L}$, $\underline{M}$ and $\underline{N} \in N$.

Lemma 8.1. For any $(\underline{L}+1) \times (\underline{M}+1) \times (\underline{N}+1)$-sized pattern $a \in F(\underline{\mathbf{L}} \times \underline{\mathbf{M}} \times \underline{\mathbf{N}}, Y)$ with no periodicity, there exists a $(\underline{L}+1) \times (\underline{M}+1) \times (\underline{N}+1)$-dimensional 3-Commutative Linear Representation System $\sigma = ((K^{(\underline{L}+1)\times(\underline{M}+1)\times(\underline{N}+1)}, F_\alpha, F_\beta, F_\gamma), x^0, h)$ which realizes it. See Figure 8.1 for the σ.

Proof. This lemma can be obtained easily by direct calculations.

Remark: According to this lemma, we can easily treat a pattern with no periodicity.

8.2. Patterns with periodicity
A pattern $a \in F(N^3, Y)$ with a period of l length in the vertical direction, m length in the horizontal direction and n length in depth is written by a three-dimensional image with a $l \times m \times n$-period.

Proposition 8.3. Any three-dimensional image with a $\ell \times m \times n$-period can be realized by a 3-Commutative Linear Representation System $\sigma_p = ((K^{\ell\times m\times n}, F_{\alpha p}, F_{\beta p}, F_{\gamma p}), x_p^0, h_p)$. See Figure 8.2 for the system σ_p.

Proof. This proposition can be obtained easily by direct calculations.

8.4. A direct sum of 3-Commutative Linear Representation Systems
We now introduce a direct sum $\sigma_1 \oplus \sigma_2$ for the 3-Commutative Linear Representation Systems σ_1 and σ_2, where $\sigma_1 = ((K^{n_1}, F_{\alpha 1}, F_{\beta 1}, F_{\gamma 1}), x_1^0, h_1)$ and $\sigma_2 = ((K^{n_2}, F_{\alpha 2}, F_{\beta 2}, F_{\gamma 2}), x_2^0, h_2)$.

The behavior of $\sigma_1 \oplus \sigma_2$ is given by $a_{\sigma_1} + a_{\sigma_2}$, namely, $a_{\sigma_1 \oplus \sigma_2} = a_{\sigma_1} + a_{\sigma_2}$ holds, where $\sigma_1 \oplus \sigma_2$ is expressed as follows:

$\sigma_1 \oplus \sigma_2 = ((K^{n_1+n_2}, F_\alpha, F_\beta, F_\gamma), x^0, h)$, where $F_\alpha = \begin{bmatrix} F_{\alpha 1} & 0 \\ 0 & F_{\alpha 2} \end{bmatrix}$, $F_\beta = \begin{bmatrix} F_{\beta 1} & 0 \\ 0 & F_{\beta 2} \end{bmatrix}$, $F_\gamma = \begin{bmatrix} F_{\gamma 1} & 0 \\ 0 & F_{\gamma 2} \end{bmatrix}$, $x^0 = \begin{bmatrix} x_1^0 \\ x_2^0 \end{bmatrix}$ and $h = \begin{bmatrix} h_1 & h_2 \end{bmatrix}$.

$$F_\alpha = \underbrace{\begin{bmatrix} F_{\alpha 0} & 0 & \cdots & 0 \\ 0 & F_{\alpha 0} & \ddots & \vdots \\ \vdots & \ddots & \ddots & 0 \\ 0 & \cdots & 0 & F_{\alpha 0} \end{bmatrix}}_{(\underline{L}+1)\times(\underline{M}+1)\times(\underline{N}+1)}, \quad F_{\alpha 0} = \underbrace{\begin{bmatrix} 0 & \cdots & 0 & 0 \\ 1 & \ddots & \vdots & \vdots \\ 0 & \ddots & 0 & \vdots \\ 0 & 0 & 1 & 0 \end{bmatrix}}_{\underline{L}+1},$$

$$F_\beta = \underbrace{\begin{bmatrix} F_{\beta 0} & 0 & \cdots & 0 \\ 0 & F_{\beta 0} & \ddots & \vdots \\ \vdots & \ddots & \ddots & 0 \\ 0 & \cdots & 0 & F_{\beta 0} \end{bmatrix}}_{(\underline{L}+1)\times(\underline{M}+1)\times(\underline{N}+1)}, \quad F_{\beta 0} = \underbrace{\begin{bmatrix} 0 & \cdots & 0 & 0 \\ I_{\underline{L}+1} & \ddots & \vdots & \vdots \\ 0 & \ddots & 0 & \vdots \\ 0 & 0 & I_{\underline{L}+1} & 0 \end{bmatrix}}_{(\underline{L}+1)\times(\underline{M}+1)},$$

$$F_\gamma = \underbrace{\begin{bmatrix} 0 & \cdots & 0 & 0 \\ F_{\gamma 0} & \ddots & \vdots & \vdots \\ 0 & \ddots & 0 & \vdots \\ 0 & 0 & F_{\gamma 0} & 0 \end{bmatrix}}_{(\underline{L}+1)\times(\underline{M}+1)\times(\underline{N}+1)}, \quad F_{\gamma 0} = \underbrace{\begin{bmatrix} I_{\underline{L}+1} & \cdots & 0 & 0 \\ 0 & \ddots & \vdots & \vdots \\ 0 & \ddots & I_{\underline{L}+1} & \vdots \\ 0 & 0 & 0 & I_{\underline{L}+1} \end{bmatrix}}_{(\underline{L}+1)\times(\underline{M}+1)},$$

$$x^0 = \left.\begin{bmatrix} 1 \\ 0 \\ \vdots \\ 0 \end{bmatrix}\right\} (\underline{L}+1)\times(\underline{M}+1)\times(\underline{N}+1), \quad I_{\underline{L}+1} \text{ is a } (\underline{L}+1)\times(\underline{L}+1) \text{ identity matrix,}$$

$$h = \underbrace{\begin{bmatrix} a_{0,0,0} \cdots a_{\underline{M},0,0}\, a_{0,1,0} \cdots a_{\underline{M},1,0} \cdots a_{0,\underline{L},0} \cdots a_{\underline{M},\underline{L},\underline{N}} \end{bmatrix}}_{(\underline{L}+1)\times(\underline{M}+1)\times(\underline{N}+1)}$$

Fig. 8.1. The 3-Commutative Linear Representation System $\sigma = ((K^{(\underline{L}+1)\times(\underline{M}+1)\times(\underline{N}+1)}, F_\alpha, F_\beta, F_\gamma), x^0, h)$ in Lemma (8.1)

To design three-dimensional images at high speed, we will use the direct sum of 3-Commutative Linear Representation Systems and number theory. Though the following lemmas are the same as the lemmas in Chapter 5, we will include them again for the sake of easy reading.

First, we state Lemma (8.5), which is a special case of Dirichlet's theorem [Hardy and Wright, 1979]:

Lemma 8.5. Let p be a prime number such that $p = L \times M + 1$ for a positive integer L and a fixed integer M. Then there are infinite primes p of the form.

Proof. This is Dirichlet's theorem itself.

Lemma 8.6. Fermat's Lemma [Hardy and Wright, 1979]
If p is prime and x is not divisible by p, then $x^{p-1} \equiv 1 \ (mod\ p)$ holds.

By virtue of Fermat's lemma (8.6), if an integer L satisfies $L < p$, then $x^L - 1 \equiv 0 \ (mod\ p)$ has L different solutions. Hence we can obtain the following lemma.

$$F_{\alpha p} = I_n \otimes I_m \otimes F_\ell = \underbrace{\begin{bmatrix} F_{\alpha 1} & 0 & \cdots & 0 \\ 0 & F_{\alpha 1} & \ddots & \vdots \\ \vdots & \ddots & \ddots & 0 \\ 0 & \cdots & 0 & F_{\alpha 1} \end{bmatrix}}_{\ell \times m \times n}, \quad F_{\alpha 1} = \underbrace{\begin{bmatrix} 0 & \cdots & 0 & 1 \\ 1 & \ddots & \vdots & 0 \\ 0 & \ddots & 0 & \vdots \\ 0 & 0 & 1 & 0 \end{bmatrix}}_{\ell}$$

$$F_{\beta p} = I_n \otimes F_m \otimes I_\ell = \underbrace{\begin{bmatrix} F_{\beta 1} & 0 & \cdots & 0 \\ 0 & F_{\beta 1} & \ddots & \vdots \\ \vdots & \ddots & \ddots & 0 \\ 0 & \cdots & 0 & F_{\beta 1} \end{bmatrix}}_{\ell \times m \times n}, \quad F_{\beta 1} = \underbrace{\begin{bmatrix} 0 & \cdots & 0 & I_\ell \\ I_\ell & \ddots & \vdots & 0 \\ 0 & \ddots & 0 & \vdots \\ 0 & 0 & I_\ell & 0 \end{bmatrix}}_{\ell \times m}$$

$$F_{\gamma p} = F_m \otimes I_n \otimes I_\ell = \underbrace{\begin{bmatrix} 0 & 0 & \cdots & F_{\gamma 1} \\ F_{\gamma 1} & 0 & \ddots & \vdots \\ \vdots & \ddots & \ddots & 0 \\ 0 & \cdots & F_{\gamma 1} & 0 \end{bmatrix}}_{\ell \times m \times n} \quad F_{\gamma 1} = \underbrace{\begin{bmatrix} I_\ell & \cdots & 0 & 0 \\ 0 & \ddots & \vdots & 0 \\ 0 & \ddots & I_\ell & \vdots \\ 0 & 0 & 0 & I_\ell \end{bmatrix}}_{\ell \times m}$$

$$x^0 = \left.\begin{bmatrix} 1 \\ 0 \\ \vdots \\ 0 \end{bmatrix}\right\} \ell \times m \times n \, ,$$

I_ℓ is a $\ell \times \ell$ identity matrix for any $\ell \in N$ and

$$h_p = \underbrace{\begin{bmatrix} a_{0,0,0} \cdots a_{\ell-1,0,0}\, a_{0,1,0} \cdots a_{\ell-1,1,0} \cdots \cdots a_{\ell-1,m-1,n-1} \end{bmatrix}}_{\ell \times m \times n}$$

Fig. 8.2. The 3-Commutative Linear Representation System $\sigma_p = ((K^{\ell \times m \times n}, F_{\alpha p}, F_{\beta p}, F_{\gamma p}), x_p^0, h_p)$ for patterns with a $\ell \times m \times n$ period

Lemma 8.7. Let p be a prime number such that $p = L \times M + 1$ for a positive integer L and a fixed integer M. Then $x^L - 1 \equiv (x - x_1)(x - x_2)(x - x_3) \cdots (x - x_L)$ $(mod\ p)$ holds.

Proof. By the condition for the selection of the prime number p, we obtain $x^L - 1 = x^{\frac{p-1}{M}} - 1 = (x^{\frac{1}{M}})^{p-1} - 1 \equiv 0$ $(mod\ p)$.

Definition 8.8. The 3-Commutative Linear Representation System $\sigma_e = ((K^{\ell \times m \times n}, F_{\alpha e}, F_{\beta e}, F_{\gamma e}), x_e^0, h_e)$ given in Figure 8.3 is called an Eigen Standard System.

Theorem 8.9. Let us consider the 3-Commutative Linear Representation System σ_p which realizes any three-dimensional image with a $\ell \times m \times n$ period, where σ_p is given by $\sigma_p = ((K^{\ell \times m \times n}, F_{\alpha p}, F_{\beta p}, F_{\gamma p}), x_p^0, h_p)$ as we discussed in Proposition (8.3).

Then σ_p is isomorphic to the Eigen Standard System σ_e.

$$F_{\alpha e}=\underbrace{\begin{bmatrix} F_{\ell e} & 0 & \cdots & 0 \\ 0 & F_{\ell e} & \ddots & \vdots \\ \vdots & \ddots & \ddots & 0 \\ 0 & \cdots & 0 & F_{\ell e} \end{bmatrix}}_{\ell \times m \times n},\ F_{\ell e}=\underbrace{\begin{bmatrix} \alpha_1 & 0 & \ldots & 0 \\ 0 & \alpha_2 & 0 & \vdots \\ \vdots & \ddots & \ddots & 0 \\ 0 & \ldots & 0 & \alpha_\ell \end{bmatrix}}_{\ell},$$

$$F_{\beta e}=\underbrace{\begin{bmatrix} F_{\beta o} & 0 & \cdots & 0 \\ 0 & F_{\beta o} & \ddots & \vdots \\ \vdots & \ddots & \ddots & 0 \\ 0 & \cdots & 0 & F_{\beta o} \end{bmatrix}}_{\ell \times m \times n},\ F_{\beta o}=\underbrace{\begin{bmatrix} F_1 & 0 & \cdots & 0 \\ 0 & F_2 & \ddots & \vdots \\ \vdots & \ddots & \ddots & 0 \\ 0 & \cdots & 0 & F_m \end{bmatrix}}_{\ell \times m},\ F_i \in K^{\ell \times \ell} \text{ for } i\ (1 \le i \le m),$$

$$F_{\gamma e}=\underbrace{\begin{bmatrix} \gamma_1 I_{\ell \times m} & 0 & \cdots & 0 \\ 0 & \gamma_2 I_{\ell \times m} & \ddots & \vdots \\ \vdots & \ddots & \ddots & 0 \\ 0 & \cdots & 0 & \gamma_n I_{\ell \times m} \end{bmatrix}}_{\ell \times m \times n},$$

$I_{\ell \times m}$ is a $(\ell \times m) \times (\ell \times m)$ identity matrix.

$$F_i=\underbrace{\begin{bmatrix} \beta_i & 0 & \ldots & 0 \\ 0 & \beta_i & 0 & \vdots \\ \vdots & \ddots & \ddots & 0 \\ 0 & \ldots & 0 & \beta_i \end{bmatrix}}_{\ell},$$

$$x^\ell - 1 \equiv (x-\alpha_1)(x-\alpha_2)(x-\alpha_3)\cdots(x-\alpha_\ell)\ (mod\ p),$$
$$y^m - 1 \equiv (y-\beta_1)(y-\beta_2)(y-\beta_3)\cdots(y-\beta_m)\ (mod\ p),$$
$$z^n - 1 \equiv (z-\gamma_1)(z-\gamma_2)(z-\gamma_3)\cdots(z-\gamma_n)\ (mod\ p),$$

$$x_e^0 = \left.\begin{bmatrix} 1 \\ 1 \\ \vdots \\ 1 \end{bmatrix}\right\} l \times m.$$

Fig. 8.3. The Eigen Standard System $\sigma_e = ((K^{\ell \times m \times n}, F_{\alpha e}, F_{\beta e}, F_{\gamma e}), x_e^0, h_e)$, as given in Definition (8.8)

Proof. Take the following $(\ell \times m \times n) \times (\ell \times m \times n)$ matrix T_e:

$$\begin{aligned} T_e = [&x_e^0\ F_{\alpha e}x_e^0 \ldots F_{\alpha e}^{\ell-1}x_e^0\ F_{\beta e}x_e^0\ F_{\alpha e}F_{\beta e}x_e^0 \\ &\ldots F_{\alpha e}^{\ell-1}F_{\beta e}x_e^0 \ldots F_{\beta e}^{m-1}x_e^0 \ldots F_{\alpha e}^{\ell-1}F_{\beta e}^{m-1}x_e^0 \\ &F_{\gamma e}x_e^0\ F_{\alpha e}F_{\gamma e}x_e^0 \ldots F_{\alpha e}^{\ell-1}F_{\gamma e}x_e^0\ F_{\beta e}F_{\gamma e}x_e^0\ F_{\alpha e}F_{\beta e}F_{\gamma e}x_e^0 \\ &\ldots F_{\alpha e}^{\ell-1}F_{\beta e}F_{\gamma e}x_e^0 \ldots F_{\beta e}^{m-1}F_{\gamma e}x_e^0 \ldots F_{\alpha e}^{\ell-1}F_{\beta e}^{m-1}F_{\gamma e}x_e^0 \\ &\ldots \ldots \ldots \ldots \ldots \ldots \\ &F_{\gamma e}^{n-1}x_e^0\ F_{\alpha e}F_{\gamma e}^{n-1}x_e^0 \ldots F_{\alpha e}^{\ell-1}F_{\gamma e}^{n-1}x_e^0\ F_{\beta e}F_{\gamma e}^{n-1}x_e^0\ F_{\alpha e}F_{\beta e}F_{\gamma e}^{n-1}x_e^0 \\ &\ldots F_{\alpha e}^{\ell-1}F_{\beta e}F_{\gamma e}^{n-1}x_e^0 \ldots F_{\beta e}^{m-1}F_{\gamma e}^{n-1}x_e^0 \ldots F_{\alpha e}^{\ell-1}F_{\beta e}^{m-1}F_{\gamma e}^{n-1}x_e^0]. \end{aligned}$$

Let h_e be $h_p := h_e T_e$. Then T_e is a 3-Commutative Linear Representation System morphism T_e:

$\sigma_p = ((K^{\ell \times m \times n}, F_{\alpha p}, F_{\beta p}, F_{\gamma p}), x_p^0, h_p) \rightarrow$
$\sigma_e = ((K^{\ell \times m \times n}, F_{\alpha e}, F_{\beta e}, F_{\gamma e}), x_e^0, h_e).$

Hence the behavior of σ_p is the same as the behavior of σ_e by Corollary (6.7).

Remark: By virtue of Theorem (8.9), when we design any three-dimensional image, the Eigen Standard System σ_e can provide that image more rapidly than it can be generated using σ_p.

8.10. Problem statement for finding a prime number
Consider the following problem in view of the need for rapid calculation in designing three-dimensional periodic geometrical patterns.

"Find a prime number p for given integers l_1, l_2 and l_3 such that $p-1 = m_1 \times l_1 = m_2 \times l_2 = m_3 \times l_3$ for some integers m_1, m_2 and m_3."

Remark: By virtue of Lemma (8.7), if such a prime number p is found for given integers l_1, l_2 and l_3, the polynomials $x^{l_1} - 1$, $x^{l_2} - 1$ and $x^{l_3} - 1$ can be factorized simultaneously via modulo p.

8.11. Algorithm for finding a prime number p

1. Calculate the greatest common divisor g of l_1, l_2 and l_3.
2. Calculate the least common multiple l of l_1, l_2 and l_3.
3. Find the minimum prime number which satisfies
   $p = m_1 \times l_1 + 1 = m_2 \times l_2 + 1 = m_3 \times l_3 + 1$ and $p \geq l_1 \times l_2 \times l_3$.

Remark: In this algorithm, the greatest common divisor can be obtained by using the well-known Euclidean algorithm, while the least common multiple is calculated by using the well-known relation $l \times g = l_1 \times l_2 \times l_3$. Upon finding the apparent minimum prime number, one only has to judge whether the given number is truly a prime or not.

8.12. Table of prime numbers
A prime number table which has been obtained by using the algorithm (8.11) is shown in Tables 8.1–8.5. In these table, the positive integers l_1, l_2 and l_3, which are the periods of three-dimensional images, are restricted from 1 through 10 for a fixed number l_3.

For given periods of a three-dimensional image l_1, l_2 and l_3, the required minimum prime number is found at the cross point of the l_2 column and the l_3 row in the table for a fixed number l_1.

Table 8.1. Prime number table for three-dimensional periodic images with period l_1,l_2 and l_3 (upper right: $l_1 = 2$, lower left: $l_1 = 1$)

-	10	9	8	7	6	5	4	3	2	1	l_2 / l_3
-	211	181	241	211	151	101	101	61	41	31	10
-	-	163	433	127	109	181	73	73	37	19	9
1	2	-	137	113	97	241	73	73	41	17	8
2	3	5	-	113	127	71	113	43	29	29	7
3	7	7	13	-	73	61	61	37	31	13	6
4	5	13	13	17	-	61	41	31	31	11	5
5	11	11	31	41	31	-	37	37	17	13	4
6	7	13	19	37	31	37	-	19	13	7	3
7	29	29	43	29	71	43	71	-	11	5	2
8	17	17	73	41	41	73	113	73	-	3	1
9	19	19	37	37	181	73	127	73	109	-	-
10	11	31	31	41	61	61	71	241	181	101	-
l_2 / l_3	1	2	3	4	5	6	7	8	9	10	-

Table 8.2. Prime number table for three-dimensional periodic images with period l_1,l_2 and l_3 (upper right: $l_1 = 4$, lower left: $l_1 = 3$)

-	10	9	8	7	6	5	4	3	2	1	l_2 / l_3
-	401	541	401	281	241	241	181	181	101	41	10
-	-	397	433	757	397	181	181	109	73	37	9
1	7	-	257	281	193	241	137	97	73	41	8
2	7	13	-	197	337	281	113	337	113	29	7
3	13	19	31	-	157	181	97	73	61	37	6
4	13	37	37	61	-	101	101	61	41	41	5
5	31	31	61	61	151	-	73	61	37	17	4
6	19	37	61	73	151	109	-	37	37	13	3
7	43	43	127	337	211	127	211	-	17	13	2
8	73	73	73	97	241	193	337	193	-	5	1
9	37	73	109	109	181	163	379	433	271	-	-
10	31	61	151	181	151	181	211	241	271	331	-
l_2 / l_3	1	2	3	4	5	6	7	8	9	10	-

Table 8.3. Prime number table for three-dimensional periodic images with period l_1,l_2 and l_3 (upper right: $l_1 = 6$, lower left: $l_1 = 5$)

-	10	9	8	7	6	5	4	3	2	1	l_2 / l_3
-	601	541	601	421	421	331	241	181	151	61	10
-	-	487	433	379	379	271	397	163	109	73	9
1	11	-	409	337	313	241	193	193	97	73	8
2	11	31	-	337	337	211	337	127	127	43	7
3	31	31	61	-	223	181	157	109	73	37	6
4	41	41	61	101	-	151	181	151	61	31	5
5	31	61	151	101	131	-	97	73	61	37	4
6	31	61	151	181	151	181	-	61	37	19	3
7	71	71	211	281	211	211	281	-	31	13	2
8	41	241	241	241	241	241	281	401	-	7	1
9	181	181	181	181	271	271	631	1801	541	-	-
10	61	101	151	241	251	331	421	401	541	521	-
l_2 / l_3	1	2	3	4	5	6	7	8	9	10	-

Table 8.4. Prime number table for three-dimensional periodic images with period l_1,l_2 and l_3 (upper right: $l_1 = 8$, lower right: $l_1 = 7$)

-	10	9	8	7	6	5	4	3	2	1	l_2 / l_3
-	881	1801	641	2521	601	401	401	241	241	241	10
-	-	937	557	1009	433	1801	433	433	433	73	9
1	29	-	521	449	409	401	257	193	137	73	8
2	29	29	-	449	337	281	281	337	113	113	7
3	43	43	127	-	313	241	193	193	97	73	6
4	29	113	337	113	-	241	241	241	241	41	5
5	71	71	211	281	211	-	137	97	73	41	4
6	43	127	127	337	211	337	-	73	73	73	3
7	71	113	211	197	281	337	379	-	41	17	2
8	113	113	337	281	281	337	449	449	-	17	1
9	127	127	379	757	631	379	631	1009	631	-	-
10	71	211	211	281	421	421	491	2521	631	701	-
l_2 / l_3	1	2	3	4	5	6	7	8	9	10	-

Table 8.5. Prime number table for three-dimensional periodic images with period l_1,l_2 and l_3 (upper right: $l_1 = 10$, lower left: $l_1 = 9$)

-	10	9	8	7	6	5	4	3	2	1	l_2 / l_3
-	1021	991	881	701	601	524	401	331	221	101	10
-	-	811	1801	631	541	541	541	271	181	181	9
1	19	-	641	2521	601	401	401	241	241	241	8
2	19	37	-	491	421	421	281	211	211	71	7
3	37	73	109	-	421	331	241	181	151	61	6
4	37	73	109	181	-	251	241	151	101	61	5
5	181	181	181	181	271	-	181	181	101	41	4
6	73	109	163	397	271	379	-	151	61	31	3
7	127	127	379	757	631	379	631	-	41	31	2
8	73	433	433	433	1801	433	1009	577	-	11	1
9	109	163	271	397	541	487	631	937	739	-	-
10	181	181	271	541	541	541	631	1801	811	991	-
l_2 / l_3	1	2	3	4	5	6	7	8	9	10	-

Example 8.13. Consider the $3 \times 3 \times 3$-periodic image in Figure 8.4.

Let K be $N/31N$, which is the quotient field modulo the prime number 31, and let the set Y of output values be K. The details of these sets are given in Figure 8.5. Given Figure 8.5, the readout map h_p represented in Figure 8.4 is as follows:

$$h_p = [1,4,4,3,0,0,3,0,0,2,0,0,0,0,0,0,0,0,2,0,0,0,0,0,0,0,0] .$$

For this image, since $l = m = n = 3$, $F_3, F_{\alpha p}, F_{\beta p}, F_{\gamma p}$ and x^0 in Figure 8.2 are written as follows:

$$F_3 = \begin{bmatrix} 0 & 0 & 1 \\ 1 & 0 & 0 \\ 0 & 1 & 0 \end{bmatrix}, \qquad x^0 = \begin{bmatrix} 1 \\ 0 \\ \vdots \\ 0 \end{bmatrix} \in K^{27},$$

$$F_{\alpha p} = I_3 \otimes I_3 \otimes F_3, \qquad F_{\beta p} = I_3 \otimes F_3 \otimes I_3, \qquad F_{\gamma p} = F_3 \otimes I_3 \otimes I_3.$$

For rapid design, we must express our image in the Eigen Standard System depicted in Figure 8.3. First, we find the prime number p which can be factorized $x^l - 1$, $x^m - 1$ and $x^n - 1$ simultaneously. Since $l = m = n = 3$, we find that $p = 31$ from Table 8.2.

Since $x^3 - 1 \equiv (x-1)(x-5)(x-25) \pmod{31}$, we can set $\alpha_1 = \beta_1 = \gamma_1 = 1$, $\alpha_2 = \beta_2 = \gamma_2 = 5$ and $\alpha_3 = \beta_3 = \gamma_3 = 25$. Then, from Figure 8.3,

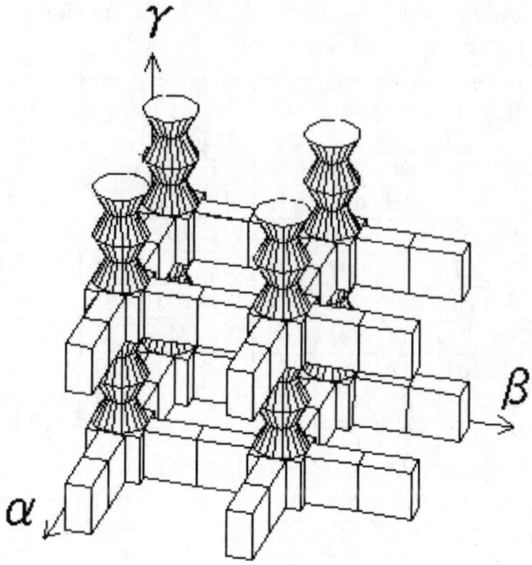

Fig. 8.4. The $3 \times 3 \times 3$-periodic image for Example (8.13)

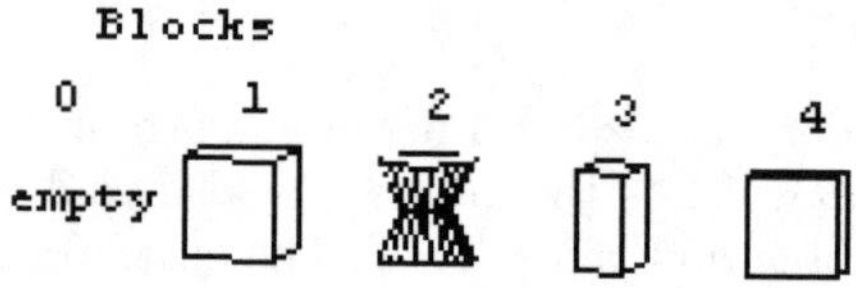

Fig. 8.5. The coding list for Example (8.13)

$F_{le}, F_1, F_2, F_3, x_e^0$ will be as follows:

$$F_{le} = \begin{bmatrix} 1 & 0 & 0 \\ 0 & 5 & 0 \\ 0 & 0 & 25 \end{bmatrix}, \qquad F_1 = \begin{bmatrix} 1 & 0 & 0 \\ 0 & 1 & 0 \\ 0 & 0 & 1 \end{bmatrix}, \qquad F_2 = \begin{bmatrix} 5 & 0 & 0 \\ 0 & 5 & 0 \\ 0 & 0 & 5 \end{bmatrix},$$

$$F_3 = \begin{bmatrix} 25 & 0 & 0 \\ 0 & 25 & 0 \\ 0 & 0 & 25 \end{bmatrix}, \qquad x_e^0 = \begin{bmatrix} 1 \\ \vdots \\ 1 \end{bmatrix} \in K^{27},$$

$F_{\alpha e}, F_{\beta_e}$ and F_{γ_e} can be constructed as in Figure 8.3.

From the proof of Theorem (8.9), the 3-Commutative Linear Representation System morphism T_e can be constructed as follows:

$T_e = [x_e^0, F_{\alpha e}x_e^0, F_{\alpha e}^2 x_e^0,$
$F_{\beta e}x_e^0, F_{\alpha e}F_{\beta e}x_e^0, F_{\alpha e}^2 F_{\beta e}x_e^0,$
$F_{\beta e}^2 x_e^0, F_{\alpha e}F_{\beta e}^2 x_e^0, F_{\alpha e}^2 F_{\beta e}^2 x_e^0,$
$F_{\gamma e}x_e^0, F_{\alpha e}F_{\gamma e}x_e^0, F_{\alpha e}^2 F_{\gamma e}x_e^0,$
$F_{\beta e}F_{\gamma e}x_e^0, F_{\alpha e}F_{\beta e}F_{\gamma e}x_e^0, F_{\alpha e}^2 F_{\beta e}F_{\gamma e}x_e^0,$
$F_{\beta e}^2 F_{\gamma e}x_e^0, F_{\alpha e}F_{\beta e}^2 F_{\gamma e}x_e^0, F_{\alpha e}^2 F_{\beta e}^2 F_{\gamma e}x_e^0,$
$F_{\gamma e}^2 x_e^0, F_{\alpha e}F_{\gamma e}^2 x_e^0, F_{\alpha e}^2 F_{\gamma e}^2 x_e^0,$
$F_{\beta e}F_{\gamma e}^2 x_e^0, F_{\alpha e}F_{\beta e}F_{\gamma e}^2 x_e^0, F_{\alpha e}^2 F_{\beta e}F_{\gamma e}^2 x_e^0,$
$F_{\beta e}^2 F_{\gamma e}^2 x_e^0, F_{\alpha e}F_{\beta e}^2 F_{\gamma e}^2 x_e^0, F_{\alpha e}^2 F_{\beta e}^2 F_{\gamma e}^2 x_e^0,] \in K^{27\times 27}.$

Then the readout map of the Eigen Standard System h_e will be as follows:

$h_e = [29, 27, 27, 12, 10, 10, 12, 10, 10, 28,$
$26, 26, 11, 9, 9, 11, 9, 9, 28, 26, 26, 11, 9, 9, 11, 9, 9].$

Example 8.14. Consider the following $3 \times 4 \times 3$-periodic image.

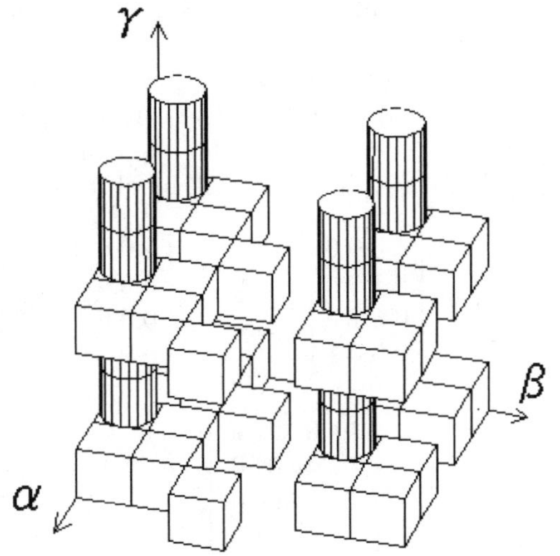

Fig. 8.6. The $3 \times 4 \times 3$ periodic image for Example (8.14)

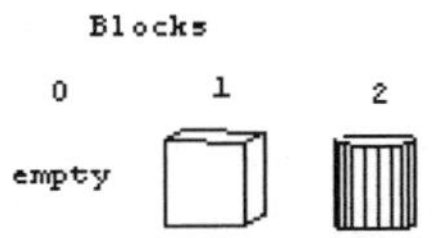

Fig. 8.7. The coding list for Example (8.14)

Let K be $N/37N$, which is the quotient field modulo the prime number 37, and let the set Y of output values be K. Given Figure 8.7, the readout map h_p represented in Figure 8.6 is as follows:

$$h_p = [1,1,0,1,1,0,0,0,1,0,0,0,2,0,\cdots,0,2,0,\cdots,0] \in K^{1\times 36}.$$

For this image, since $l = 3$, $m = 4$ and $n = 3$, $F_3, F_4, F_{\alpha p}, F_{\beta p}$ and x^0 in Figure 8.3 are written as follows:

$$F_3 = \begin{bmatrix} 0&0&1 \\ 1&0&0 \\ 0&1&0 \end{bmatrix}, \quad F_4 = \begin{bmatrix} 0&0&0&1 \\ 1&0&0&0 \\ 0&1&0&0 \\ 0&0&1&0 \end{bmatrix}, \quad x^0 = \begin{bmatrix} 1 \\ 0 \\ \vdots \\ 0 \end{bmatrix} \in K^{36},$$

$$F_{\alpha p} = I_3 \otimes I_4 \otimes F_3, \qquad F_{\beta p} = I_3 \otimes F_4 \otimes I_3, \qquad F_{\gamma p} = F_3 \otimes I_4 \otimes I_3.$$

For rapid design, we must express our image in the Eigen Standard System depicted in Figure 8.4. First, find the prime number p which can be factorized $x^l - 1$, $x^m - 1$ and $x^n - 1$ simultaneously. Since $l = 3$, $m = 4$ and $n = 3$, we find that $p = 37$ from Table 8.2.

Since $x^3 - 1 \equiv (x-1)(x-10)(x-26) \pmod{37}$ and $x^4 - 1 \equiv (x-1)(x-6)(x-31)(x-36) \pmod{37}$, we can set $\alpha_1 = \gamma_1 = 1$, $\alpha_2 = \gamma_2 = 10$, $\alpha_3 = \gamma_3 = 26$, $\beta_1 = 1$, $\beta_2 = 6$, $\beta_3 = 31$ and $\beta_4 = 36$. Then, from Figure 8.3, $F_{le}, F_1, F_2, F_3, F_4$ and x_e^0 will be obtained as follows:

$$F_{le} = \begin{bmatrix} 1&0&0 \\ 0&10&0 \\ 0&0&26 \end{bmatrix}, \quad F_1 = \begin{bmatrix} 1&0&0&0 \\ 0&1&0&0 \\ 0&0&1&0 \\ 0&0&0&1 \end{bmatrix}, \quad F_2 = \begin{bmatrix} 6&0&0&0 \\ 0&6&0&0 \\ 0&0&6&0 \\ 0&0&0&6 \end{bmatrix},$$

$$F_3 = \begin{bmatrix} 31&0&0&0 \\ 0&31&0&0 \\ 0&0&31&0 \\ 0&0&0&31 \end{bmatrix}, \quad F_4 = \begin{bmatrix} 36&0&0&0 \\ 0&36&0&0 \\ 0&0&36&0 \\ 0&0&0&36 \end{bmatrix},$$

$$x_e^0 = \begin{bmatrix} 1 \\ \vdots \\ 1 \end{bmatrix} \in K^{36}.$$

$F_{\alpha e}, F_{\beta_e}$ and F_{γ_e} can be constructed as in Figure 8.3.

From the proof of Theorem (8.9), the 3-Commutative Linear Representation System morphism T_e can be constructed as follows:

$$
\begin{aligned}
T_e = [&x_e^0, F_{\alpha e}x_e^0, F_{\alpha e}^2 x_e^0,\\
&F_{\beta e}x_e^0, F_{\alpha e}F_{\beta e}x_e^0, F_{\alpha e}^2 F_{\beta e}x_e^0,\\
&F_{\beta e}^2 x_e^0, F_{\alpha e}F_{\beta e}^2 x_e^0, F_{\alpha e}^2 F_{\beta e}^2 x_e^0,\\
&F_{\beta e}^3 x_e^0, F_{\alpha e}F_{\beta e}^3 x_e^0, F_{\alpha e}^2 F_{\beta e}^3 x_e^0,\\
&F_{\gamma e}x_e^0, F_{\alpha e}F_{\gamma e}x_e^0, F_{\alpha e}^2 F_{\gamma e}x_e^0,\\
&F_{\beta e}F_{\gamma e}x_e^0, F_{\alpha e}F_{\beta e}F_{\gamma e}x_e^0, F_{\alpha e}^2 F_{\beta e}F_{\gamma e}x_e^0,\\
&F_{\beta e}^2 F_{\gamma e}x_e^0, F_{\alpha e}F_{\beta e}^2 F_{\gamma e}x_e^0, F_{\alpha e}^2 F_{\beta e}^2 F_{\gamma e}x_e^0,\\
&F_{\beta e}^3 F_{\gamma e}x_e^0, F_{\alpha e}F_{\beta e}^3 F_{\gamma e}x_e^0, F_{\alpha e}^2 F_{\beta e}^3 F_{\gamma e}x_e^0,\\
&F_{\gamma e}^2 x_e^0, F_{\alpha e}F_{\gamma e}^2 x_e^0, F_{\alpha e}^2 F_{\gamma e}^2 x_e^0,\\
&F_{\beta e}F_{\gamma e}^2 x_e^0, F_{\alpha e}F_{\beta e}F_{\gamma e}^2 x_e^0, F_{\alpha e}^2 F_{\beta e}F_{\gamma e}^2 x_e^0,\\
&F_{\beta e}^2 F_{\gamma e}^2 x_e^0, F_{\alpha e}F_{\beta e}^2 F_{\gamma e}^2 x_e^0, F_{\alpha e}^2 F_{\beta e}^2 F_{\gamma e}^2 x_e^0,\\
&F_{\beta e}^3 F_{\gamma e}^2 x_e^0, F_{\alpha e}F_{\beta e}^3 F_{\gamma e}^2 x_e^0, F_{\alpha e}^2 F_{\beta e}^3 F_{\gamma e}^2 x_e^0,] \in K^{36\times 36}.
\end{aligned}
$$

Then the readout map of the Eigen Standard System h_e will be as follows:

$$
\begin{aligned}
h_e = [&14, 3, 11, 0, 17, 3, 17, 3, 11, 31, 26, 19, 17, 6, 14, 3, 20, 6,\\
&20, 6, 14, 34, 29, 22, 17, 6, 14, 3, 20, 6, 20, 6, 14, 34, 29, 22].
\end{aligned}
$$

8.2 Design Methods for Geometrical Patterns

To make a geometrical pattern like Figure 8.4 or 8.6 on a computer screen, the usual procedure or design technique is as follows:

1) A pattern for the screen is imagined and decided on.
2) It is determined whether the desired pattern is periodic or non-periodic.
3) The design elements are produced manually at the place where they are to appear in the finished pattern. For a periodic pattern, one whole period is made. In the case of a non-periodic design, every design elements is set manually at each place, and the geometrical pattern is now complete.
4) In the case of a periodic pattern, the period made in Step 3) is copied manually into every desired direction as need arises.

These steps are very laborious, tedious and monotonous; moreover, the error rate is high. If a given geometrical pattern is complex, the work of generating the pattern becomes even harder. Thus the usual method of transferring a design fro mind or from nature to a computer screen is not very efficient for getting a desired geometrical pattern.

Examples (8.13) and (8.14) show that our method is performed mainly by computation using a computer program. Once the program based on the σ_e discussed in Theorem (8.9) is correctly made, any complex geometrical pattern can be generated on the screen automatically and with no error which is caused by executing the design process. The design procedure is as follows:

First, we make a table in which numerals are assigned to each of the picture elements and their thickness, after which a three-dimensional image is made with the numerals replacing the corresponding elements of the design which lies in our minds. Then, this new design may be executed by a program composed of the following procedures:

1) Input data for a pattern to be displayed on a screen.
2) Determine whether the desired pattern is non-periodic or periodic.
2-a) For a non-periodic pattern, determine the 3-Commutative Linear Representation System σ (Lemma (8.1)).
2-b) For a periodic pattern, determine the 3-Commutative Linear Representation System σ_e (Theorem (8.9)).
3) According to a calculation of the behavior of the determined 3-Commutative Linear Representation System, show the design pattern on the screen by using the corresponding table of colors and numerals.

This program will allow any designer to see the designed pattern on the screen directly and confirm easily whether the pattern is good or not.

8.3 Historical Notes and Concluding Remarks

In this chapter, we have proposed a new method of designing three-dimensional patterns on everyday artifacts such as fabrics without actual (visual) patterns. We showed that geometrical patterns can be concisely generated by a mathematical model that is called a 3-Commutative Linear Representation System.

It is easy to understand that this method is an extension of two-dimensional geometrical patterns in Chapter 5. As the concept of design is the same as two-dimensional geometrical patterns, see the comments in Section 5.3 for details.

References

V.M. Adamjan, D.Z. Arov and M.G. Krein

1971 Analytic properties of Schmidt pairs for a Hankel operatpor and the generarized Schur-Takagi problem, Math. USSR Sbornik Vol. 15, No. 1, 31–73.

1978 Infinite Hankel block matrices and related extension problems, Amer. Math. Soc. Transl. (2) Vol. 111, 133–156.

M. Aono and T.L. Kunii

1984 Botanical Tree Image Generation, IEEE Computer Graphics, May, 10–34.

J.S. Bonet and P. Viola

1997 A Non-parametric Multi-Scale Statiscal Model for Natural Images, Neural Information Processing.

N. Bourbaki

1958 *Elements de Mathematique*, Algebre, Herman, Paris.

1968 *Elements of Mathematics, Theory of Sets*, English Translation, Hermann, Paris.

1974 *Elements of Mathematics, Algebra Part 1*, English Translation, Hermann, Paris.

C. Caratheodory and L. Fejer

1911 Über den Zusammenhang der Extremen von harmonischen Funktionen mit ihren Koeffizienten und über den Picard-Landauschen Satz, Rend Circ. Mat. Palermo, 32, 218–239.

R.W. Brockett

1976 Nonlinear systems and differential geometry, Proc. of IEEE, 64–1, 61–72.

C. Chevalley

1956 *Fundamental concept of algebra*, Academic Press.

K. Culik II and V. Valenta

1997 Finite automata based compressed of bi-level and simple color images, Computers & Graphics, 21, 61–68.

D. D'alessandro, A. Isidori and A. Ruberti

1974 Realization and structure theory of bilinear dynamical systems, SIAM J. Contr., 12, 517–534.

E.J. Davison
1966 A method for simplifying linear dynamic systems, IEEE Trans. on AC 11-1, 93–101.

J.C. Doyle, K. Glover, P.P. Kargonekar and B.A. Francis
1989 State space solutions to standard H^2 and H^∞ control problems, IEEE Trans. on AC 34-8, 831–847.

C. Eckart and G. Young
1936 The approximation of one matrix by another of lower rank, Psychometrika, 1, 211–218.

S. Eilenberg
1974 *Automata, Languages and Machines, vol.A*, Academic Press, New York.

R. Eising
1974 Separability of 2-D Transfer Matrices, IEEE Trans. on AC 24-3, 508–510.

M. Fliess
1970 Series reconnaissables, rationnelles et algebriques, Bull. Math., 94, 231–239.
1973 Sur la realisation des systemes dynamiques bilineaires, C.R.Acad. Sc. Paris, t. 277, serie A, 923–926.
1974 Matrices de Hankel, J.Math, pures et al appl,. 53, 197–224.
1978 Un codge non commutatif pour certains systemes echantillonnes nonlineaires, Inf. and Contr., 39, 82.
1979 A remark on the transfer functions and realization of homogeneous continuous-time nonlinear systems, IEEE Trans. on AC 24-3, 507–508.
1980 A note on volterra series expansion for nonlinear differntial systems, IEEE Trans. on AC 25-1, 116–117.

M. Fliess, M. Lamnabhi and F. Lamnabhi-Lagarrigue
1983 An algebra approach to nonlinear functional expansions, IEEE Trans. on Circ. and Syst., 30–8, 554–570.

E. Fornasini and G. Marchesini
1976 State-space realization theory of two-dimensional filters, IEEE Trans. on Automatic Control, AC-21, 484–492.

F.R. Gantmacher
1959 *The theory of matrices, vol.2*, Chelsea, NewYork.

K. Glover
1984 All optimal Hankel-norm approximation of linear multivariable systems and their L^∞-error bounds, Int. J. Control, Vol. 39, No. 6, 1115–1193.

P.R. Halmos
1958 *Finite dimensional vector spaces*, D. Van. Nos. Com.

G.H. Hardy and E.M. Wright
1979 *An Introduction to the Theory of Numbers*, Fifth Edition, Oxford University Press.

S. Haruyama and B.A. Barsky
1984 Using Stochastic Modeling for Texture Generation, IEEE Computer Graphics and Applications, March, 7–19.

Y. Hasegawa, K. Hamada and T. Matsuo
2000 A Realization Algorithm for Discrete-Time Linear Systems, Problems in Modern Applied Mathematics, World Scientifics, 192–197.

Y. Hasegawa and T. Matsuo
1979a Realization Theory of discrete-time linear representation systems, SICE Transaction, 15-3, 298–305, (in Japanese).
1979b On the discrete time finite-dimensional linear representation systems, SICE Transaction, 15-4, 443–450, (in Japanese).
1992 Realization theory of discrete-time Pseudo-Linear Systems, SICE Transaction, 28-2, 199/207, (in Japanese).
1993 Realization theory of Discrete-Time finite-dimensional Pseudo-Linear systems, SICE Transaction, 29-9, 1071–1080, (in Japanese).
1994a Realization theory of discrete-time Almost-Linear Systems, SICE Transaction, 30-2, 150/157, (in Japanese).
1994b Realization theory of Discrete-Time finite-dimensional Almost-Linear systems, SICE Transaction, 30-11, 1325–1333, (in Japanese).
1995 Partial Realization & Real-Time, Partial Realization Theory of Discrete-Time Pseudo-Linear System, SICE Transaction, 31-4, 471–480, (in Japanese).
1996a Modeling for Hysteresis Characterestic by Affine Dynamical Systems, International Conference on Power Electronics, Drivers and Energy Systems, New-Delhi, 1006/1011.
1996b Real-Time Partial Realization Theory of Discrete-Time Non-Linear Systems, SICE Transaction, 32-3, 345–354, (in Japanese).
1996c A relation between Discrete-Time Almost-Linear Systems and "So-called" Linear Systems, SICE Transaction, 32-5, 782–784, (in Japanese).
1996d Partial realization and real-time partial realization theory of discrete-time almost linear systems, SICE Transaction, 32-5, 653–662.

Y. Hasegawa, T. Matsuo and T. Hirano
1982 On the Partial Realization Problem of discrete-time linear representation systems, SICE Transaction, 18–2, 152/159, (in Japanese).

Y. Hasegawa, S. Niinomi and T. Matsuo
1996 Realization Theory of Continuous-Time Pseudo-Linear Systems, IEEE International Symposium on Circuits and Systems, USA, Vol.3, 186/189.
1998 Realization Theory of Continuous-Time Finite Dimensional Pseudo-Linear Systems, Recent Advances in Circuits and Systems World Scientific, 109–114.

Y. Hasegawa, K. Takeichi and T. Matsuo
1997 Realization Theory of 3-Dimensional Arrays, Trans. IECE, 80-A, 3, 542–551 (in Japanese).
1999 Three-Dimensional Arrays and Finite-Dimensional 3-Commutative Linear Representation Systems, Trans. IECE, 82-A, 7, 1101–1114 (in Japanese).

B. Hassibi, A.H. Sayed and T. Kailath

1986 Linear Estimation in Krein Spaces (Part I: Theory, Part II:Application), IEEE Trans. on AC 41-1, 18–49.

D.J. Heeger and J.R. Bergen

1995 Pyramid-Based Texture Analysis/Synthesis, In Computer Graphics Proceedings, 229–238.

B.L. Ho and R.E. Kalman

1966 Effective construction of linear state-variable models from input/output functions, Regelungstechnik, 14, 545–548.

G.M. Hunter and K. Steiglitz

1979 Operations on Images Using Quad Trees, IEEE Trans. pattern analysis Machine Intelligence, PAM1-1, 2, 145–153.

A. Isidori

1973 Direct construction of minimal bilinear realization from nonlinear input-output maps, IEEE Trans. on Autom. Contr., AC-18, 626–631.

C.L. Jackins and S.L. Tanimoto

1980 Oct-Trees and their use in Representing Three-Dimensional Objects, Computer Graphics and Image Processing 14, 249–270.

R.E. Kalman

1960 On the general theory of control systems, Proc. 1st IFAC Congress, Moscow, Butterworths London.

1963 Mathematical description of linear dynamical systems, SIAM J. Contr., 1, 152–192.

1965 Algebraic structure of linear dynamical systems. I, The module of Σ, Proc. Nat. Acad. Sci. (USA), 54, 1503–1508.

1967 *Algebraic aspects of the theory of dynamical systems*, in Differential Equations and Dynamical systems, J. K. Hale and J. P. Lasale (eds), 133–143, Academic Press, New York. Lectures on controllability and observability, note for a course held at C.I.M.E., Bologna, Italy. Cremonese, Roma.

R.E. Kalman, P.L. Falb and M.A. Arbib

1969 *Topics in mathematical system theory*, McGraw-Hill, New York.

L.M. Kaplan and C.-C.J. Kuo

1992 An Improved Method for 2-D Self-Similar Image Synthesis, IEEE Trans. on Image Processing, 5-5, 754–761.

A. Kawakami

1991 A realization of three-dimensional digital filters, Proc. IEEE International Symposium on Circuits and Systems, Vol. 1, 456–459.

C. Machover

1995 Computer Art, IEEE Computer Graphics and Applications, May, 19–23.

T. Matsuo

1968 System theory of linear continuous-time systems using generalized function theory, Res. Rept. of Auto. Contr. Lab., 15, Nagoya, Japan. (in Japanese).

1969 Mathematical theory of linear continuous-time systems, Res. Rept. of Auto. Contr. Lab., 16, 11–17, Nagoya, Japan.
1975 On the realization theory linear infinite dimensional systems, Res. Rept. No. SC-75-9, The Institute of Electrical Engineer of Japan, (in Japanese).
1977 Foundations of mathematical system theory, Journal of SICE, 16, 648–654.
1980 On-line partial realization of a discrete-time non-linear black-box by linear representation systems, Proc. of 18th Annual Allerton Con. on Com. Contr. and Computing. Monticello, Illinois.
1981 *Realization theory of continuous-time dynamical systems*, Lecture Notes in Control and Information Science, 32, Springer.

T. Matsuo and Y.Hasegawa
1977 On linear representation systems and affine dynamical systems, Res. Rept. No. SC-77-27, The institute of Electrical Engineers of Japan. (in Japanese).
1979 Realization theory of discrete-time systems, SICE Transaction, 152, 178–185.
1981 Two-Dimensional Arrays and Finite-Dimensional Commutative Linear Representation Systems, Electronics and Communications in Japan, 64-A, 5, 11–19.
1996 Partial Realization & Real-Time, Partial Realization Theory of Discrete-Time Almost-Linear System, SICE Transaction, 32-5, 653–662, (in Japanese).
2003 *Realization theory of discrete-time dynamical systems*, Lecture Notes in Control and Information Science, 296, Springer.

T. Matsuo, Y. Hasegawa, S. Niinomi and M.Togo
1981 Foundations on the Realization Theory of Two-Dimensional Arrays, Electronics and Communications in Japan, 64-A, 5, 1–10.

T. Matsuo, Y. Hasegawa and Y. Okada
1981 The Partial Realization Theory of Finite Size Two-Dimensional Arrays, Trans. IECE, 64-A, 10, 811–818, (in Japanese).
1994 A Partial Realization Algorithm for Finite-Sized Two-Dimensional Arrays, Trans. IECE, 77-A, 10, 1383–1389 (in Japanese).
1996 Structures of Commutative Linear Representation Systems and an Efficient Coding of Two-Dimensional Arrays, Trans. IECE, 79-A, 3, 719–727 (in Japanese).

T. Matsuo and S. Niinomi
1981 The realization theory of continuous-time affine dynamical systems, SICE Trans., 17-1, 56–63, (in Japanese).

B.H. Mcormick and S.N. Jayaramamurthy
1974 Time Series Model for Texture Synthesis, Int. J. Computer and Information Science, 13-4, 329–343.

D. Meagher
1982 Geometric Modeling Using Octtree Encoding, Computer Graphics and Image Processing, 19, 129–147.

M. Meiszner and B. Eberhardt
1998 The Art of Knitted Fabrics, Realistic and Physically Based Modeling of Knitted patterns, Computer Graphics Forum, 17-3, 354–362.

B.C. Moore
1981 Principal component analysis in linear systems: Controllability, observability and model reduction, IEEE Trans. on AC 26-1, 17–32.

A. Nerode
1958 Linear automaton transformation, American Math. Soc., 2, 541–544.

R. Nevatia
1982 *Machine Perception*, Academic Press.

C.T. Millis and R.A. Roberts
1976 Synthesis of minimum roundoff noise fixed point digital filters, IEEE Circuit and Systems CAS 23-9, 551–562.

S. Niinomi and T. Matsuo
1981 Relation between discrete-time linear representation systems and affine dynamical systems, SICE Trans., 17-3, 350–357, (in Japanese).

A. Netravali and J.O. Limb
1980 Picture Coding: A Review, Proceeding of IEEE, 68-3, 366–406.

H.E. Pade
1892 Sur la representation approchee d'une fonction par des fractions rationelles, Annales Scientifique de l'Ecole Normale Superieure, Vol. 9, no. 3, (supplement) 1-93.

B. Pareigis
1970 *Categories and functors*, Academic Press, New York.

A. Paz
1966 Some aspects of Probabilistic Automata, Inform. and Contr., 9.

J.G. Perlman
1980 Realizability of multilinear input/output maps, Int. J. Cont., 32-2, 271–283.

G.R.B. Prony
1795 Essai experimental et analytique sur les lois de la dilatabilite de fluides elastiques et sur cells de la force expansion de la vapeur de l'alcool, a differentes temperatures, Journel de l'Ecole Polytechnique (Paris), Vol. 1, no. 2, 24-76.

A. Rosenfeld, R.A. Hummel and S.W. Zucker
1976 Scene Labeling by Relaxation Operations, IEEE, Trans. on Systems, Man and Cyb. SMC-6, 420–433.

A. Rosenfeld and A.C. Kak
1976 *Digital picture processing*, Academic Press, New York.

J. Serra
1982 *Image Analysis and Mathematical Morphology*, Academic Press.

B.S. Schahter and N. Ahuja
1975 Random Pattern Generation Process, Computer Graphics and Image processing, 10, 95–114.

M.P. Schutzenberger
1961 On the definition of family of automata, Information and Control, 4, 245–270.

L. M. Silverman
1971 Realization of linear dynamical systems, IEEE Trans. on Automatic Control, AC-16, 554–567.

E.D. Sontag
1979 Realization theory of discrete-time nonlinear systems: part1, IEEE Trans. on Circuit and Systems, CAS-26, 342–356.

H.J. Sussmann
1976 *Semigroup representations, bilinear approximation of input/output maps and generalized inputs*, In Mathematical System Theory (Proc. Internat. Sympos., Internat. Centre Mech. Sci., Udine 1975 / G.Marchesini and S.K.Mitter Eds.), Lecture Notes in Economics and Mathematical Systems, **131**, 172–192, Springer-Verlag.
1977 Existence and uniqueness of minimal realization of non-linear systems, Mathematical System Theory, 10, No.3, 263–284.

R. Szeliski
1991 Fast Shape from Shading, Vision Graphics, Image Processing, Understanding, 53, 2, 129–153.

T.J. Tarn and S. Nonoyama
1976 Realization of discrete-time internally bilinear systems, Proc.IEEE Conf. D.C., 1–3.

K. Vijay and D.G. Madisetti
1998 The digital signal processing handbook, CRC Press.

J.C. Willems
1986 From Time Series to Linear System Part II. Exact Modelling, Automatica, Vol.22, No.6, 675–694.

P. Zeiger
1967 Ho's algorithm, commutative diagram, and the uniqueness of minimal linear systems, Information and Control, 11, 71–79.

Y. Zang and L.T. Bruton
1994 Application of 3-D LCR networks in the design of 3-D recursive filters for processing image sequences, IEEE Trans. on Circuits and Systems for Video Technology, Vol. 4, No. 4, 369–382.

S.C. Zhu, Y. Wu and D. Mumford
1998 Filters, Random fields And Maximum Entropy(FRAME), Int'l Journal of Computer Vision, 27-2, 1–20, March/April.

Index

Lecture Notes in Control and Information Sciences

Edited by M. Thoma and M. Morari

Further volumes of this series can be found on our homepage: springer.com

Vol. 314: Gil', M.I.
Explicit Stability Conditions for Continuous Systems
193 p. 2005 [3-540-23984-7]

Vol. 313: Li, Z.; Soh, Y.; Wen, C.
Switched and Impulsive Systems
277 p. 2005 [3-540-23952-9]

Vol. 312: Henrion, D.; Garulli, A. (Eds.)
Positive Polynomials in Control
313 p. 2005 [3-540-23948-0]

Vol. 311: Lamnabhi-Lagarrigue, F.; Loría, A.;
Panteley, E. (Eds.)
Advanced Topics in Control Systems Theory
294 p. 2005 [1-85233-923-3]

Vol. 310: Janczak, A.
Identification of Nonlinear Systems Using Neural
Networks and Polynomial Models
197 p. 2005 [3-540-23185-4]

Vol. 309: Kumar, V.; Leonard, N.; Morse, A.S. (Eds.)
Cooperative Control
301 p. 2005 [3-540-22861-6]

Vol. 308: Tarbouriech, S.; Abdallah, C.T.; Chiasson, J. (Eds.)
Advances in Communication Control Networks
358 p. 2005 [3-540-22819-5]

Vol. 307: Kwon, S.J.; Chung, W.K.
Perturbation Compensator based Robust Tracking
Control and State Estimation of Mechanical Systems
158 p. 2004 [3-540-22077-1]

Vol. 306: Bien, Z.Z.; Stefanov, D. (Eds.)
Advances in Rehabilitation
472 p. 2004 [3-540-21986-2]

Vol. 305: Nebylov, A.
Ensuring Control Accuracy
256 p. 2004 [3-540-21876-9]

Vol. 304: Margaris, N.I.
Theory of the Non-linear Analog Phase Locked Loop
303 p. 2004 [3-540-21339-2]

Vol. 303: Mahmoud, M.S.
Resilient Control of Uncertain Dynamical Systems
278 p. 2004 [3-540-21351-1]

Vol. 302: Filatov, N.M.; Unbehauen, H.
Adaptive Dual Control: Theory and Applications
237 p. 2004 [3-540-21373-2]

Vol. 301: de Queiroz, M.; Malisoff, M.; Wolenski, P. (Eds.)
Optimal Control, Stabilization and Nonsmooth Analysis
373 p. 2004 [3-540-21330-9]

Vol. 300: Nakamura, M.; Goto, S.; Kyura, N.; Zhang, T.
Mechatronic Servo System Control
Problems in Industries and their Theoretical Solutions
212 p. 2004 [3-540-21096-2]

Vol. 299: Tarn, T.-J.; Chen, S.-B.; Zhou, C. (Eds.)
Robotic Welding, Intelligence and Automation
214 p. 2004 [3-540-20804-6]

Vol. 298: Choi, Y.; Chung, W.K.
PID Trajectory Tracking Control for Mechanical Systems
127 p. 2004 [3-540-20567-5]

Vol. 297: Damm, T.
Rational Matrix Equations in Stochastic Control
219 p. 2004 [3-540-20516-0]

Vol. 296: Matsuo, T.; Hasegawa, Y.
Realization Theory of Discrete-Time Dynamical Systems
235 p. 2003 [3-540-40675-1]

Vol. 295: Kang, W.; Xiao, M.; Borges, C. (Eds)
New Trends in Nonlinear Dynamics and Control,
and their Applications
365 p. 2003 [3-540-10474-0]

Vol. 294: Benvenuti, L.; De Santis, A.; Farina, L. (Eds)
Positive Systems: Theory and Applications (POSTA 2003)
414 p. 2003 [3-540-40342-6]

Vol. 293: Chen, G. and Hill, D.J.
Bifurcation Control
320 p. 2003 [3-540-40341-8]

Vol. 292: Chen, G. and Yu, X.
Chaos Control
380 p. 2003 [3-540-40405-8]

Vol. 291: Xu, J.-X. and Tan, Y.
Linear and Nonlinear Iterative Learning Control
189 p. 2003 [3-540-40173-3]

Vol. 290: Borrelli, F.
Constrained Optimal Control
of Linear and Hybrid Systems
237 p. 2003 [3-540-00257-X]

Vol. 289: Giarré, L. and Bamieh, B.
Multidisciplinary Research in Control
237 p. 2003 [3-540-00917-5]

Vol. 288: Taware, A. and Tao, G.
Control of Sandwich Nonlinear Systems
393 p. 2003 [3-540-44115-8]

Vol. 287: Mahmoud, M.M.; Jiang, J.; Zhang, Y.
Active Fault Tolerant Control Systems
239 p. 2003 [3-540-00318-5]

Vol. 286: Rantzer, A. and Byrnes C.I. (Eds)
Directions in Mathematical Systems
Theory and Optimization
399 p. 2003 [3-540-00065-8]

Vol. 285: Wang, Q.-G.
Decoupling Control
373 p. 2003 [3-540-44128-X]

Vol. 284: Johansson, M.
Piecewise Linear Control Systems
216 p. 2003 [3-540-44124-7]

Vol. 283: Fielding, Ch. et al. (Eds)
Advanced Techniques for Clearance of
Flight Control Laws
480 p. 2003 [3-540-44054-2]